Synthesis, Design, and Resource Optimization in Batch Chemical Plants

Synthesis, Design, and Resource Optimization in Batch Chemical Plants

Edited by Thokozani Majozi
Esmael Reshid Seid • Jui-Yuan Lee

CRC Press
Taylor & Francis Group
Boca Raton London New York

CRC Press is an imprint of the
Taylor & Francis Group, an **informa** business

CRC Press
Taylor & Francis Group
6000 Broken Sound Parkway NW, Suite 300
Boca Raton, FL 33487-2742

First issued in paperback 2017

© 2015 by Taylor & Francis Group, LLC
CRC Press is an imprint of Taylor & Francis Group, an Informa business

No claim to original U.S. Government works

ISBN-13: 978-1-4822-5241-5 (hbk)
ISBN-13: 978-1-138-89330-6 (pbk)

Visit the Taylor & Francis Web site at
http://www.taylorandfrancis.com

and the CRC Press Web site at
http://www.crcpress.com

Contents

PART III   *Resource Conservation*

Preface

The book presents state-of-the-art techniques for synthesis, design, and resource optimization in batch chemical plants. In all the aspects addressed in the book, emphasis is placed on the rigor and essence of a chosen scheduling framework. The manner in which time is captured is of utmost importance and forms the foundation for synthesis, design, and optimization in batch chemical plants. However, there are still serious challenges with handling time in batch plants, as evinced in published literature. Most techniques tend to assume either a fixed time dimension or adopt time average models in order to tame the time dimension, thereby simplifying the resultant mathematical models. A direct consequence of this simplification is a suboptimal process. The material presented in this textbook aims to close this scientific gap.

The book is arranged in three parts according to focus of content. The first part of the book focuses on scheduling techniques. This part entails five chapters that present different ways to represent and capture time in the optimal allocation of tasks to various units with the objective of maximizing throughput or minimizing makespan. These cover unit-specific event points, global event points, slot-based scheduling methods as well as graph-theoretic frameworks in the scheduling of batch plants. Work on long-term scheduling is also presented. Chapter 1 gives a broad background and overview of batch process scheduling, whereas Chapter 5 is more specific to the pharmaceuticals industry. Chapters 2 through 4 present some of the most recent and proven state-of-the-art techniques for scheduling of multipurpose batch plants, including pipeless batch plants, which are arguably the most complex type in this category of chemical processes.

The second part of the book covers synthesis and design where the objective is mainly to yield a chemical facility, which satisfies all the targets with minimum capital cost investment. Mainly continuous-time formulations are presented in this part of the book, since most contributions in this area have been based on continuous rather than discrete time representation.

Lastly, the book deals with resource conservation aspects in batch plants, where water and energy take the center stage. The main objective in this part of the book is to synthesize and design chemical plants that optimally utilize water and energy resources, while optimally achieving the desired throughput.

Professor Thokozani Majozi
School of Chemical and Metallurgical Engineering
University of the Witwatersrand
Republic of South Africa

Acknowledgments

The publication of this book was made possible by researchers who contributed chapters to the three themes that comprise the content contained herein. We are forever indebted to them for all their effort. We also thank the publishers for all the editorial work and finally assembling the material in the most presentable format. Moreover, paramount in the success of this book is the kind gesture of the Stellenbosch Institute for Advanced Study (STIAS), who provided their facilities for a month to the lead editor, Professor Thokozani Majozi, to spend time in formulating and consolidating the ideas shared in this book. Indeed, that was the time well spent.

Editors

Thokozani Majozi is a full professor in the School of Chemical and Metallurgical Engineering, University of the Witwatersrand, South Africa, where he also holds an NRF/DST chair in sustainable process engineering. He completed his PhD in process integration at the University of Manchester Institute of Science and Technology (UMIST) in the United Kingdom. He was also an associate professor in computer science at the University of Pannonia in Hungary from 2005 to 2009. He is a member of Academy of Sciences of South Africa and a Fellow for the Academy of Engineering of SA. He has received numerous awards for his research including S2A3 British Association Medal (Silver) and the South African Institution of Chemical Engineers Bill Neal-May Gold Medal. Majozi is author and coauthor of more than 150 scientific publications, including a book *Batch Chemical Process Integration* published by Springer in January 2010.

Esmael Reshid Seid obtained his BSc in chemical engineering from Bhair Dar University, Ethiopia. He then worked in process industry for 3 years after his BSc before joining the University of Pretoria, South Africa, in 2009, where he obtained his MScEng and PhD in chemical engineering. Dr. Seid has several publications in international refereed journals on design, synthesis, scheduling, and resource conservation, with particular emphasis on water and energy for multipurpose batch plants.

Jui-Yuan Lee is currently a postdoctoral fellow at Carbon Cycle Research Centre, National Taiwan University (NTU). He received his BS in chemical engineering from National Cheng Kung University, Taiwan in 2006, and his PhD in chemical engineering from NTU in 2011. Afterwards he spent two-and-a-half more years at NTU doing postdoctoral research (until September 2013), before joining the School of Chemical and Metallurgical Engineering, University of the Witwatersrand, Johannesburg, where he stayed 8 months as a postdoctoral fellow (until May 2014). Lee's research centers on process integration for energy savings and waste reduction using mathematical programming. He has published dozens of research papers in international peer-reviewed journals, and is working closely with several collaborators in Asia and Africa.

Contributors

Vincentius Surya Kurnia Adi is currently a postdoctoral researcher in the Chemical Engineering Department, National Cheng Kung University, Tainan, Taiwan. He received his BS in chemical engineering from Institut Teknologi Bandung, Indonesia, in 2005. He worked as process engineer for PT. Arianto Darmawan until 2006, and in 2008, he received his MS in chemical engineering from National Cheng Kung University, and continued with his studies to receive his PhD in chemical engineering from National Cheng Kung University in 2013.

Kathleen B. Aviso is an associate professor in the Chemical Engineering Department of De La Salle University, Manila, the Philippines. She received her PhD in industrial engineering and her MS in environmental engineering and management from De La Salle University, Manila, the Philippines. She obtained her BS in chemical engineering from the University of the Philippines, Diliman. She received the Paterno and Natividad Bacani Professorial Chair in chemical engineering in 2013 from the Gokongwei College of Engineering at De La Salle University. She is a recipient of multiple awards from the Philippine National Academy of Science and Technology. Her research interest is in the development of decision support tools in applications to process systems engineering, life-cycle assessment, and industrial ecology.

Santanu Bandyopadhyay obtained his BTech (Hons) in energy engineering from Indian Institute of Technology, Kharagpur in 1992, and MTech and PhD in energy systems engineering from Indian Institute of Technology, Mumbai, in 1995 and 1999, respectively. He then joined the Heat and Mass Transfer Division of M/s Engineers India Limited, New Delhi. In 2001, he joined Department of Energy Science and Engineering (formerly, Energy Systems Engineering), Indian Institute of Technology, Mumbai. His research interest includes process integration, pinch analysis, industrial energy conservation, modeling, and simulation of energy systems, design and optimization of renewable energy systems, and so on. Since 1994, Dr. Bandyopadhyay has been associated with and contributed toward various developmental, industrial, and research activities involving different structured approaches to process design, energy integration, and conservation as well as renewable energy systems design.

Ana Paula Barbosa-Póvoa obtained her PhD in engineering from Imperial College of Science Technology and Medicine, London. She is currently a full professor of operations and logistics at the Department of Management and Engineering of Instituto Superior Técnico (IST), University of Lisbon, Portugal, where she is the director of the bachelor's and master's programs in engineering and management. Ana is vice-president of the Portuguese Association for Operational Research. She has been acting as reviewer to several national and international research scientific boards on research projects. Her research interests are on the supply chain management, where both forward and reserve structures are included and on the design, planning, and scheduling of flexible production systems. Ana has published widely in these areas and has supervised several master and PhD students.

Chuei-Tin Chang received his BS in chemical engineering from National Taiwan University, and his PhD in chemical engineering from Columbia University, New York City, USA, in 1976 and 1982, respectively. He worked as a process engineer in FMC Corporation (Princeton, New Jersey, USA) from 1982 to 1985, and also as an assistant professor at the Department of Chemical Engineering in University of Nebraska (Lincoln, Nebraska, USA) from 1985 to 1989. He later joined the faculty of the Chemical Engineering Department of National Cheng Kung University (Tainan, Taiwan) in 1989, and became a full professor in 1993.

Nitin Dutt Chaturvedi obtained his BTech in chemical engineering, Malaviya National Institute of Technology, Jaipur, and PhD from Department of Energy Science and Engineering, Indian Institute of Technology, Mumbai. After receiving his BTech, he worked in process industries for a couple of years. His research interests include water and energy targets for batch process and development of process integration techniques to address various resource conservation problems.

Cheng-Liang Chen is a professor at the Department of Chemical Engineering, National Taiwan University (NTU). His research centers on process integration, optimization, and control. Dr. Chen has actively conducted professional training for practicing engineers in the field of process design and control. He served as a board member of the Chinese Petroleum Corporation, Taiwan (2008–2013) as well as deputy executive director of the National Science and Technology Program on Energy (2011–2013). Dr. Chen is currently serving as director of the Petro-Chemistry Research Center of NTU (since 2010).

Joel L. Cuello is a professor of biosystems engineering in the Department of Agricultural and Biosystems Engineering at the University of Arizona, USA. Dr. Cuello has designed, constructed, and operated varied types of engineered agricultural or biological systems, including environmentally controlled micropropagation systems for plant tissue culture, scalable plant cell, and organ bioreactors for biochemical production, integrated hybrid solar-and-electric lighting system for crop production for space and earth applications, a UV-ozone bioreactor for wastewater treatment, integrated plant hydroponic and aquaculture production system, and an electrical elicitation system for overproduction of chemicals from plant cells and tissues, among others. Over the last 10 years, Dr. Cuello has focused on the economical scale up and system integration of algae production for nutraceuticals, biofuels, animal feed, and other bioproducts. He is co-inventor of a patented algae photobioreactor design, the Accordion photobioreactor. Dr. Cuello has consulted for numerous companies globally, and recently coauthored the report "Sustainability of Algal Biofuels in the United States" published by the US National Academies.

Alvin B. Culaba is an academician of the National Academy of Science and Technology (NAST), the Philippines. He is also a university fellow and distinguished professor of mechanical engineering at De La Salle University, Manila. A multi-awarded scientist, science administrator, and teacher, he is an internationally acknowledged energy expert. His research interests include life-cycle assessment studies of energy and environment as well as bioenergy systems, and science and technology policy. He is currently a member of the Presidential Coordinating Council for Research and Development, the Congressional Commission on Science & Technology and Engineering (COMSTE), technical panels of the Commission on Higher Education (CHED) and the Department of Science and Technology (DOST), the Philippines. He was a former president of the National Research Council of the Philippines (NRCP) and the Philippine–American Academy of Science and Engineering (PAASE).

Hong-Guang Dong is an associate professor in the Department of Chemical Engineering at the Dalian University of Technology (DUT). His main research interest is chemical process synthesis and integration. He earned his BSE (1985), MS (1993), and PhD (2005) in chemical engineering from DUT, China. Dr. Dong joined DUT as a faculty member in 1993 and is currently deputy director of the Institute of Chemical Process Engineering. Dr. Dong is author or coauthor of more than 50 refereed scientific publications and two undergraduate textbooks. He is a member of the American Institute of Chemical Engineers (AIChE).

Mahmoud M. El-Halwagi is the McFerrin Professor at Artie McFerrin Department of Chemical Engineering, Texas A&M University. Dr. El-Halwagi's main areas of expertise are process integration, synthesis, design, operation, and optimization. Specifically, Dr. El-Halwagi's research focuses on sustainable design. In addition to the theoretical foundations he assisted in these areas, he has been active in education, technology transfer, and industrial applications. He has served as a consultant to a wide variety of chemical, petrochemical, petroleum, gas processing, pharmaceutical, and metal finishing industries. He is the coauthor of more than 200 papers and book chapters, the coeditor of five books, and the author of three textbooks. Dr. El-Halwagi is the recipient of several awards including the American Institute of Chemical Engineers Sustainable Engineering Forum (AIChE SEF) Research Excellence Award, the National Science Foundation's National Young Investigator Award, the Lockheed Martin Excellence in Engineering Teaching Award, the Celanese Excellence in Teaching Award, and the Fluor Distinguished Teaching Award. Dr. El-Halwagi received his PhD in chemical engineering from the University of California, Los Angeles, and his MS and BS from Cairo University.

Dominic Chwan Yee Foo is a professor of process design and integration at the University of Nottingham Malaysia Campus, and is the founding director for the Centre of Excellence for Green Technologies. He is a fellow of the Institution of Chemical Engineers UK (IChemE), a chartered engineer with the UK Engineering Council, a professional engineer with the Board of Engineer Malaysia (BEM), as well as the 2012/2013 and 2013/2014 sessions chairman for the Chemical Engineering Technical Division of the Institution of Engineers Malaysia (IEM). He served on International Scientific Committees for many important international conferences [International Congress of Chemical and Process Engineering/ Conference on Process Integration, Modelling and Optimization for Energy Saving and Pollution Reduction (CHISA/PRES), Foundations of Computer-Aided Process Design (FOCAPD), European Symposium on Computer-Aided Process Engineering (ESCAPE), Process Systems Engineering Conference (PSE)]. Professor Foo is the chief editor for *IEM Journal*, subject editor for *Trans IChemE Part B*, editorial board member for *Clean Technology and Environmental Policy*, and *Chemical Engineering Transactions*. He is the winner of the Innovator of the Year Award 2009 of IChemE, Young Engineer Award 2010 of IEM, Outstanding Young Malaysian Award 2012 of Junior Chamber International (JCI), as well as the SCEJ (Society of Chemical Engineers, Japan) Award for Outstanding Asian Researcher and Engineer.

Iskandar Halim is a scientist at ICES (Institute of Chemical Engineering Sciences), Singapore. His expertise includes chemical process sustainability and process modeling, simulation, and optimization. Prior to joining ICES, he worked at the Environmental Technology Institute (ETI) and Centre for Advanced Water Technology (CAWT) in a variety of water and wastewater treatment projects. He holds a BEng in chemical engineering from Monash University (Australia), an MSc in environmental technology from UMIST (UK) and an MEng and a PhD in chemical engineering from the National University of Singapore.

Tânia Pinto-Varela has been an assistant professor in the Engineering and Management Department in Instituto Superior Técnico (IST), University of Lisbon, since 2011. Previously, she was a researcher at the National Laboratory of Energy and Geology. She is currently the scientific coordinator of the Mobility Program in Engineering and Management, and integrates the Center for Management Studies board. She has a PhD in engineering and management from IST, and develops her research interests in process systems engineering, more specifically in: design, planning and scheduling, mixed integer optimization, process integration and design, and planning supply chain network, considering not only economic but also environmental aspects. She has authored several papers in refereed international journals and conferences and has been involved in several projects.

Munawar A. Shaik has about 15 years of experience in modeling and optimization. He has developed several novel mathematical programming models for solving a variety of problems in process systems engineering including: planning and scheduling of batch and continuous operations, process synthesis & design, and evolutionary computation. He obtained his PhD from IIT Bombay in 2005 and did his postdoctoral studies at Princeton University, New Jersey, USA, during 2005–2007, before joining IIT Delhi. He is currently serving as associate professor in the Department of Chemical Engineering at IIT Delhi.

Rajagopalan Srinivasan is a professor of chemical engineering at Indian Institute of Technology Gandhinagar. His research interests are in process systems engineering, inherent safety, sustainability, process design and operations, and supply chain management. He earned a PhD in chemical engineering from Purdue University,

Indiana, USA, and a BTech in chemical engineering from the Indian Institute of Technology Madras. His research has received best paper awards from various journals and conferences.

Raymond R. Tan is a professor of chemical engineering, University Fellow and current vice chancellor for Research and Innovation at De La Salle University, Manila, the Philippines. His main areas of research are process systems engineering and process integration, with applications to low-carbon energy systems and sustainable industrial operations. Professor Tan received his BS and MS in chemical engineering and PhD in mechanical engineering from De La Salle University, and is the author of more than 100 published and forthcoming articles in ISI-indexed journals in the fields of chemical, environmental and energy engineering. He has over 130 Scopus-indexed publications with an h-index of 27, is member of the editorial board of the journal *Clean Technologies and Environmental Policy* and editor of the book *Recent Advances in Sustainable Process Design and Optimization*. He is also the recipient of multiple scientific awards from the Commission on Higher Education (CHED), the National Academy of Science and Technology (NAST) and the National Research Council of the Philippines (NRCP).

Hella Tokos is a lecturer at the Department of Chemical and Process Engineering, University of Surrey. She was awarded a PhD in chemical engineering from University of Maribor, Slovenia, in 2009. Her doctoral thesis was in the area of process systems engineering with a focus on the industrial application of Computer-Aided Process Engineering (CAPE) Tools, addressing process integration problems and assessment of environmental impact. In 2010, Dr. Tokos joined the Department of Chemical and Biochemical Engineering at Zhejiang University, China, as a postdoctoral research associate. She worked primarily on problems in multilevel and bi-objective optimization of large-scale industrial water networks. In 2012, she was appointed as assistant professor at the Department of Chemical Engineering, School of Engineering, Nazarbayev University, Kazakhstan.

Aristotle T. Ubando is an assistant professor in the Mechanical Engineering Department of De La Salle University, Manila, the Philippines, with research interest in energy systems optimization. A recipient of the Fulbright Foundation Fellowship for Doctoral Dissertation Program, he is a PhD candidate in the mechanical engineering at De La Salle University with research attachment at the University of Arizona and Texas A&M University studying microalgal multifunctional bioenergy systems. He obtained his MS in mechanical engineering from University of the Philippines, Diliman, in 2009 and received his BS in mechanical engineering at De La Salle University, Manila, in 2000.

Miguel Vieira obtained his MS in chemical engineering from Instituto Superior Técnico, University of Lisbon. Recently, he received the Executive Master's degree in technology management enterprise at the same host institution. He is currently a PhD student in Leaders for Technical Industries—Engineering Design and

Advanced Manufacturing, a doctoral program that is part of the MIT Portugal initiative. Miguel's research interests cover the fields of planning and scheduling optimization in industrial processes.

Yi Zhang is a PhD candidate at School of Chemical Engineering in Dalian University of Technology (DUT). He earned his BS in chemical Engineering from DUT, China in 2013. His main research focuses on computer-aided molecular design and product design.

Xiong Zou is a PhD candidate at School of Chemical Engineering in Dalian University of Technology (DUT). He earned his BS in chemical Engineering from DUT, China in 2010. His main research focuses on heat-integrated distillation sequence synthesis and membrane-based separation system optimization.

Part I

Scheduling

Part I

1 Introduction to Batch Chemical Processes

*Thokozani Majozi, Esmael Reshid Seid,
and Jui-Yuan Lee*

CONTENTS

1.1 INTRODUCTION

By definition, a batch process is any process that is a consequence of discrete tasks that follow a predefined sequence from raw materials to final products. The prescribed sequence is known as a recipe. These processes are mostly suited to production of low volume, high value-added products, where equipment sharing is common. In most instances, equipment sharing is a result of similar recipes. The following sections are aimed at giving a brief account on the very key areas of research covered in this book. These pertain to scheduling in batch plants, which deals solely with handling of time, followed by energy and water minimization. The role of time is fundamental in the operation of batch facilities, since most batch unit operations are time-constrained. Only the essential elements of batch plants are captured with references, where necessary, to further sources of information for the benefit of the reader.

1.1.1 SCHEDULING OF BATCH PLANTS

Much research has been done on developing mathematical models to improve batch plant efficiency. The substantial advancement in modern computers allows the possibility of handling large and more complex problems by using optimization techniques. Excellent reviews of current scheduling techniques based on different time representations and associated challenges have been conducted (Floudas and Lin, 2004; Méndez et al., 2006; Shaik et al., 2006). In the reviews, with regard to time representation, the models are classified as slot based, event based, and precedence based (sequence based). In the slot-based models (Pinto and Grossmann, 1994; Lim and Karimi, 2003; Liu and Karimi, 2008), the time horizon is divided into "nonuniform unknown slots" and tasks start and finish in the same slot. On the other hand, slot models exist that use nonuniform unknown slots where tasks are allowed to continue to the next slots (Schilling and Pantelides, 1996; Karimi and McDonald 1997; Reddy et al., 2004; Sundaramoorthy and Karimi, 2005; Erdirik-Dogan and Grossmann, 2008; Susarla et al., 2010). The event-based models can also be categorized into those that use uniform unknown events, where the time associated with the events is common across all units (Maravelias and Grossmann, 2003; Castro et al., 2004), and those that use unit-specific events where the time associated with the events can be different across the units (Ierapetritou and Floudas, 1998; Majozi and Zhu, 2001; Shaik et al., 2006; Janak and Floudas, 2008; Shaik and Floudas, 2009; Li et al., 2010). The heterogeneous location of events across the units gives fewer event points as compared to both the global event-based and slot-based models. As a result, unit-specific event-based models are computationally superior. The sequence-based or precedence-based representation uses either direct precedence (Hui and Gupta, 2000; Liu and Karimi, 2007) or indirect precedence sequencing of pairs of tasks in units (Méndez and Cerdá, 2000; Méndez et al., 2001; Méndez and Cerdá, 2003; Ferrer-Nadal et al., 2008). The models do not require prepostulation of events and slots. Seid and Majozi (2012) presented a mixed integer linear programming (MILP) formulation based on the state sequence network and unit-specific time points, which can handle proper sequencing of tasks and fixed intermediate storage (FIS) policy. The model yields a reduced number of events or time points and gives a better performance index and CPU time in comparison with models in published literature.

1.1.2 ENERGY INTEGRATION IN BATCH PLANTS

Many heat integration techniques are applied to predefined schedules, which is inherently suboptimal, since the time dimension is suppressed as an optimization variable. Vaklieva-Bancheva et al. (1996) considered direct heat integration with the objective of minimizing total costs. The resulting overall formulation was an MILP problem, solved to global optimality, although only specific pairs of units were allowed to undergo heat integration. Uhlenbruck et al. (2000) improved OMNIUM, which is a tool developed for the heat exchanger network (HEN) synthesis by Hellwig and Thöne (1994). The improved OMNIUM tool increased the energy recovery by 20%. Bozan et al. (2001) developed a single step, interactive computer program (BatcHEN) used for the determination of campaigns, as well as the heat exchange areas of all possible heat exchangers in the campaigns. Production campaign in this context refers to the set of products that can be produced simultaneously. The interactive computer program is also capable of generating feasible HEN. This work addressed the limitation of the graph theory method for the determination of the campaign by Bancheva et al. (1996). Krummenacher and Favrat (2001) proposed a new systematic procedure, supported by graphics, which made it possible to determine the minimum number of heat storage units. Chew et al. (2005) applied cascade analysis, proposed by Kemp and Macdonald (1987), to reduce the utility requirement for the production of oleic acid from palm olein using immobilized lipase. The result obtained showed a savings of 71.4% and 62.5% for hot and cold utilities, respectively. Pires et al. (2005)

developed the BatchHeat software, which was aimed at highlighting the energy inefficiencies in the process, thereby enabling the scope for possible heat recovery to be established through direct heat exchange or storage.

Boer et al. (2006) evaluated the technical and economic feasibility of an industrial heat storage system for an existing production facility of organic surfactants. Fritzson and Berntsson (2006) applied process integration methods to investigate the potential of decreasing energy usage in the slaughtering and meat processing industry. The result obtained illustrates that 30% of the external heat demand and more than 10% of the shaft work used can be saved. Morrison et al. (2007) developed a user friendly software package known as optimal batch integration (OBI) wherein a pseudo-continuous behavior was assumed, thereby adapting the batch aspect of the processes to continuous heat integration techniques. Chen and Ciou (2008) formulated a method to design an optimization of indirect energy storage systems for batch process. Their work aimed at simultaneously solving the problem of indirect heat exchange network synthesis and its associated thermal storage policy for recirculated hot/cold heat storage medium. Most of the prior contributions solved this sequentially. Foo et al. (2008) extended the minimum units targeting and network evolution techniques that were developed for batch mass exchange network (MEN) into batch HEN. They applied the technique for energy integration of oleic acid production from palm olein using immobilized lipase. Halim and Srinivasan (2009) discussed a sequential method using direct heat integration. A number of optimal schedules with minimum makespan were found, and heat integration analysis was performed on each. The schedule with minimum utility requirement was chosen as the best. Later, Halim and Srinivasan (2011) extended their technique to synthesize a network that considers water reuse and heat integration simultaneously. One key feature of this method is its ability to find the heat integration and water reuse solution without sacrificing the quality of the scheduling solution.

Atkins et al. (2010) applied indirect heat integration using heat storage for a milk powder plant in New Zealand. The traditional composite curves were used to estimate the maximum heat recovery and to determine the optimal temperatures of the stratified tank. Tokos et al. (2010) applied a batch heat integration technique to a large beverage plant. The opportunities of heat integration between batch operations were analyzed by an MILP model, which was slightly modified by considering specific industrial circumstances. Muster-Slawitsch et al. (2011) came up with the Green Brewery concept to demonstrate the potential for reducing thermal energy consumption in breweries. Three detailed case studies were investigated. The "Green Brewery" concept has shown a savings potential of over 5000 t/y fossil CO_2 emissions from thermal energy supply for the three breweries that were closely considered. Becker et al. (2012) applied the time average energy integration approach to a real case study of a cheese factory with nonsimultaneous process operations. Their work addressed appropriate heat pump integration. A cost saving of more than 40% was reported.

For a more optimal solution, scheduling and heat integration may be combined into an overall problem. Papageorgiou et al. (1994) embedded a heat integration model within the scheduling formulation of Kondili et al. (1993). Opportunities for both direct and indirect heat integration were considered as well as possible heat losses from a heat storage tank. The operating policy, in terms of heat integrated or standalone operating modes, was predefined for tasks. Adonyi et al. (2003) used the "S-Graph" scheduling approach and incorporated one-to-one direct heat integration. Barbosa-Póvoa et al. (2001) presented a mathematical formulation for the detailed design of multipurpose batch process facilities with heat integration. Pinto et al. (2003) extended the work of Barbosa-Póvoa et al. (2001) by considering the economic savings in utility requirements, while considering cost of auxiliary structures, that is, heat transfer area and the design of the utility circuits and associated piping costs. Majozi (2006) presented a direct heat integration formulation based on the state sequence network of Majozi and Zhu (2001), which uses an unevenly discretized time horizon. The direct heat integration model developed by Majozi (2006) was later extended to incorporate heat storage for more flexible schedules and utility savings (Majozi, 2009). However, the storage size was considered a parameter in this formulation. This was addressed later by Stamp and Majozi (2011), where the storage size was treated as an optimization variable. Chen and Chang (2009) extended the work of Majozi (2006) to periodic scheduling, based on the resource task network (RTN) scheduling framework. The reader can get a more comprehensive and detailed review on energy recovery for batch processes in the paper by Fernández et al. (2012).

1.1.3 WASTEWATER MINIMIZATION IN BATCH PLANTS

In most instances, a vast amount of wastewater in batch plants is generated during cleaning of multipurpose equipment, which is mandatory when changing over from one product to another. Significant amounts of wastewater are also generated when water is used as a solvent that is dispensed at the end of the process. Indeed, water is also a consequence of floor washing operations. Tight environmental regulations and increased public awareness demand that batch plants consider rational use of water during their operation. Many researchers have developed methodologies for the efficient use of water through direct reuse, indirect reuse, and regeneration of wastewater. Direct reuse refers to the use of an outlet wastewater stream from one processing unit into another processing unit. Indirect reuse is when wastewater is temporarily stored in a storage vessel and later reused in a processing unit requiring water. Water can also be directly or indirectly recycled to the same unit if time and concentration constraints allow. Worthy of mention, however, is the fact that this option is seldom preferred in practice.

Based on the analogy between heat and mass integration, several methodologies for synthesizing water reuse networks in batch processes have also been developed. Gouws et al. (2010) presented a review of water minimization techniques by focusing on graphical and mathematical optimization approaches. The graphical techniques bear resemblance to the so-called pinch analysis, which was developed in the late 1970s for energy optimization through heat integration in continuous processes. The seminal work on pinch analysis application to batch water network was presented by Wang and Smith (1994). Foo et al. (2005) proposed a time-dependent water cascade analysis to obtain minimum water flows required in a process. While these graphical techniques are useful, they share a common drawback in that their application is limited to single contaminant cases. Moreover, they all cannot handle time as a variable. This implies that time has to be fixed *a priori*.

The mathematical optimization techniques, which are capable of solving multiple contaminant problems, can be differentiated into two groups, namely, those based on a predefined schedule and those in which time is treated as an optimization variable. Typical examples of the former include the work of Almató et al. (1997), which addressed the problem of water reuse through storage tank allocation based on a predefined optimal schedule and the work of Kim and Smith (2004), which proposed a more generalized method for optimal design of discontinuous water reuse network. Their approach assumed a fixed production schedule, and the direct reuse of water between operations within the same time interval was allowed without passing through storage tanks. On the other hand, methods in which time is treated as a variable include the work of Cheng and Chang (2007), which considered optimization of the batch production schedule, water reuse schedule, and wastewater treatment schedule in a single problem based on the discrete time scheduling framework. At the end of optimization, the production schedule, the number and sizes of buffer tanks, and the physical configuration of the pipeline network were obtained. Majozi and Gouws (2009) proposed a continuous-time scheduling framework to simultaneously optimize the schedule and water reuse while addressing both single and multiple contaminants. Adekola and Majozi (2011) extended the work of Majozi and Gouws (2009) by incorporating wastewater regeneration for further improvement of water utilization.

The foregoing review makes it apparent that wastewater minimization and heat integration in batch plants are generally addressed separately, except in a very few cases, like in the contributions of Halim and Srinivasan (2011) and Adekola et al. (2013). In the work of Halim and Srinivasan (2011), the overall problem is decomposed into three parts: scheduling, heat integration, and water reuse optimization, and solved sequentially. Batch scheduling is solved first to meet an economic objective function. Next, alternate schedules are generated through an integer cut procedure based on a stochastic search. For each resulting schedule, minimum energy and water reuse targets are established and networks are identified. Adekola et al. (2013) also addressed this problem by developing a model that simultaneously optimized energy, water, and production throughput. They demonstrated that the unified approach where all resources are optimized simultaneously gives a better economic performance compared to the common sequential techniques for wastewater and energy integration techniques developed for multipurpose batch plants. However, the limitation of this contribution is that it forces the heat-integrated units to start simultaneously, thereby restricting the flexibility of the schedule. Consequently, this is likely to yield suboptimal results.

REFERENCES

Adekola, O., Stamp, J., Majozi, T., Garg, A., Bandyopadhyay, S., 2013. Unified approach for the optimization of energy and water in multipurpose batch plants using a flexible scheduling framework. *Ind. Eng. Chem. Res.* 52, 8488–8506.

Adekola, O., Majozi, T., 2011. Wastewater minimization in multipurpose batch plants with a regeneration unit: Multiple contaminants. *Comput. Chem. Eng.* 35, 2824–2836.

Adonyi, R., Romero, J., Puigjaner, L., Friedler, F., 2003. Incorporating heat integration in batch process scheduling. *Appl. Therm. Eng.* 23, 1743–1762.

Almató, M., Sanmartí, E., Espuña, A., Puigjaner, L., 1997. Rationalizing the water use in the batch process industry. *Comput. Chem. Eng.* 21, S971–S976.

Atkins, M.J., Walmsley, M.R.W., Neale, J.R., 2010. The challenge of integrating non-continuous processes—Milk powder plant case study. *J. Clean. Prod.* 18, 927–934.

Bancheva, N., Ivanov, B., Shah, N., Pantelides, C.C., 1996. Heat exchanger network design for multipurpose batch plants. *Comput. Chem. Eng.* 20, 989–1001.

Barbosa-Póvoa, A.P.F.D., Pinto, T., Novais, A.Q., 2001. Optimal design of heat-integrated multipurpose batch facilities: A mixed-integer mathematical formulation. *Comput. Chem. Eng.* 25, 547–559.

Becker, H., Vuillermoz, A., Maréchal, F., 2012. Heat pump integration in a cheese factory. *Appl. Therm. Eng.* 43, 118–127.

Boer, R., Smeding, S.F., Bach, P.W., 2006. Heat storage systems for use in an industrial batch process (Results of) a case study. In: *10th International Conference on Thermal Energy Storage ECOSTOCK*. Stockton, USA.

Bozan, M., Borak, F., Or, I., 2001. A computerized and integrated approach for heat exchanger network design in multipurpose batch plants. *Chem. Eng. Process.* 40, 511–524.

Castro, P.M., Barbosa-Povóa, A.P., Matos, H.A., Novais, A.Q., 2004. Simple continuous-time formulation for short-term scheduling of batch and continuous processes. *Ind. Eng. Chem. Res.* 43, 105–118.

Chen, C.L., Chang, C.Y., 2009. A resource-task network approach for optimal short-term/periodic scheduling and heat integration in multipurpose batch plants. *Appl. Therm. Eng.* 29, 1195–1208.

Chen, C.L., Ciou, Y.J., 2008. Design and optimization of indirect energy storage systems for batch process plants. *Ind. Eng. Chem. Res.* 47, 4817–4829.

Cheng, K.F., Chang, C.T., 2007. Integrated water network designs for batch processes. *Ind. Eng. Chem. Res.* 46, 1241–1253.

Chew, Y.H., Lee, C.T., Foo, C.Y., 2005. Evaluating heat integration scheme for batch production of oleic acid. *Malaysian Science and Technology Congress (MSTC)*. Kuala Lumpur, Malaysia, pp. 18–20.

Erdirik-Dogan, M., Grossmann, I.E., 2008. Slot-based formulation for the short-term scheduling of multistage, multiproduct batch plants with sequence-dependent changeovers. *Ind. Eng. Chem. Res.* 47, 1159–1163.

Fernández, I., Renedo, C.J., Pérez, S.F., Ortiz, A., Mañana, M., 2012. A review: Energy recovery in batch processes. *Renewable Sustainable Energy Rev.* 16, 2260–2277.

Ferrer-Nadal, S., Capón-Garćia, E., Méndez, C.A., Puigjaner, L., 2008. Material transfer operations in batch scheduling. A critical modeling issue. *Ind. Eng. Chem. Res.* 47, 7721–7732.

Floudas, C.A., Lin, X., 2004. Continuous-time versus discrete-time approaches for scheduling of chemical processes: A review. *Comput. Chem. Eng.* 28, 2109–2129.

Foo, D.C.Y., Chew, Y.H., Lee, C.T., 2008. Minimum units targeting and network evolution for batch heat exchanger network. *Appl. Therm. Eng.* 28, 2089–2099.

Foo, D.C.Y., Manan, Z.A., Tan, Y.L., 2005. Synthesis of maximum water recovery network for batch process systems. *J. Clean. Prod.* 13, 1381–1394.

Fritzson, A., Berntsson, T., 2006. Efficient energy use in a slaughter and meat processing plant—Opportunities for process integration. *J. Food Eng.* 76, 594–604.

Gouws, J.F., Majozi, T., Foo., D.C.Y., Chen., C.L., Lee, J Y., 2010. Water minimization techniques for batch processes. *Ind. Eng. Chem. Res.* 48 (19), 8877–8893.

Halim, I., Srinivasan, R., 2009. Sequential methodology for scheduling of heat-integrated batch plants. *Ind. Eng. Chem. Res.* 48, 8551–8565.

Halim, I., Srinivasan, R., 2011. Sequential methodology for integrated optimization of energy and water use during batch process scheduling, *Comput. Chem. Eng.* 35, 1575–1597.

Hellwig, T., Thöne, E., 1994. Omnium: ein verfahren zur optimierung der abwarmenutzung. *BWK (Brennstoff, Warme, Kraft)*. 46, 393–397 [in German].

Hui, C.W., Gupta, A., 2000. A novel MILP formulation for short-term scheduling of multi-stage multi-product batch plants. *Comput. Chem. Eng*. 24, 2705–2717.

Ierapetritou, M.G., Floudas, C.A., 1998. Effective continuous-time formulation for short-term scheduling: 1. Multipurpose batch processes. *Ind. Eng. Chem. Res*. 37, 4341–4359.

Janak, S.L., Floudas, C.A., 2008. Improving unit-specific event based continuous time approaches for batch processes: Integrality gap and task splitting. *Comput. Chem. Eng*. 32, 913–955.

Karimi, I.A., McDonald, C.M., 1997. Planning and scheduling of parallel semicontinuous processes. II. Short-term scheduling. *Ind. Eng. Chem. Res*. 36, 2701–2714.

Kemp, I.C., Macdonald, E.K., 1987. Energy and process integration in continuous and batch processes. Innovation in process energy utilization. *IChemE Symp Series*. 105, 185–200.

Kim, J.K., Smith, R., 2004. Automated design of discontinuous water systems. *Process. Saf. Environ*. 82, 238–248.

Kondili, E., Pantelides, C.C., Sargent, R.W.H., 1993. A general algorithm for short-term scheduling of batch operations. I. MILP formulation. *Comput. Chem. Eng*. 17, 211–227.

Krummenacher, P., Favrat, D., 2001. Indirect and mixed direct–indirect heat integration of batch processes based on Pinch Analysis. *Int. J. Appl. Thermodyn*. 4, 135–143.

Li, J., Susarla, N., Karimi, I.A., Shaik, M., Floudas, C.A., 2010. An analysis of some unit-specific event-based models for the short-term scheduling of non-continuous processes. *Ind. Eng. Chem. Res*. 49, 633–647.

Lim, M.F., Karimi, I.A., 2003. Resource-constrained scheduling of parallel production lines using asynchronous slots. *Ind. Eng. Chem. Res*. 42, 6832–6842.

Liu, Y., Karimi, I.A., 2007. Scheduling multistage, multiproduct batch plants with non identical parallel units and unlimited intermediate storage. *Chem. Eng. Sci*. 62, 1549–1566.

Liu, Y., Karimi, I.A., 2008. Scheduling multistage batch plants with parallel units and no interstage storage. *Comput. Chem. Eng*. 32, 671–693.

Majozi, T., 2006. Heat integration of multipurpose batch plants using a continuous-time framework. *Appl. Therm. Eng*. 26, 1369–1377.

Majozi, T., 2009. Minimization of energy use in multipurpose batch plants using heat storage: An aspect of cleaner production. *J. Clean. Prod*. 17, 945–950.

Majozi, T., Gouws, J.F., 2009. A mathematical optimisation approach for wastewater minimization in multipurpose batch plants: Multiple contaminants. *Comp. Chem. Eng*. 33, 1826–1840.

Majozi, T., Zhu, X.X., 2001. A novel continuous-time MILP Formulation for multipurpose batch plants. *Ind. Eng. Chem. Res*. 40, 5935–5949.

Maravelias, C.T., Grossmann, I.E., 2003. New general continuous-time state-task network formulation for short-term scheduling of multipurpose batch plants. *Ind. Eng. Chem. Res*. 42, 3056–3074.

Méndez, C.A., Cerdá, J., 2000. Optimal scheduling of a resource-constrained multiproduct batch plant supplying intermediates to nearby end product facilities. *Comput. Chem. Eng*. 24, 369–376.

Méndez, C.A., Cerdá, J., 2003. An MILP continuous-time framework for short-term scheduling of multipurpose batch processes under different operation strategies. *Opt. Eng*. 4, 7–22.

Méndez, C.A., Cerdá, J., Grossmann, I.E., Harjunkoski, I., Fahl, M., 2006. State-of-the-art review of optimization methods for short-term scheduling of batch processes. *Comput. Chem. Eng*. 30, 913–946.

Méndez, C.A., Henning, G.P., Cerdá, J., 2001. An MILP continuous-time approach to short-term scheduling of resource-constrained multistage flowshop batch facilities. *Comput. Chem. Eng*. 25, 701–711.

Morrison, A.S., Walmsley, M.R.W., Neale, J.R., Burrell, C.P., Kamp, P.J.J., 2007. Non-continuous and variable rate processes: Optimisation for energy use. *Asia-Pac. J. Chem. Eng.* 5, 380–387.

Muster-Slawitsch, B., Weiss, W., Schnitzer, H., Brunner, C., 2011. The green brewery concept Energy efficiency and the use of renewable energy sources in breweries. *Appl. Therm. Eng.* 31, 2123–2134.

Papageorgiou, L.G., Shah, N., Pantelides, C.C., 1994. Optimal scheduling of heat-integrated multipurpose plants. *Ind. Eng. Chem. Res.* 33, 3168–3186.

Pinto, J.M., Grossmann, I.E., 1994. Optimal cyclic scheduling of multistage continuous multiproduct plants. *Comput. Chem. Eng.* 1994, 18, 797–816.

Pinto, T., Novais, A.Q., Barbosa-Póvoa, A.P.F.D., 2003. Optimal design of heat-integrated multipurpose batch facilities with economic savings in utilities: A mixed integer mathematical formulation. *Ann. Operat. Res.* 120, 201–30.

Pires, A.C., Fernandes, C.M., Nunes, C.P., 2005. An energy integration tool for batch process, sustainable development of energy, water and environment systems. In: *Proceedings of the 3rd Dubrovnik Conference.* Dubrovnik, Croatia, pp. 5–10.

Reddy, P.C.P., Karimi, I.A., Srinivasan R., 2004. A new continuous-time formulation for scheduling crude oil operations. *Chem. Eng. Sci.* 59, 1325–1341.

Schilling, G., Pantelides, C., 1996. A simple continuous-time process scheduling formulation and a novel solution algorithm. *Comput. Chem. Eng.* 20, 1221–1226.

Seid, R., Majozi, T., 2012. A robust mathematical formulation for multipurpose batch plants. *Chem. Eng. Sci.* 68, 36–53.

Shaik, M., Floudas, C., 2009. Novel unified modeling approach for short term scheduling. *Ind. Eng. Chem. Res.* 48, 2947–2964.

Shaik, M.A., Janak, S.L., Floudas, C.A., 2006. Continuous-time models for short-term scheduling of multipurpose batch plants: A comparative study. *Ind. Eng. Chem. Res.* 45, 6190–4209.

Stamp, J., Majozi, T., 2011. Optimal heat storage design for heat integrated multipurpose batch plants. *Energy.* 36(8), 1–13.

Sundaramoorthy, A., Karimi I.A., 2005. A simpler better slot-based continuous-time formulation for short-term scheduling in multipurpose batch plants. *Chem. Eng. Sci.* 60, 2679–2702.

Susarla, N., Li, J., Karimi, I.A., 2010. A novel approach to scheduling of multipurpose batch plants using unit slots. *AICHE J.* 56, 1859–1879.

Tokos, H., Pintarič, Z.N., Glavič, P., 2010. Energy saving opportunities in heat integrated plant retrofit. *Appl. Therm. Eng.* 30, 36–44.

Uhlenbruck, S., Vogel, R., Lucas, K., 2000. Heat integration of batch processes. *Chem. Eng. Technol.* 23, 226–229.

Vaklieva-Bancheva, N., Ivanov, B.B., Shah, N., Pantelides, C.C., 1996. Heat exchanger network design for multipurpose batch plants. *Comput. Chem. Eng.* 20, 989–1001.

Wang, Y. P., Smith, R., 1994. Wastewater minimization. *Chem. Eng. Sci.* 49(7), 981–1006.

2 Effective Technique for Scheduling in Multipurpose Batch Plants

Esmael Reshid Seid, Jui-Yuan Lee, and Thokozani Majozi

CONTENTS

Several scheduling techniques exist in literature based on either even or uneven discretizations of the time horizon. The latter are commonly referred to as continuous time representation due to their exactness in handling time. Over the years, most work in the area of batch process scheduling has adopted the continuous time representation over even discretization of time. This choice is attributable to the fact that methods based on continuous time representation tend to result in much fewer binary variables compared to even discretization of time. Continuous time formulations are also divided into unit specific and global time point representations. The models based on unit-specific time points have shown better solution efficiency by reducing the number of time points and problem size. In this chapter, novel scheduling techniques based on unit-specific time point continuous time representation are presented. The proposed models allow nonsimultaneous material transfer into a unit. Nonsimultaneous transfer is encountered when a task requires more than one intermediate state. In this situation, it is possible for one state to be transferred and stored in a unit that is dedicated to processing it and wait for a while for the other intermediates to come together to commence the task. This approach gives a better schedule as compared to most published formulations.

In this chapter, novel scheduling techniques that use unit-specific time points are presented. The proposed models can handle the different storage operational policies that do not violate real-time storage constraints. Paramount among these is the finite intermediate storage (FIS) operational policy, which has been inadvertently violated in most published models. The developed models are compared to the different models in the literature based on bench mark examples. Through literature examples, it is demonstrated that the proposed formulations are simpler, tighter, and give less number of time points as compared to other formulations in the literature. Moreover, the resulting formulations give better optimal objective value.

2.1   MOTIVATING EXAMPLE

A case study taken from Shaik and Floudas (2009) discussed as Example 3 and a case study taken from Susarla et al. (2010) discussed as Example 5 are often used in literature to check the efficiency of models in terms of optimal objective value and CPU time required to get optimal objective value, and are presented as motivating examples. The plant accommodates many common features of multipurpose batch plants such as unit performing multiple tasks, multiple units suitable for a task, and dedicated units for specific tasks. The state task network (STN) representation for this motive example is depicted in Figure 2.1 and the corresponding data are shown in Table 2.1.

The processing time of a task i on unit j is assumed to be linearly dependent, $\alpha_i + \beta_i B$ of its batch size B. Where α_i is constant term of processing time of task i and

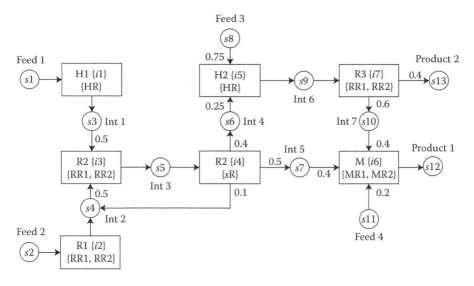

FIGURE 2.1   STN representation of motivating example.

β_i is the coefficient of the variable term of processing time task i. The batch size B of a task should be in the range of V_i^L and V_i^U representing the minimum and maximum batch size, respectively. The initial stock levels for all intermediates are assumed to be zero except for state $s6$ and $s7$ with an initial stock of 50 (mu). The stock is available as required for states $s1$, $s2$, $s3$, and $s11$. The price of product 1 and 2 is \$5/mu. This motivating example is solved for unlimited intermediate storage (UIS) and FIS operational policy. In the case of FIS, the storage capacity for states $s3$, $s4$, $s5$, $s6$, $s7$, $s9$, and $s10$ are 100, 100, 300, 150, 150, 150, and 150 mu, respectively. The storage capacity for raw material and product states $s1$, $s2$, $s8$, $s11$, $s12$, and $s13$ are assumed unlimited.

TABLE 2.1

Data Coefficient of Processing Time of Tasks, Limits on Batch of Units for the Motivating Example

Task	Label i	Unit	Label j	α_i	β_i	$V_i^L - V_i^U$
Heating 1	($i1$)	Heater	HR	0.667	0.00667	0–100
Heating 2	($i5$)	Heater	HR	1	0.01	0–100
Reaction 1	($i2$)	Reactor 1	RR-1	1.333	0.01333	0–100
	($i2$)	Reactor 2	RR-2	1.333	0.00889	0–150
Reaction 2	($i3$)	Reactor 1	RR-1	0.667	0.00667	0–100
	($i3$)	Reactor 2	RR-2	0.667	0.00445	0–150
Reaction 3	($i7$)	Reactor 1	RR-1	1.333	0.0133	0–100
	($i7$)	Reactor 2	RR-2	1.333	0.00889	0–150
Separation	($i4$)	Separator	SR	2	0.00667	0–300
Mixing	($i6$)	Mixer 1	MR-1	1.333	0.00667	20–200
	($i6$)	Mixer 2	MR-2	1.333	0.0067	20–200

This motivating example is solved by Shaik and Floudas (2009) for UIS policy using different literature models to maximize the profit using a time horizon (H) of 10 h. These models are unit-specific models of Ierapetritou and Floudas (1998a, b) (I&F), Shaik and Floudas (2008) (S&F), and Shaik and Floudas (2009) (S&F), the Global-event-based models of Castro et al. (2004) (CBMN) and Maravelias and Grossmann (2003) (M&G), and the slot-based model of Sundaramoorthy and Karimi (2005) (S&K). Table 2.2 provides a comparison of different models in terms of the problem statistics such as the number of binary and continuous variables, number of constraints, CPU time taken to solve to the specified integrality gap, the number of nodes taken to reach the optimal solution, the relaxed mixed integer linear programming (RMILP) objective value, and the mixed integer linear programming (MILP) objective value for 10 h time horizon. For fair comparison, n event points for the model of S&K (2005) represents $n - 1$ slots when compared to the global and unit-specific event-based models. The parameters (Δt) in CBMN and (Δn) in S&F (2009) represent the maximum number of events over which a task can occur.

The slot-based model of S&K (2005) and the global-event-based models of MG and CBMN required 10 event points and took excessive computational time giving a suboptimal objective value of $2249.6, $2108.7, and $2283.8, respectively. This suboptimality is due to single grid time representation, which forces all tasks starting at the same slot to have an identical starting time. The unit-specific models by I&F and S&F (2008) required seven event points, giving a suboptimal objective value of $2305.6 when compared to the global optimal solution $2345.3 obtained by the model of S&F (2009). The better optimal objective value by the model of S&F (2009) is due to the fact that the model allows the task to span over multiple event points; for this case it needs the parameter $(\Delta n = 1)$ to get the global optimal solution. The material processed by each task, the type of task each unit is conducting, and the starting and finishing time for each task for the global optimal solution obtained by the model of S&F (2009) are shown in Figure 2.2. The storage profile for the global optimal solution is also shown in Figure 2.3.

Recently, this example is solved for FIS operational philosophy by Susarla et al. (2010) for the objective of maximizing profit for a time horizon of 10 h by using different literature models. These models are the unit-specific slot-based models of Susarla et al. (2010) (SLK2&SLK1), S&F (2009), S&K, and M&G. Table 2.3 provides a comparison of different models statistics for a 10 h time horizon.

The models SLK2, SLK1, and S&F (2009) $(\Delta n = 1)$ required nine event points to get the optimal objective value of $2337.36. The models S&K and M&G needed 10 event points to give an inferior optimal objective value of $2260.9 and $2137.1, respectively. The better objective value by SLK2, SLK1, and S&F (2009) $(\Delta n = 1)$ is attributed to fact that the models are based on multigrid time representation and allow a task to span over multiple event points.

The optimal objective value obtained by the UIS, which is $2345.3, is better than the optimal objective value $2337.36 obtained by FIS operational philosophy. The number of even points required by the model of S&F (2009) to get the global optimal solution for the UIS operational policy is seven which is two event points lower than nine event points required by models SLK2, SLK1, and S&F (2009) for FIS operational policy. From Figure 2.3, the storage profile obtained for UIS indicates that the

TABLE 2.2

Computational Results for Motivating Example under Maximization of Profit with UIS

Model	Events	CPU Time (s)	Nodes	RMILP ($)	MILP ($)	BV	CV	Constraints	Nonzeros	Relative Gap (%)
S&K	10	>71,000	11,152,876	3474	2249.6	153	876	896	3153	19.6
M&G	10	>71,000	7,020,465	3474	2108.7	198	1056	2459	9492	37.36
CBMN ($\Delta t = 3$)	10	>71,001	46,171,236	3873.1	2283.8	264	467	647	3092	3.5
I&F	7	83.87	127,697	3465.6	2305.6	52	225	451	1403	—
S&F (2008)	7	110.68	201,885	3369.7	2305.6	52	267	465	1633	—
S&F (2009)										
($\Delta n = 0$)	7	222.74	344,568	3369.7	2305.6	52	302	523	1450	—
($\Delta n = 1$)	7	4319.08	2,902,108	3369.7	2345.3	100	350	1147	3535	—

BV = binary variables.
CV = continuous variables.

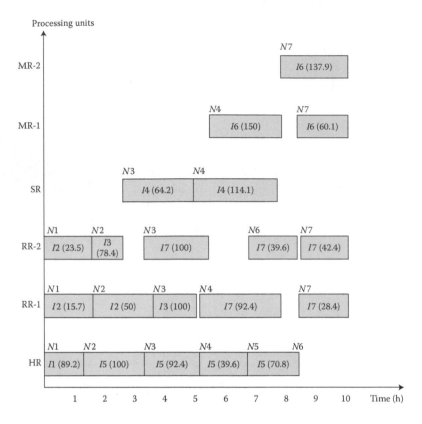

FIGURE 2.2 Gantt chart for the time horizon of 10 h. (Adapted from Shaik, M., Floudas, C., 2009. *Industrial Engineering Chemistry Research*, 48, 2947–2964.)

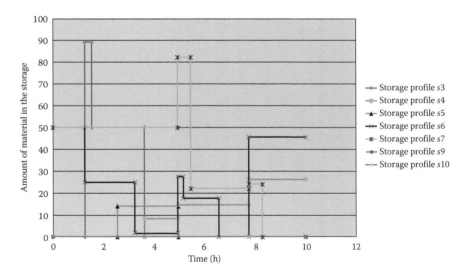

FIGURE 2.3 Storage profile for the time horizon of 10 h.

TABLE 2.3

Computational Results for Motivating Example under Maximization of Profit with FIS

Model	p	CPU Time (s)	Nodes	RMILP	MILP	BV	CV	Constraints	Nonzeros	Relative Gap (%)
SLK2	9	10,000	872,204	3618.6	2337.36	136	853	1428	4970	5.07
SLK1	9	10,000	711,493	3618.6	2337.36	136	799	1380	4886	11.6
SF										
($\Delta n = 0$)	8	115	1,160,988	3618.6	2292.5	88	382	884	2847	–
($\Delta n = 1$)	8	10,000	2,256,858	3618.6	2337.36	165	459	1658	6187	1.68
SK	10	10,000	2,666,604	3473.9	2260.9	153	874	1121	3380	10.7
MG	10	10,000	1,415,516	3473.9	2137.1	220	1151	2691	10,710	23.9

maximum storage profiles for the time horizon of 10 h for states $s3$, $s4$, $s5$, $s6$, $s7$, $s9$, and $s10$ are 89.2, 17.64, 14.2, 50, 82.12, 0, and 42.75 mu, respectively, which is less than the capacity of the storage for each state given for FIS operational policy. This indicates that the optimal objective value obtained for FIS is suboptimal and the developed formulations to handle FIS operational policy are unable to find the global optimal solutions, besides the models need increased event points that lead to increased computational time. The developed mathematical formulation in this chapter addresses the suboptimality and computational efficiency by regress representation of the scheduling constraints.

2.2   STATE SEQUENCE NETWORK

Majozi and Zhu (2001) introduced a new concept called the state sequence network (SSN) representation for scheduling of batch plants. The SSN was a graphical network representation of all the states that exist in batch plants and the network was formulated based on the production recipe. A state changes from one state to another state when it undergoes a unit operation such as mixing, separation, or reaction. The building blocks of the SSN are shown in Figure 2.4. From these building blocks it is easy to construct an SSN for any process recipe. In the SSN representation, only states are considered while tasks are implicitly incorporated.

For better understanding, a common literature problem is considered. Figure 2.5 shows the flowsheet of this batch plant that involves a mixer reactor and purificator. The state task and the SSN representation of this batch plant are depicted in Figure 2.6.

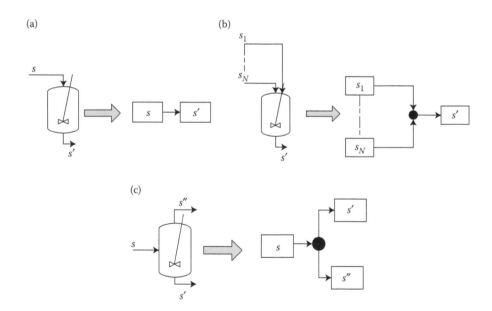

FIGURE 2.4   Building blocks of SSN. (a) Simple unit operation, (b) unit operation with mixing/reaction, and (c) unit operation with splitting.

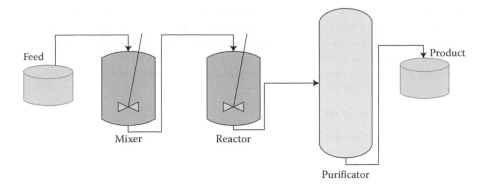

FIGURE 2.5 Flowsheet for a simple batch plant.

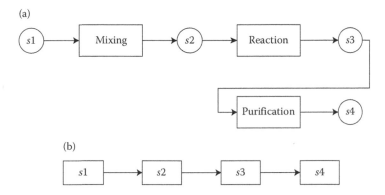

FIGURE 2.6 (a) STN representation, (b) SSN representation of the flowsheet given in Figure 2.5.

2.3 PROBLEM STATEMENT

In scheduling of multipurpose batch plants, the following are given: (i) the production recipe that indicates the sequence of unit operations in which raw materials are converted into products, (ii) the capacity of units and the type of tasks each unit can perform, (iii) the maximum storage capacity for each material, and (iv) the time horizon of interest.

Using the given data, it is required to determine (i) the maximum achievable profit of the plant, (ii) the minimum makespan if throughput is given, and (iii) a production schedule related to the optimal resource utilization.

2.4 MATHEMATICAL MODEL USING SSN

It is important to explain the effective state in the SSN representation because it renders the opportunity to reduce the number of binary variables. If a process requires multiple raw materials to make a particular product, then it is a fact that if one of the raw materials is fed then all the other feed materials also exist to make the product.

By noting this, it is easy to see that only one of the states needs to be defined as an effective state. The choice of effective state is not unique, however, once the choice of effective state has been made, it should remain consistent throughout the formulation. Effective states are considered in defining the binary variables. The mathematical model presented in this chapter entails the following constraints.

2.4.1 ALLOCATION CONSTRAINTS

Constraint (2.1) implies that at time point p only one task is allowed to be performed in unit j.

$$\sum_{s_{in,j} \in S^*_{in,j}} y(s_{in,j}, p) \leq 1, \quad \forall j \in J, p \in P \tag{2.1}$$

2.4.2 CAPACITY CONSTRAINTS

Constraint (2.2) implies that the total amount of all the states consumed at time point p is limited by the capacity of the unit which consumes the states. $V^L_{s_{in,j}}$ and $V^U_{s_{in,j}}$ represent lower and upper bounds in the capacity of a given unit that processes the effective state $s_{in,j}$.

$$V^L_{s_{in,j}} y(s_{in,j}, p) \leq mu(s_{in,j}, p) \leq V^U_{s_{in,j}} y(s_{in,j}, p), \quad \forall p \in P, \ j \in J, \ s_{in,j} \in S_{in,J} \tag{2.2}$$

2.4.3 MATERIAL BALANCE FOR STORAGE

Constraint (2.3) states that the amount of material stored at each time point p is the amount stored at the previous time point adjusted to some amount resulting from the difference between state s produced by tasks at the previous time point $(p - 1)$ and used by tasks at the current time point p. This constraint is used for a state other than a product, since the latter is not consumed, but only produced within the process.

$$q_s(s, p) = q_s(s, p - 1) - \sum_{s_{in,j} \in S^{sc}_{in,J}} \rho^{sc}_{s_{in,j}} mu(s_{in,j}, p) + \sum_{s_{in,j} \in S^{sp}_{in,J}} \rho^{sp}_{s_{in,j}} mu(s_{in,j}, p - 1),$$

$$\forall p \in P, s \in S \tag{2.3}$$

Constraint (2.4) states that the amount of product stored at time point p is the amount stored at the previous time point and the amount of product produced at the time point p.

$$q_s(s^p, p) = q_s(s^p, p - 1) + \sum_{s_{in,j} \in s^{sp}_{in,J}} \rho^{sp}_{s_{in,j}} mu(s_{in,j}, p), \quad \forall p \in P, s^p \in S^p \tag{2.4}$$

2.4.4 DURATION CONSTRAINTS (BATCH TIME AS A FUNCTION OF BATCH SIZE)

Constraint (2.5) describes the duration constraint modeled as a function of batch size, where the processing time is a linear function of the batch size. For zero-wait (ZW), only the equality sign is used.

$$t_p(s_{in,j}, p) \geq t_u(s_{in,j}, p) + \tau(s_{in,j})\, y\, (s_{in,j}, p) + \beta(s_{in,j}) mu(s_{in,j}, p),$$

$$\forall\ j \in J,\ p \in P,\ s_{in,j} \in S_{in,J} \tag{2.5}$$

2.4.5 SEQUENCE CONSTRAINTS

The two subsections address the proper allocation of tasks in a given unit that ensures the starting time of a new task to be later than the finishing time of the previous task.

2.4.5.1 Same Task in Same Unit

Constraint (2.6) states that a state can only be used in a unit, at any time point, after all the previous tasks are complete. In essence, this implies that a unit must be available before it can be used. It is worth noting that all the tasks referred to in constraint (2.6) pertains to the same state, that is, same task for different batches in the same unit.

$$t_u(s_{in,j}, p) \geq t_p(s_{in,j}, p - 1), \quad \forall\ j \in J,\ p \in P,\ s_{in,j} \in S^*_{in,j} \tag{2.6}$$

2.4.5.2 Different Tasks in Same Unit

Constraint (2.7) states that a task can start in the unit after the completion of all the previous tasks that can be performed in the unit. In the context of constraint (2.7), tasks pertain to different states, hence different tasks in the same unit.

$$t_u\left(s_{in,j}, p\right) \geq t_p\left(s'_{in,j}, p\right), \quad \forall\ j \in J,\ p \in P,\ s_{in,j} \neq s'_{in,j},\ s_{in,j},\ s'_{in,j} \in S^*_{in,j} \tag{2.7}$$

If the state is consumed and produced in the same unit, where the produced state is unstable, then in addition to constraint (2.7), constraints (2.8) and (2.9) are used.

$$t_p\left(s^{usp}_{in,j}, p - 1\right) \geq t_u\left(s^{usc}_{in,j}, p\right) - H\left(1 - y\left(s^{usp}_{in,j}, p - 1\right)\right),$$

$$\forall\ j \in J,\ p \in P,\ s^{usc}_{in,j} \in S^{usc}_{in,j},\ s^{usp}_{in,j} \in S^{usp}_{in,j} \tag{2.8}$$

2.4.6 SEQUENCE CONSTRAINTS DIFFERENT TASKS IN DIFFERENT UNITS

These constraints state that for different tasks that consume and produce the same state, the starting time of the consuming task at time point p must be later than the finishing time of any task at the previous time point $p - 1$ provided that the state is used.

2.4.6.1 If an Intermediate State S Is Produced from One Unit

Constraints (2.9) and (2.10) work together in the following manner:

$$\rho(s_{in,j}^{sp})mu(s_{in,j}, p-1) \leq q_s(s,p) + V_j^U t(j,p), \quad \forall\, j \in J,\ p \in P,\ s_{in,j} \in S_{in,J}^{sp} \qquad (2.9)$$

$$tu(s_{in,j'}, p) \geq tp(s_{in,j}, p-1) - H\left(\left(2 - y\ (s_{in,j}, p-1) - t\ (j,p)\right)\right),$$
$$\forall\, j \in J,\ p \in P,\ s_{in,j} \in S_{in,J}^{sp},\ s_{in,j'} \in S_{in,J}^{sc} \qquad (2.10)$$

Constraint (2.9) states that if the state s is produced from unit j at time point $p - 1$, but is not consumed at time point p by another unit j', that is, $t(j,p) = 0$, then the amount produced cannot exceed the allowed storage, that is, $q_s(s,p)$. On the other hand, if state s produced from unit j at time point $p - 1$ is used by another unit j' then the amount of state s stored at time point p, that is, $q_s(s,p)$ is less than the amount of state s produced at time point $p - 1$. The outcome is that the binary variable $t(j,p)$ becomes 1 in order for constraint (2.9) to hold. If the unit performs tasks such as separation, distillation, and other tasks that produce more than one intermediate at time point p, then the binary variable $t(j,p)$ becomes $t(j,s,p)$. This allows us at the same time point for constraint (2.10) to be relaxed for the unit that is not using the state produced by unit j at time point p. Simultaneously, for the other unit that uses the state produced by unit j at time point p the sequence constraint (2.10) holds.

Constraint (2.10) states that the starting time of a task-consuming state s at time point p must be later than the finishing time of a task that produces state s at the previous time point $p - 1$, provided that state s is used. Otherwise, the sequence constraint is relaxed.

2.4.6.2 If an Intermediate State Is Produced from More than One Unit

Constraint (2.11) states that the amount of state s used at time point p can either come from storage, or from other units that produce the same state depending on the binary variable $t(j,p)$. If the binary variable $t(j,p)$ is 0, which means that state s produced from unit j at time point $p - 1$ is not used at time point p, then constraint (2.10) is relaxed. If $t(j,p)$ is 1, state s produced from unit j at time point $p - 1$ is used, as a result constraint (2.10) holds. Although constraint (2.11) is nonlinear, it can be linearized exactly using Glover transformation developed by Glover (1975). It is highly imperative to realize that constraint (2.10) plays a pivotal role in both instances when a state is produced from one unit and when a state is produced from many units.

$$\sum_{s_{in,j} \in S_{in,J}^{sc}} \rho_{s_{in,j}}^{sc} mu(s_{in,j}, p) \leq qs(s, p-1)$$

$$+ \sum_{s_{in,j} \in S_{in,J}^{sp}} \rho_{s_{in,j}}^{sp} mu(s_{in,j}, p-1)\, t(j,p) \quad \forall\, j \in J,\ p \in P \qquad (2.11)$$

Constraint (2.12) states that a consuming task can start after the completion of the previous task. Constraint (2.12) takes care of the proper sequencing time when a unit uses material that is previously stored is when the producing task is active at time point $(p - 2)$, and later produces and transfers the material to the storage at time point $(p - 1)$. This available material in the storage at time point $(p - 1)$ is then used by the consuming task in the next time point; this necessitates the starting time of the consuming task to be later than the finishing time of the producing task at time point $p - 2$. Constraint (2.10) is used for proper sequencing if the amount of material used by the consuming task at time point p is from that which is currently produced by the producing units. Consequently, constraint (2.12) together with constraint (2.10) results in a feasible sequencing time when the consuming task uses material which is previously stored or/and material which is currently produced by the producing units.

$$t_u(s_{in,j'}, p) \geq t_p(s_{in,j}, p - 2) - H(1 - y(s_{in,j}, p - 2)),$$

$$\forall\, j \in J,\ p \in P,\ s_{in,j} \in S_{in,J}^{sp},\ s_{in,j'} \in S_{in,j'}^{sc} \tag{2.12}$$

2.4.7 SEQUENCE CONSTRAINT FOR FIS POLICY

As aforementioned, suboptimality in previous formulations pertains to modeling for FIS. In most formulations, that is, Majozi and Zhu (2001), Ierapetritou and Floudas (1998a,b), and Lin and Floudas (2001), this is overlooked and results in UIS behavior in the final schedule. The following constraints are aimed at addressing this drawback. A new binary variable $x(s,p)$ is introduced as shown in constraints (2.13) and (2.14). This binary variable indicates the availability ($x[s,p] = 1$) and absence of storage ($x[s,p] = 0$). According to constraint (2.13), any state s can only be stored if the capacity of available storage will not be exceeded. Otherwise, state s will either be produced and consumed immediately or not produced at all. Constraint (2.14) enforces this condition.

$$\sum_{s_{in,j} \in S_{in,J}^{sp}} \rho_{s_{in,j}}^{ps} mu(s_{in,j}, p - 1) + qs(s, p - 1) \leq QS^U$$

$$+ \sum_{j \in J_s} V_j^U \left(1 - x(s,p)\right)\quad \forall\, j \in J,\ p \in P,\ s \in S \tag{2.13}$$

According to constraint (2.14), the starting time of a task that consumes state s at time point p must be equal to the finishing time of a task that produces state s at time point $p - 1$, if both consuming and producing tasks are active at time point p and time point $p - 1$, respectively, and if there is no storage to store the amount of state s produced at time point p. In a case when storage is available to store a state s at time point p, then the starting time of the consuming task at time point p is not necessarily equal to the finishing time of a task producing state s at time point $p - 1$. Constraints

(2.13) and (2.14) reduce the number of time points and give a better objective value as can be seen in the case study.

$$tp(s_{in,j'}, p-1) \leq tp(s_{in,j}, p-1) + H\left(2 - y\left(s_{in,j'}, p\right) - y(s_{in,j}, p-1)\right) + H\left(x(s,p)\right)$$
$$\forall\, j \in J,\ p \in P,\ s_{in,j} \in S_{in,J}^{sp},\ s_{in,j'} \in S_{in,J}^{sc}$$

(2.14)

2.4.8 Storage Constraints

Constraint (2.15) indicates that the amount of state s stored at any time point must not exceed the maximum capacity of the storage. The state s that is produced at time point $p-1$ can be stored for a while in a unit that is producing it in the next time point until it is used, if the unit is not performing tasks.

$$q_s(s,p) \leq QS^U + \sum_{s_{in,j}\, \in\, s_{in,J}^{sp}} u(s_{in,j}, p) \quad \forall\, s \in S,\ p \in P,\ j \in J \qquad (2.15)$$

Constraint (2.16) states that the portion of the state that is produced at time point p can be stored in the unit for consecutive time points, if the unit is not active in those consecutive time points. This is a similar concept to that of slot-based formulations that allow a task to continue in the next consecutive slot, indicating that the unit stores the states in those consecutive time points.

$$u\,(s_{in,j}, p) \leq \rho_{s_{in,j}}^{sp} mu(s_{in,j}, p-1) + u(s_{in,j}, p-1) \quad \forall\, p \in P,\ j \in J \qquad (2.16)$$

Constraint (2.17) ensures that if a state is stored at time point p in the unit then the unit should not be active to start any other task.

$$u(s_{in,j}, p) \leq V_j^U \left(1 - \sum_{s_{in,j}\in S_{in,j}^*} y(s_{in,j}, p)\right) \quad \forall\, p \in P,\ j \in J \qquad (2.17)$$

2.4.9 Time Horizon Constraints

The usage and the production of states should be within the time horizon of interest. These conditions are expressed in

$$t_u(s_{in,j}, p) \leq H, \quad \forall\, s_{in,j} \in S_{in,J},\ p \in P,\ j \in J \qquad (2.18)$$

$$t_p(s_{in,j}, p) \leq H, \quad \forall\, s_{in,j} \in S_{in,J},\ p \in P,\ j \in J \qquad (2.19)$$

2.4.10 TIGHTENING CONSTRAINTS

Constraint (2.20) is used to tighten the model. The sum of the duration of all tasks in a unit must be within the time horizon.

$$\sum_{s_{in,j} \in S^*_{in,j}} \sum_{p} \left(\tau(s_{in,j}) y(s_{in,j}, p) + \beta(s_{in,j}) mu(s_{in,j}, p) \right) \leq H, \quad \forall \, p \in P, \, j \in J \qquad (2.20)$$

2.4.11 OBJECTIVE FUNCTION

The objective of the scheduling problem is to maximize the product throughput if the target is not known *a priori* or to minimize the makespan, if the target is known beforehand, as given in constraint (2.21):

$$\text{Maximize} \sum_{s} price(s^p) \, qs(s^p, p), \quad \forall \, p = P, \, s^p \in S^p$$

or

$$\text{Minimize } H \qquad (2.21)$$

The model ML1 consists of constraints (2.1) through (2.8), (2.10), (2.13) through (2.15), and (2.18) through (2.21). This model does not cater to a situation where the sequence constraint for a different task in different units is to be relaxed in a case where the state is not used. For this model the sequences constraints (2.10) and (2.15) can be written as

$$tu(s_{in,j'}, p) \geq tp(s_{in,j}, p) - H\left(1 - y(s_{in,j}, p - 1)\right),$$
$$\forall \, j \in J, \, p \in P, \, s_{in,j} \in S^{sp}_{in,J}, \, s_{in,j'} \in S^{sc}_{in,J} \qquad (2.22)$$

$$q_s(s,p) \leq QS^U \quad \forall \, s \in S, \, p \in P \qquad (2.23)$$

The model ML2 consists of constraints (2.1) through (2.23).

2.5 CASE STUDIES

To demonstrate the application of the proposed models, four case studies found in literature are presented and discussed below. The results in all the case studies for the proposed model ML1 and ML2 were obtained using CPLEX 9.1.2/GAMS 22.0 on a 2.4 GHz, 4 GB of RAM, Acer TravelMate 5740G computer. All the case studies are taken from published literature. The sequence of constraints in the model statement can significantly affect the CPU time and it is fair to report the sequence of

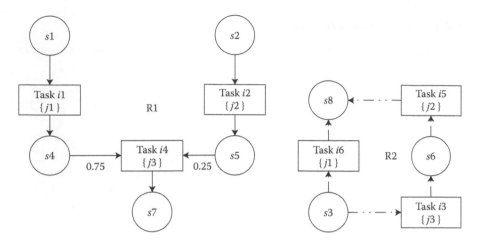

FIGURE 2.7   STN representation for Case I.

constraints in the model statement. For all case studies, the sequence of constraints in the model statement is as that of the sequence in the mathematical model presented in Section 2.4.

2.5.1   CASE I

A case study carried out by Susarla et al. (2010) indicates the necessity of the nonsimultaneous transfer of intermediates into a unit to get a better objective value. This batch plant constitutes six tasks performed in three units and three storages for the intermediates. The plant produces two products using two different production paths (R1 and R2) as depicted in the STN representation (Figure 2.7). The processing time for a task, the capacity of the unit to process a task, the storage capacity, and the initial inventories for each state are given in Tables 2.4 and 2.5.

2.5.2   RESULTS AND DISCUSSIONS

The computational results for literature models are taken from Susarla et al. (2010). For revenue maximization of the time horizon of 6 h, the models by MG, SK, and SF give an optimal solution of $560. The models by Susarla et al. (2010), that is, SLK1 and SLK2 and the proposed models, that is, ML1 and ML2, give a better objective value of $650. The better objective value by SLK1&2 and ML1&2 is because the models allow nonsimultaneous material transfers, which means for a task that uses more than one intermediate state, it is possible for one state to be stored in a unit that is processing it for a while and wait for the other intermediates to come together to start the task.

In this case, task 4 needs state $s4$ and $s5$ to produce $s7$. The 10 kg of $s5$ is produced at 3 h and 30 kg of $s4$ is produced at 4 h (Figure 2.8a). Since $s5$ is produced earlier when compared to $s4$ and there is not enough storage to store $s5$ and later to use together with $s4$, it is required that state $s5$ is stored in unit 3 for 1 h and to start

TABLE 2.4

Batch Size Data for Cases I–III

Task	Label i	Unit	Label j	$\tau(s_{in,j})$	$\beta(s_{in,j})$	$V^L_{s_{in,j}} - V^U_{s_{in,j'}}$
			Case I			
Task 1	1	Unit 1	Unit-$j1$	1.666	0.0778	0–30
Task 2	2	Unit 2	Unit-$j2$	2.333	0.0667	0–10
Task 3	3	Unit 3	Unit-$j3$	0.669	0.0777	0–30
Task 4	4	Unit 3	Unit-$j3$	0.667	0.033325	0–40
Task 5	5	Unit 2	Unit-$j2$	1.332	0.0556	0–30
Task 6	6	Unit 1	Unit-$j1$	1.5	0.025	0–20
			Case II			
Task 1	1	Unit 1	Unit-1	1.333	0.01333	0–100
		Unit 2	Unit-2	1.333	0.01333	0–150
Task 2	2	Unit 3	Unit-3	1	0.005	0–200
Task 3	3	Unit 4	Unit-4	0.667	0.00445	0–150
		Unit 5	Unit-5	0.667	0.00445	0–150
			Case III			
Heating	$i1$	Heater	HR	0.667	0.00667	0–100
Reaction-1	$i2$	Reactor 1	RR-1	1.334	0.02664	0–50
	$i2$	Reactor 2	RR-2	1.334	0.01665	0–80
Reaction-2	$i3$	Reactor 1	RR-1	1.334	0.02664	0–50
	$i3$	Reactor 2	RR-2	1.334	0.01665	0–80
Reaction-3	$i4$	Reactor 1	RR-1	0.667	0.01332	0–50
	$i4$	Reactor 2	RR-2	0.667	0.00833	0–80
Separation	$i5$	Separator	SR	1.3342	0.00666	0–200

processing together with $s4$ at 4 h. By doing so a better revenue is obtained since unit 2 transfers the state $s5$ to unit 3 at 3 h and starts processing $s6$ at the same time to produce the product state $s8$. The models by SF, SK, and MG give suboptimal results since the models do not allow nonsimultaneous material transfer. In this case, state $s5$ produced at 3 h stayed in unit 2 for 1 h until used at 4 h by unit 3; as a result unit 2 is inactive for 1 h (Figure 2.8b). The amount of material processed by each unit, the type of task each unit is conducting, and the starting and finishing time of each task are shown in Figure 2.8.

2.5.3 CASE II

This case study has been studied extensively in published literature. It is a simple batch plant requiring only one raw material to get a product as depicted in the STN representation (Figure 2.9).

The plant encompasses five units and two intermediate storages. The conversion of the raw material into a product is achieved through three sequential processes. The first task can be performed in two units ($j1$ and $j2$), the second task can be

TABLE 2.5

Storage Capacities, Initial Inventories, and Price of Products for Cases I–III

	Case I		Case II		Case III	
States	**Storage Capacity (mu)**	**Initial Inventory (mu)**	**Storage Capacity (mu)**	**Initial Inventory (mu)**	**Storage Capacity (mu)**	**Initial Inventory (mu)**
1	UL	AA	UL	AA	UL	AA
2	UL	AA	200	0	UL	AA
3	UL	AA	250	0	UL	AA
4	10	0	UL	0	UL	AA
5	5	0	–	–	100	0
6	10	0	–	–	150	0
7	UL	0	–	–	UL	0
8	UL	0	–	–	200	–
9	–	0	–	–	200	–
10	–	–	–	–	UL	–
11	–	–	–	–	–	–
12	–	–	–	–	–	–
13	–	–	–	–	–	–

UL = Unlimited; AA = Available as and when required.
Case I: Price for $s7$ is 10 \$/mu and $s8$ is 5 \$/mu.
Case II: Price for $s4$ is 5 \$/mu.
Case III: Price for $s8$ and $s9$ is 10 \$/mu.

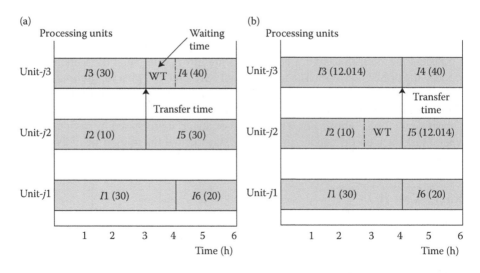

FIGURE 2.8 (a) Schedule from SLK1&2 and ML1&2. (b) Schedule from MG, SK, and SF.

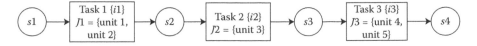

FIGURE 2.9 STN representation for Case II.

performed only in unit $j3$, and the third task is suitable for units $j4$ and $j5$. The required data to solve the case study is given in Tables 2.4 and 2.5. The computational statistics of the different models showing the number of continuous and binary variables, constraints, number of time points, nonzero elements, CPU time required, and the optimal objective value obtained by each model are shown in Table 2.6.

The computational results for literature models are taken from Susarla et al. (2010). For the time horizon of 10 h, the proposed models ML1&2 require five slots ($p = 6$) to get the optimal objective value of $2628.2. Almost all the models get the same optimal objective value in a similar CPU time. For the time horizon of 12 h, the proposed models ML1&2 require six slots ($p = 7$) to get the optimal objective value of 3463.6. The models ML1&2 require two slots less when compared to the other models. The models ML1&2 outperform both single grid and multigrid models in terms of CPU time required to get the optimal objective value of $2628.2 (0.25 s for ML1 and 4.5 s for ML2 vs. 781 s for SLK2, 1492 s for SLK1, 585 s SF [$\Delta n = 1$], 11.6 s for SK and 10.3 s for MG). In a case where a task need not span over on multiple time points, SF ($\Delta n = 0$) and ML1 perform better—1.88 s for SF ($\Delta n = 0$) and 0.25 s for ML1. While SF solves faster for ($\Delta n = 0$), one would need to solve it several times to get the best solution which makes it difficult to compare the solution time with those of other models. For the time horizon of 16 h the proposed model ML2 requires eight slots ($p = 9$), which is three slots less when compared to other models to get the optimal objective value of $5038.2. The models ML1&2 outperform the multigrid models SF and SLK1&2 in terms of the CPU time required (75.8 s for ML1 and 60 s for ML2 vs. specified CPU time of 10,000 s for SLKs and SF [$\Delta n = 1$]). In this case also, the models ML1&2 give a better CPU time when compared to the single grid models (377 s for SK and 2431 s for MG).

For makespan minimization, two scenarios are taken with product demands of 2000 and 4000 mu. In the first scenario, the model ML2 gives a better objective value of 27.98 h in a specified CPU time of 10,000 s when compared to other models. The model ML1&2 required 12 slots ($p = 13$) to get an optimal objective value, which is four slots less when compared to the other models. For the second scenario the model ML2, again, gives a better objective value of 53.8 h when compared to other models in literature that give an objective value of 56.432 h. The superior performance of the proposed models is due to the reduction of the required time points and does not allow the task to span over multiple time points to get a better objective value, like in the models of SF and SLK1&2. The Gantt chart for the time horizon of 12 h is shown in Figure 2.10.

TABLE 2.6

Computational Statistics of the Proposed Model and Other Models in Case II

Model	p	CPU Time (s)	Nodes	RMILP	MILP	BV	CV	Constraints	Nonzeros	Relative Gap (%)
					Case IIa ($H = 10$)					
SLK2	7	1.95	2764	4000	2628.2	60	359	491	1556	—
SLK1	7	3.2	6341	4000	2628.2	60	324	461	1496	—
SF										
($\Delta n = 0$)	6	0.17	285	3973.9	2628.2	30	119	209	639	—
($\Delta n = 1$)	6	1.33	1607	4000	2628.2	55	144	479	1539	—
SK	7	1.06	1090	3384.3	2628.2	60	316	300	1001	—
MG	7	0.88	770	3548.4	2628.2	70	309	846	2766	—
ML1	6	0.2	543	3973.9	2628.2	42	157	272	724	—
ML2	6	1.64	6215	4000	2628.2	60	205	368	989	—
					Case IIb ($H = 12$)					
SLK2	9	781	284,342	4951.2	3463.6	80	467	657	2082	—
SLK1	9	1492	600,476	4951.2	3463.6	80	422	617	2002	—
SF										
($\Delta n = 0$)	8	1.88	8136	4951.2	3463.6	40	157	281	865	—
($\Delta n = 1$)	8	585	254,877	4951.2	3463.6	75	192	657	2117	—
SK	9	11.6	12,480	4481	3463.6	80	416	408	1359	—
MG	9	10.3	32,092	4563.8	3463.6	90	397	1084	3879	—
ML1	7	0.25	807	4937.3	3463.6	49	183	317	852	—
ML2	7	4.5	15,568	4951.2	3463.6	70	239	429	1166	—
					Case IIc ($H = 16$)					
SLK2	12	10,000	1,628,804	6601.7	5038.1	110	629	906	2871	18
SLK1	12	10,000	1,294,912	6601.7	5038.1	110	569	851	2761	19.1

SF									
(Δn = 0)	11	113	484,764	6601.7	5038.1	55	214	389	1204	–
(Δn = 1)	11	10,000	1,727,132	6601.7	5038.1	105	264	909	2984	15.87
SK	12	377	461,037	6312.6	5038.1	110	566	570	1896	–
MG	12	2431	1,974,025	6332.8	5038.1	120	529	1441	5811	–
ML1	10	60	278,759	6601.7	5038.1	70	261	452	1239	–
ML2	9	75.8	278,047	6601.7	5038.1	90	307	551	1520	–
Case IId: d(s4) = 2000 mu and H = 50									
SLK2	17	10,000	328,879	24.2	28.772	160	901	1330	4203	15.8
SLK1	17	10,000	380,619	24.2	28.772	160	816	1250	4043	15.8
SF									
(Δn = 0)	16	10,000	1,093,172	24.2	28.884	80	309	574	1783	6.6
(Δn = 1)	16	10,000	769,022	24.2	28.772	155	384	1344	4443	15.8
SK	17	5403	3,214,852	24.72	28.772	160	816	843	2794	–
MG	17	10,000	1,210,125	24.7	29.5	170	750	2045	9879	8.88
ML1	14	10,000	11,453,748	24.2	28.74	98	366	633	1897	2.2
ML2	13	10,000	4,804,688	24.2	27.98	130	444	796	2368	10.8
Case IIe: d(s4) = 4000 mu and H = 100									
SLK2	23	4944	1,522,250	48.5	56.432	220	1225	1828	5781	–
SLK1	23	10,000	1,880,663	48.5	56.432	220	1110	1718	5561	1.89
SF									
(Δn = 0)	22	35	42,758	48.5	56.432	110	423	790	2461	–
(Δn = 1)	22	8586	1,957,756	48.5	56.432	215	528	1860	6177	–
SK	26	10,000	3,115,485	49.11	56.432	250	1266	1329	4405	6.04
MG	26	10,000	562,110	49.01	56.432	260	1146	3116	19,212	10.25
ML1	22	10,000	5,126,760	48.5	55.43	154	574	993	3001	2.4
ML2	22	10,000	8,201,001	48.5	53.8	220	750	1345	4051	4.2

Source: Adapted from Susarla, N., Li, J., Karimi, I.A., 2010. *AIChE Journal*, 56, 1859–1879.

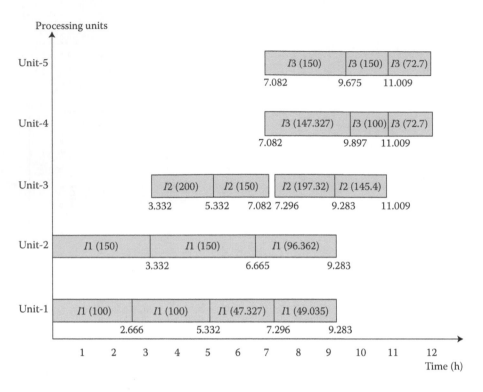

FIGURE 2.10 Gantt chart for the time horizon of 12 h for Case II.

2.5.4 CASE III

This case study was first studied by Kondili et al. (1993) and has become one of the most common examples that appeared in literature. This batch plant produces two different products sharing the same processing units; Figure 2.11 shows the plant flowsheet. The unit operations consist of preheating, three different reactions, and separation. The STN and SSN representation of the flowsheet are shown in Figure 2.12. Tables 2.4 and 2.5 give the required data to solve the scheduling problem for this case.

Choice of Effective States. Because this problem involves more than one state entering some units, that is, reactors 1 and 2, it is necessary to choose effective states. The set indicated below is one of the choices of effective states which represent tasks.

$$S_{in,J} = \left\{ s_{1in,j}, s_{2in,j}, s_{6in,j}, s_{8in,j}, s_{9in,j} \right\}, \quad j = 1, 2, 3, 4$$

There are eight effective states, leading to $8p$ binary variables. It should be mentioned that the only requirement for the choice of the effective states is that only one of the input states that are used simultaneously in a particular unit should be chosen. For example, reaction 2 requires state $s3$ and $s2$ as an input state. We can choose

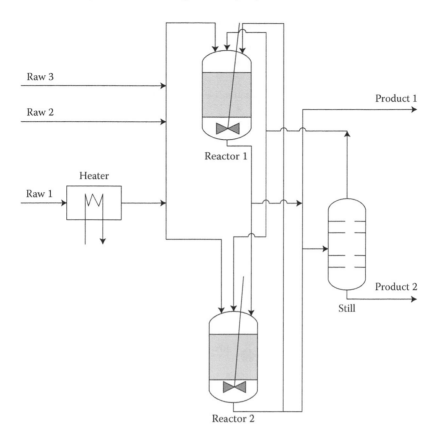

FIGURE 2.11   Flowsheet for Case III.

an effective state between the two; once the choice of the effective state is made, it should be consistent throughout the mathematical formulation. For more information, the reader can see the contribution of Majozi and Zhu (2001).

2.5.5   RESULTS AND DISCUSSIONS

The computational statistics of the different models based on the continuous time representation for this example are shown in Table 2.7. The computational results for literature models are taken from Susarla et al. (2010). For the time horizon of 10 h, both models in this chapter require two time points less when compared to SK and MG, and one event point less when compared to SF, SLK1, and SLK2. The proposed model ML2 gives a better objective $1962.7 value when compared to $1943.2 by the model ML1, because when the state produced at time point $p - 1$ by unit j is not used at time point p by another unit j', that is, the sequence constraints for different tasks in different units is relaxed. Single grid models of SK and MG give a better RMILP objective of $2730.7. All the multigrid models solve faster than the single grid models, because they need a lower number of time points.

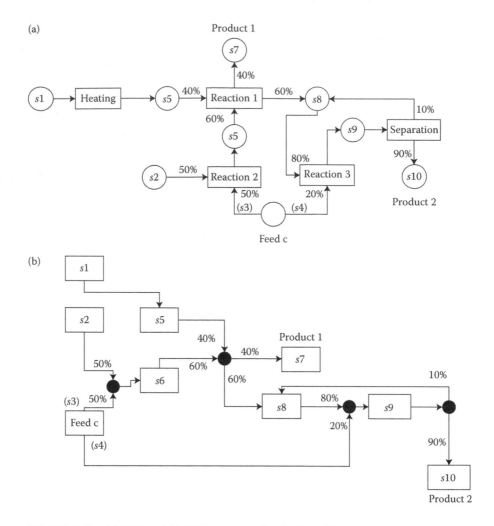

FIGURE 2.12   (a) STN and (b) SSN representation for Case III.

For the time horizon of 12 h, the proposed models require six slots ($p = 7$), which is four slots less when compared to the single grid models. For this reason, the model solves faster (35.4 s for ML2 and 3.9 s for ML1 vs. 3330 s for SK and 9124.5 s for MG). When compared to multigrid models, the proposed models require two slots less to SF and one slot less to SLK1 and SLK2. ML1&2 solve faster for this time horizon when compared to SLK1&2 and SF in order to get the optimal objective value. In this case, both ML1 and ML2 give the same optimal objective value of $2658.5 as other formulations. For a case where ML1 and ML2 give the same objective value, ML1 solves faster than ML2 (3.9 s for ML1 vs. 35.4 s for ML2) since ML1 requires less binary variables when compared to ML2. For the time horizon of 16 h, the proposed model ML2 requires seven slots ($p = 8$) which is one slot less when compared

TABLE 2.7

Computational Statistics of the Proposed Model and Other Models in Case III

Model	p	CPU Time (s)	Nodes	RMILP	MILP	BV	CV	Constraints	Nonzeros	Relative Gap (%)
					Case IIIa (H = 10)					
SLK2	7	12	19,043	2730.7	1962.7	72	449	645	2319	–
SLK1	7	28.2	44,015	2730.7	1962.7	72	421	621	2315	–
SF										
(Δn = 0)	6	3.19	7978	2730.7	1931.92	48	208	439	1415	–
(Δn = 1)	6	8.13	10,915	2730.7	1962.7	88	248	847	3150	–
SK	8	57.13	65,587	2690.6	1962.7	84	489	458	1686	–
MG	8	126	54,753	2690.6	1962.7	112	617	1468	5464	–
ML1	6	2.4	4520	2730.7	1943.2	68	273	562	1580	–
ML2	6	11.3	23,837	2730.7	1962.7	88	383	779	2306	–
					Case IIIb (H = 12)					
SLK2	8	38	58,065	3301	2658.5	84	517	753	2710	–
SLK1	8	62	80,093	3301	2658.5	84	485	725	2706	–
SF										
(Δn = 0)	8	53.2	137,217	3350.5	2658.5	64	274	597	1925	–
(Δn = 1)	8	825	899,959	3350.5	2658.5	120	330	1157	4342	–
SK	11	3330	2,614,049	3343.4	2658.5	120	687	665	2442	–
MG	11	9125	3,174,288	3343.4	2658.5	160	842	2020	8467	–
ML1	7	3.9	11,575	3301	2658.5	80	319	664	1873	–
ML2	7	18	39,563	3301	2658.5	104	448	923	2739	–
					Case IIIc (H = 16)					
SLK2	9	30	32,351	4291.7	3738.38	96	585	861	3101	–
SLK1	9	76	62,786	4291.7	3738.38	96	549	829	3097	–
SF										
(Δn = 0)	9	140	315,573	4438.9	3738.38	72	307	676	2180	–

(Continued)

TABLE 2.7 (Continued)
Computational Statistics of the Proposed Model and Other Models in Case III

Model	p	CPU Time (s)	Nodes	RMILP	MILP	BV	CV	Constraints	Nonzeros	Relative Gap (%)
(Δn = 1)	9	3903	3,554,870	4438.9	3738.38	136	371	1312	4938	—
SK	10	156	96,734	4318.8	3738.38	108	621	596	2190	—
MG	10	703	200,592	4318.8	3738.38	144	767	1836	7410	—
ML1	8	20	31,792	4291.7	3738.38	92	365	664	1873	—
ML2	8	79.7	138,508	4291.7	3738.38	120	513	1067	3172	—
Case IIId: d(s7) = d(s10) = 200 mu and H = 50										
SLK2	10	821	175,107	18.7	19.34	108	658	983	3517	—
SLK1	10	1276	152,939	18.7	19.34	108	618	947	3513	—
SF										
(Δn = 0)	9	1.14	507	18.7	19.34	72	307	685	2199	—
(Δn = 1)	9	1331	203,829	18.7	19.34	136	371	1321	4957	—
SK	10	171	55,349	18.7	19.34	108	621	604	2197	—
MG	10	314	36,146	18.7	19.34	160	842	1962	7767	—
ML1	9	500	345,789	18.7	19.34	104	412	870	2607	—
ML2	9	2978	1,452,700	18.7	19.34	136	579	1213	3753	—
Case IIIe: d(s7) = 500 mu, d(s10) = 400 mu and H = 100										
SLK2	22	10,000	302,206	47.4	47.6835	252	1474	2279	8209	0.64
SLK1	22	10,000	260,603	47.4	47.6835	252	1186	2195	8205	0.64
SF										
(Δn = 0)	21	10,000	986,563	47.5	47.754	168	703	1633	5259	0.49
(Δn = 1)	21	10,000	340,927	47.4	49.012	328	863	3181	12,109	3.34
SK	23	10,000	398,979	48.78	49.05	264	1479	1501	5473	0.55
MG	23	10,000	59,431	48.78	49.05	368	1734	4471	26,279	0.55
ML1	22	10,000	401,658	47.4	47.75	260	1010	2196	6624	0.05
ML2	21	10,000	1,545,601	46.4	47.6835	328	1359	2941	9141	1.8

Source: Adapted from Susarla, N., Li, J., Karimi, I.A., 2010. *AIChE Journal*, 56, 1859–1879.

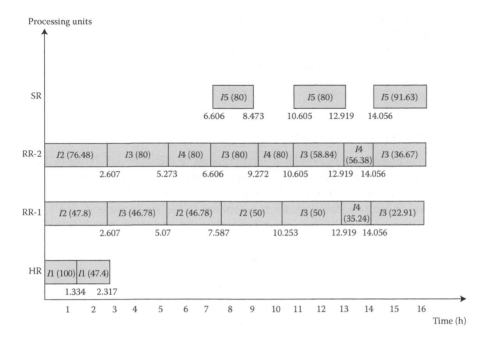

FIGURE 2.13   Gantt chart for the time horizon of 16 h for Case III.

to SLK1&2 and two slots less when compared to SF, MG, and SK to get an optimal objective value of $3738.38. The multigrid models ML1&2 and SLK1&2 solve much faster than SF, MG, and SK (79.7 s for ML2, 20 s for ML1 and 30 s for SKL2 vs. 3903 s for SF [$\Delta n = 1$], 140 s for SF [$\Delta n = 0$], 156 s for SK, 703 s for MG). For this case also, the models ML1 and ML2 require the same optimal number of time points to get the optimal objective value. As a result, the model ML1 performs better than the model ML2 in terms of CPU time due to the lesser number of binary variables.

With regard to makespan minimization, the case study is solved for two demand scenarios Case III d (d7 = d10 = 200 mu) and Case III e (d7 = 500 mu, d10 = 400 mu). For Example 3d, the single grid models of MG and SK perform well in obtaining the optimal objective value of 19.34 h. In a case where there is no need for a task to span over multiple time points to get optimal objective value, the model SF ($\Delta n = 0$) performs better since the models require few binary variables. For Example 3e, all the models do not close the relative gap in the specified CPU time of 10,000 s. All the multigrid models require the same time points and one time point less when compared to the single grid models. The models ML2 and SLK1&2 give a better objective value of 47.6835 h when compared to other models in literature. The Gantt chart for the time horizon of 16 h is shown in Figure 2.13.

2.5.6   CASE IV

At first, Sundaramoorthy and Karimi (2005) studied this case study which is a relatively more complex problem and later often used in the literature to check the efficiency of models in terms of optimal objective value and CPU time required to get to optimal objective value. This case study is exactly the same as the motivating example. This case study is solved for the FIS policy using the presented model ML1 and ML2. The STN representation of this case study is depicted in Figure 2.1. Data required for this case study is the same as that of the motivating example for the FIS policy.

2.5.7 RESULTS AND DISCUSSION

The computational statistics of the different models in literature are shown in Table 2.8. The computational results for literature models are taken from Susarla et al. (2010). For the time horizon of 8 h, the proposed model ML2 requires four slots ($p = 5$) which are two slots less when compared to the multigrid and single grid models. The model ML2 solves much faster than the multigrid models of SF and SLK1&2 (1.25 s for ML2 vs. 284 s for SLK2 and 56.8 s for SF [$\Delta n = 1$]). The proposed model ML2 gives the lowest RMILP value of 2100. All models obtained the same optimal objective value of $1583.4.

For the time horizon of 10 h, again, the proposed model ML2 requires five slots ($p = 6$), which is four slots less when compared to single grid models of MG and SK and three slots less when compared to the multigrid models of SLK1&2 and SF. The model ML2 gives a better objective value ($2345.3 for ML2 vs. $2337.36 for SF [$\Delta n = 1$] and SLK1&2, $2260.9 for MG and $2137.1 for SK). The proposed model ML2 again gives the lowest RMILP value of $2871.9. A CPU time of 8 s is required by ML2 to solve the problem to a 0% relative gap, which is much less when compared to the specified CPU time of 10,000 s for all models in literature. Figure 2.14 shows the Gantt chart for the time horizon of 10 h obtained by the proposed model ML2.

As observed in the Gantt chart shown in Figure 2.14, the starting time (6.58 h) of heating task 2 at time point $p5$ is less than the finishing time (7.75 h) of separation task at $p4$ because the intermediate state s6 produced by the separator at time point $p4$ is not used by heating task 2 at time point $p5$. This schedule is not feasible for the models by Ierapetritou and Floudas (1998a,b), Majozi and Zhu (2001), and Shaik and Floudas (2008). Consider the sequence constraint for different tasks in different units in the model presented by Shaik and Floudas (2008) as shown below.

$$T^s(i, n + 1) \geq T^s(i', n) + \alpha_{i'} w(i', n) + \beta_{i'} b(i', n) - H(1 - w(i', n)), \forall \quad s \in S,$$

$$i' \in I_s^p, i \in I_s^c, i \in I_j, i' \in I_{j'}, j, j' \in J, j' \neq j, n \in N, n < N \tag{i}$$

The above constraint states that the starting time of the consuming task at the current event point n should be after the end time of the producing task at the previous event point $n - 1$. This need not be true if there is sufficient material for the consuming task to start production, as a result the above constraint leads to suboptimal results. This constraint in the proposed model ML2 is relaxed without spanning a task over multiple time points.

TABLE 2.8

Computational Statistics of the Proposed Model and Other Models in Case IV

Model	p	CPU Time (s)	Nodes	RMILP	MILP	BV	CV	Constraints	Nonzeros	Relative Gap (%)
					Case IVa ($H = 8$)					
SLK2	7	284	388,832	2751	1583.4	102	655	1070	3671	—
SLK1	7	2659	3,674,795	2751	1583.4	102	613	1034	3654	—
SF										
($\Delta n = 0$)	6	9.37	18,574	2751	1583.4	66	290	716	2169	—
($\Delta n = 1$)	6	56.8	80,223	2751	1583.4	121	345	1336	4628	—
SK	7	45.5	39,305	2560.6	1583.4	102	595	728	2207	—
MG	7	81.2	55,146	2560.6	1583.4	154	806	1893	6630	—
ML1	5	0.9	653	2751	1583.4	83	314	648	1863	—
ML2	5	1.25	3662	2100	1583.4	107	418	889	2680	—
					Case IVb ($H = 10$)					
SLK2	9	10,000	872,204	3618.6	2337.36	136	853	1428	4970	5.07
SLK1	9	10,000	711,493	3618.6	2337.36	136	799	1380	4886	11.6
SF										
($\Delta n = 0$)	8	115	1,160,988	3618.6	2292.5	88	382	884	2847	—
($\Delta n = 1$)	8	10,000	2,256,858	3618.6	2337.36	165	459	1658	6187	1.68
SK	10	10,000	2,666,604	3473.9	2260.9	153	874	1121	3380	10.7
MG	10	10,000	1,415,516	3473.9	2137.1	220	1151	2691	10,710	23.9
ML1	6	1.34	4378	2871.9	2292.5	101	378	792	2289	—
ML2	6	8	9252	2871.9	2345.3	131	516	1093	3308	—

(Continued)

TABLE 2.8 (Continued)
Computational Statistics of the Proposed Model and Other Models in Case IV

Model	p	CPU Time (s)	Nodes	RMILP	MILP	BV	CV	Constraints	Nonzeros	Relative Gap (%)
Case IVc (H = 12)										
SLK2	8	10	3096	3465.6	3041.3	119	754	1249	4289	–
SLK1	8	422	236,847	3465.6	3041.3	119	706	1207	4270	–
SF										
(Δn = 0)	7	2	1262	3465.6	3041.3	77	336	844	2560	–
(Δn = 1)	7	4	881	3465.6	3041.3	143	402	1579	5505	–
SK	9	95	43,951	3867.3	3041.3	136	781	990	2989	–
MG	9	296	71,877	3867.3	3041.3	198	1036	2425	9269	–
ML1	7	10	4847	3465.6	3041.3	119	442	936	2715	–
ML2	7	128	103,886	3465.6	3041.3	155	604	1297	3936	–
Case IVd (H = 16)										
SLK2	11	10,000	705,090	5225.9	4241.5	170	1051	1786	6143	0.29
SLK1	11	10,000	3,042,257	5225.9	4237.61	170	985	1726	6118	1.64
SF										
(Δn = 0)	10	1475	515,144	5225.9	4241.5	110	474	1228	3733	–
(Δn = 1)	10	10,000	1,511,596	5225.9	4241.5	209	573	2311	8136	0.01
SK	11	1687	283,938	5125.9	4240.83	170	967	1252	3771	–
MG	11	10,000	739,728	5125.9	4240.83	242	1266	2957	12,232	1.69
ML1	9	318	253,967	4653.1	4240.83	155	570	1224	3567	–
ML2	10	10,000	4,455,151	4653.1	4261.9	227	868	1909	5820	2

Case IVe: d(12) = 100 mu, d(s13) = 200 mu and H = 50

SLK2	9	249	42,218	11.3	13.367	136	859	1448	4944	—
SLK1	9	2538	502,461	11.3	13.367	136	805	1400	4923	—
SF										
($\Delta n = 0$)	9	71	43,967	11.3	13.367	99	428	1112	3369	—
($\Delta n = 1$)	9	1867	770,883	11.3	13.367	187	516	2079	7286	—
SK	11	727	233,212	11.417	13.367	170	967	1264	3782	—
MG	11	1930	162,982	11.417	13.367	242	1267	2970	12,427	—
ML1	7	0.48	372	11.4	13.367	119	443	938	2848	—
ML2	7	1.5	260	11.4	13.367	155	605	1299	4069	—

Case IVf: d(12) = d(s13) = 250 mu and H = 100

SLK2	11	10	820	14.3	17.025	170	1057	1806	6180	—
SLK1	11	6	663	14.3	17.025	170	991	1746	6155	—
SF										
($\Delta n = 0$)	10	5	2236	14.3	17.199	110	474	1240	3760	—
($\Delta n = 1$)	10	7	683	14.3	17.025	209	573	2323	8163	—
SK	12	107	12,992	15.001	13.306	187	1060	1395	4173	—
MG	12	247	14,683	15.001	13.306	264	1382	3236	14,047	—
ML1	10	30.7	41,687	14.3	17.199	173	635	1370	4180	—
ML2	10	15	15,060	14.3	17.025	227	869	1911	6007	—

Source: Adapted from Susarla, N., Li, J., Karimi, I.A., 2010. *AIChE Journal*, 56, 1859–1879.

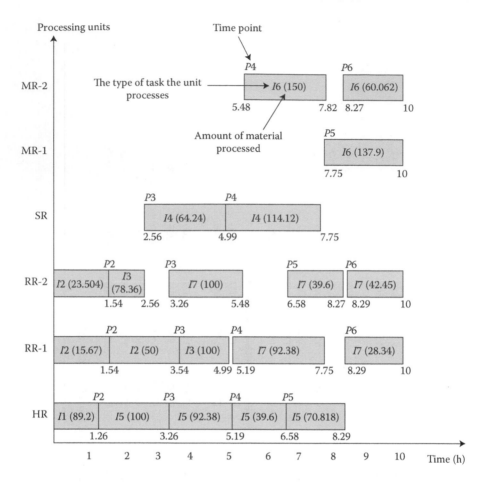

FIGURE 2.14   Gantt chart for the time horizon of 10 h for Case IV.

To understand the above case of relaxing the sequence constraints different task in different units by the proposed model without spanning a task over multiple time points, the following constraint for different tasks in different units that are used in the model presented by Shaik and Floudas (2009) is examined.

$$T^s(i, n+1) \geq T^f(i', n) - H\left(1 - \sum_{\substack{n' \in N \\ n - \Delta n \leq n' \leq n}} w(i', n', n)\right),$$

$$\forall s \in S, \; i' \in I_s^p, \; i \in I_s^c, \; i \in I_j, \; i' \in I_{j'}, \; j, j' \in J, \; j' \neq j, \; n \in N, \; n < N \qquad \text{(ii)}$$

In order for the starting time (6.58 h) of heating task 2 at time point $p5$ to be less than the finishing time (7.75 h) of the separation task at time point $p4$, the separation task must span over multiple time points. Instead of the separation task to end at time point $p4$, it continues to the next time point so that the binary variable w (i',n,n) becomes 0 to relax the sequence constraint. This method of modeling a scheduling problem requires an increased number of time points as shown in Table 2.8, which results in a large RMILP objective value that necessitate a high CPU time.

Again as observed from the Gantt chart in Figure 2.14, the starting time of reaction 2 in reactor 2 and 3 begins at 1.54 h at time point $p2$ which is later than the finishing time (1.26 h) of heating 1 at time point $p1$. This is possible since the amount of material produced by the heater at time point $p1$ can be stored and later used by reaction 2 at time point $p2$. This can be achieved through constraints (2.13) and (2.14). The binary variable $x(s3,p2)$ is 1 since state $s3$ produced at time point $p1$ is less than the capacity of the storage as a result constraint (2.14) is to be relaxed. The starting time of reaction 2 at time point $p2$ is not necessarily equal to the finishing time of heating 1 at time point $p1$. Similarly, the starting time of mixer 2 (MR-2) at time point $p5$ is 5.48 h is not equal to the finishing time of the separator at time point $p3$ which is 4.99 h, since there is enough storage for the material produced by the separator. This approach is the novelty of models ML1&2 which give a better objective value and less number of time points by eliminating the drawback of the sequence constraint pertaining to different tasks in different units for a FIS case, where the starting time of reaction 2 at time point $p2$ must be equal to the finishing time of heating 1 at time point $p1$.

For the time horizon of 16 h, the proposed model ML2 requires nine slots ($p = 10$) which is one slot less than other models in the literature. Again for this case, the model ML2 gives a better objective value of 4261.9 using 10 time points within a specified CPU time of 10,000 s.

With regard to makespan minimization, the case study is solved for two demand scenarios, Case IV e ($d_{12} = 100$ and $d_{13} = 250$ mu) and Case IV f ($d_{12} = 250$, mu = d_{13}). For Case IV e, the proposed models ML1&2 require up to three slots less when compared to the multigrid model of SF (7 time points for ML1&2 vs. 9 for SLK1&2), and four slots less when compared to the single grid models of SK and MG. The models ML1&2 require the lowest CPU time of (0.48 s for ML1 and 1.5 s for ML2) to give the optimal objective value of 13.367 h. In Example 4f, the models SLK1&2 and SF perform better than ML2 and the single grid model of MG and SK. The model ML2 and SLK1&2 and SF ($\Delta n = 1$) give a better objective value of 17.025. The Gantt chart for makespan minimization for Case IV f is shown in Figure 2.15.

2.6   COMPARISON WITH CYCLIC SCHEDULING

Case III was also solved for a time horizon of 168 h to check the proposed model efficiency, and the results are compared to the cyclic scheduling model of Wu and Ierapetritou (2004).

As it is seen from Table 2.9, an optimal objective value of 46,448.5 is obtained by the model ML2 in the specified CPU time of 40,000 s. The objective values obtained by the proposed models are much better than 45,698.9 obtained by using the cyclic scheduling model of Wu and Ierapetritou (2004). This indicates that the proposed models are tighter and capable of solving the scheduling problem even for the long time horizon. To the knowledge of the authors, there are no mathematical techniques

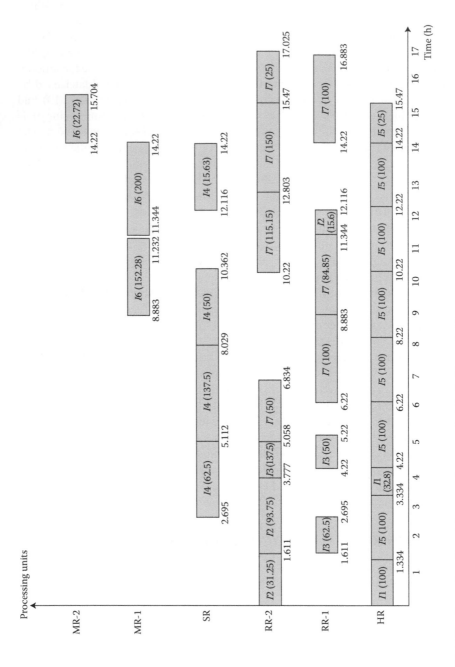

FIGURE 2.15   Gantt chart for minimum makespan schedule for Example 4f.

TABLE 2.9

Computational Results for Case III Maximization of Revenue for the Time Horizon of 168 h

Model	p	CPU Time (s)	RMILP	MILP	BV	CV	Constraints	Nonzeros	Relative Gap (%)
				$H = 168$ h					
ML1	77	40,000	47,064.50	46,303	924	4313	8114	23,931	1.7
ML2	77	40,000	47,064.50	46,448.5	1232	5312	11,186	33,857	1.26

in published literature that solved scheduling problem for this time horizon and get a better objective value as compared to cyclic scheduling. Cyclic scheduling is proposed as a way of scheduling for long-time horizon, due to the computational complexity of scheduling models as the time horizon increases. It is evident that the proposed models can solve scheduling problems even for the long-time horizon, which take an advantage over cyclic scheduling where the objective value is compromised for computational efficiency.

2.7 CONCLUSIONS

In this chapter, a novel scheduling technique for multipurpose batch plants has been presented. The technique involves two mathematical formulations that are based on SSN representation. Both presented models exhibit MILP structure for which global optimality can be proven. The proposed models allow nonsimultaneous transfer of material to get a better objective value. In most of the cases, the models in this chapter require a lower number of time points, as a result giving a better CPU time as compared to existing models from the literature. Improved objective value is also obtained for some of the case studies as compared to values obtained from models in literature.

In the published literature, the sequence constraint that pertains to tasks that consume and produce the same state, the starting time of the consuming task at time point p must be later than the finishing time of the task at the previous time point $p - 1$. This constraint is relaxed by the model ML2 if the state is not used at the current time point p. This relaxation gives a better objective value. In a case where the number of time points required and optimal objective value is the same by both models ML1 and ML2, the model ML1 outperforms ML2 in terms of CPU time, number of binary variables, continuous variables, and constraints. The proposed models can also solve scheduling problems for the long-time horizon, which take an advantage over cyclic scheduling where the objective value is compromised for computational efficiency.

NOTATION

SETS

I {$i|i$ is a task}

J {$j|j$ is a unit}

J_s $\{j_s | j_s$ is a unit producing state $s\}$
P $\{p | p$ is a time point$\}$
S $\{s | s$ any state other than a product$\}$
$S_{in,J}^{sc}$ $\{s_{in,j}^{sc} | s_{in,j}^{sc}$ task which consumes state $s\}$
$S_{in,j}^{*}$ $\{s_{in,j}^{*} | s_{in,j}^{*}$ tasks performed in unit $j\}$
$S_{in,J}$ $\{s_{in,j} | s_{in,j}$ is an effective state representing a task$\}$
$S_{in,J}^{usc}$ $\{s_{in,j}^{usc} | s_{in,j}^{usc}$ task which consumes unstable state $s\}$
$S_{in,J}^{sp}$ $\{s_{in,j}^{sp} | s_{in,j}^{sp}$ task which produces state s other than a product$\}$
$S_{in,J}^{usp}$ $\{s_{in,j}^{usp} | s_{in,j}^{usp}$ task which produces unstable state $s\}$
$S_{in,J}^{s^p p}$ $\{s_{in,j}^{s^p p} | s_{in,j}^{s^p p}$ task which produces state s, which is a product$\}$
S^p $s^p | s^p$ a state which is a product$\}$

VARIABLES

$mu(s_{in,j}, p)$ amount of material processed by a task at time point p
$q_s(s,p)$ amount of state s stored at time point p
$t(j,p)$ binary variable associated with usage of state produced by unit j at time point p
$t(j,s,p)$ binary variable associated with usage of state s produced by unit j at time point p if the unit produces more than one intermediate at time point p
$t_p(s_{in,j}, p)$ time at which task ends at time point p, $s_{in,j} \in S_{in,j}$
$t_u(s_{in,j}, p)$ time at which task starts at time point p, $s_{in,j} \in S_{in,j}$
$u(s_{in,j}, p)$ amount of material stored in unit j at time point p
$x(s,p)$ binary variable associated with availability of storage for state s at time point p
$y(s_{in,j}, p)$ binary variable for assignment of task at time point p

PARAMETERS

$\beta(s_{in,j})$ coefficient of variable term of processing time of a task $S_{in,j}$
$\rho(s_{in,j}^{sp})$ portion of state s produced by a task
$\rho(s_{in,j}^{sc})$ portion of state s consumed by a task
$\tau(s_{in,j})$ coefficient of constant term of processing time of a task
H time horizon of interest
V_j^U maximum capacity of unit j
$V_{s_{in,j}}^U$ maximum capacity of unit j to process a particular task
$V_{s_{in,j}}^L$ minimum capacity of unit j to process a particular task
QS^o initial amount of state s stored
QS^U maximum capacity of storage to store a state s

REFERENCES

Castro, P.M., Barbosa-Póvoa, A.P., Matos, H.A., Novais, A.Q., 2004. Simple continuous-time formulation for short-term scheduling of batch and continuous processes. *Industrial and Engineering Chemistry Research*, 43, 105–118.

Glover, F., 1975. Improved linear integer programming formulations of nonlinear integer problems. *Management Science*, 22, 455–460.

Ierapetritou, M.G., Floudas, C.A., 1998a. Effective continuous-time formulation for short-term scheduling: 1. Multipurpose batch processes. *Industrial and Engineering Chemistry Research*, 37, 4341–4359.

Ierapetritou, M.G., Floudas, C.A., 1998b. Effective continuous-time formulation for short-term scheduling: 2. Continuous and semi-continuous processes. *Industrial and Engineering Chemistry Research*, 37, 4360–4374.

Kondili, E., Pantelides, C.C., Sargent, R.W.H., 1993. A general algorithm for short-term scheduling of batch operations. I. *MILP formulation. Computer and Chemical Engineering*, 17, 211–227.

Lin, X., Floudas, C.A., 2001. Design, synthesis and scheduling of multipurpose batch plants via an effective continuous-time formulation. *Computer and Chemical Engineering*, 25, 665–674.

Majozi, T., Zhu, X.X., 2001. A novel continuous-time MILP formulation for multipurpose bach plants. *Industrial and Engineering Chemistry Research*, 40, 5935–5949.

Maravelias, C.T., Grossmann, I.E., 2003. New general continuous-time state-task network formulation for short-term scheduling of multipurpose batch plants. *Industrial and Engineering Chemistry Research*, 42, 3056–3074.

Shaik, M., Floudas, C., 2009. Novel unified modeling approach for short term scheduling. *Industrial Engineering Chemistry Research*, 48, 2947–2964.

Shaik, M.A., Floudas, C.A., 2008. Unit-specific event-based continuous time approach for short-term scheduling of batch plants using RTN framework. *Computer and Chemical Engineering*, 32, 260–274.

Sundaramoorthy, A., Karimi, I.A., 2005. A simpler better slot-based continuous-time formulation for short-term scheduling in multipurpose batch plants. *Chemical Engineering Science*, 60, 2679–2702.

Susarla, N., Li, J., Karimi, I.A., 2010. A novel approach to scheduling of multipurpose batch plants using unit slots. *AIChE Journal*, 56, 1859–1879.

Wu, D., Ierapetritou, M., 2004. Cyclic short-term scheduling of multiproduct batch plants using continuous-time representation. *Computer and Chemical Engineering*, 28, 2271–2286.

3 Short-Term Scheduling of Multipurpose Pipeless Plants

Munawar A. Shaik

CONTENTS

3.1 INTRODUCTION

The manufacturing sector is facing constant new challenges due to the global economy and is quickly adapting to new technologies to improve its profitability by producing multiple products in multipurpose machines. In this quest, the concept of pipeless plants has evolved which offers a potential alternative for small and medium scale industries through flexible production recipes. In contrast to traditional plants, in pipeless plants material is transferred and held in mobile vessels moving between

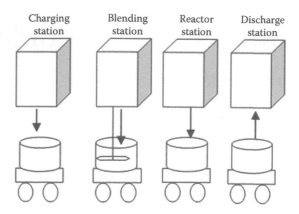

Charging station   Blending station   Reactor station   Discharge station

FIGURE 3.1 Schematic of a pipeless batch plant.

different processing stations meant for performing typical operations such as charging, mixing, heating, cooling, discharging, cleaning and so on as shown in Figure 3.1. Pipeless plants offer several advantages compared to traditional plants in terms of convenience in cleaning and maintenance. Pipeless plants can be used in many applications such as manufacturing of lubricants, paints, adhesives, food products, pharmaceuticals, and fine chemicals.

The major problems concerning pipeless plants are: scheduling and resource optimization, design and layout aspects, and vehicle routing. In the design and layout problem for pipeless plants, the objective is to determine the number of vessels required, the number of different processing stations required, and their locations or layouts. The processing stations can be arranged in different layouts such as linear, circular, or herring-bone layouts, each leading to different vessel transfer times between different processing stations. In the vehicle routing problem, the objective is to ensure conflict-free movement of vessels on automated guided vehicles (AGVs) between different processing stations. Scheduling in the context of pipeless plants involves optimal allocation of different resources such as time, processing stations, and vessels; and sequencing of various processing tasks over a given time horizon in order to meet the specified objectives. The operating efficiency of pipeless plants depends on an efficient layout, design, scheduling, and vehicle routing. In this work, we focus on scheduling aspects in pipeless plants for a given layout and design, where optimum production schedules involve the allocation of moveable vessels to different processing stations, and the sequencing of production tasks at each processing station.

A generalized framework for operation strategy of pipeless plants was presented by Liu and McGreavy (1996). Mixed-integer linear programming (MILP) formulations considering vessel dispatch rules and certain heuristics have been developed (Gonzalez and Realff 1998a,b) in addition to constraint programming approaches (Huang and Chung 2000; Zeballos and Mendez 2010). Hybrid evolutionary optimization techniques have been applied for optimal operation of pipeless plants (Piana and Engell 2009, 2010). The modeling of layout and design aspects in pipeless plants were addressed (Patsiatzis et al. 2005; Yoo et al. 2005) apart from scheduling and optimal operation aspects. Integration techniques have been proposed for simultaneous design, layout, scheduling, and routing for optimal operation of pipeless plants (Realff et al. 1996; Huang and Chung 2005).

Similar to batch scheduling literature, scheduling approaches for pipeless plants can be broadly classified into discrete-time (Pantelides et al. 1995; Realff et al. 1996) and continuous-time formulations (Bok and Park 1998; Huang and Chung 2000; Ferrer-Nadal et al. 2007). Both slot-based (Bok and Park 1998) and precedence-based (Huang and Chung 2000; Ferrer-Nadal et al. 2007) models have been applied for pipeless plants. However, there are no applications of unit-specific event-based models for scheduling of pipeless plants, although they have been extensively used for batch scheduling (Floudas and Lin 2004; Shaik et al. 2006; Shaik and Floudas 2008, 2009; Seid and Majozi 2012; Vooradi and Shaik 2012). Owing to nonuniform locations of events, unit-specific event-based models require fewer events to solve a scheduling problem compared to other continuous-time based models. In this work, we present an extension of the three-index unit-specific event-based model of Vooradi and Shaik (2012) for the scheduling of pipeless plants assuming a fixed design and layout of processing stations without considering detailed vehicle routing. Two scheduling models are presented based on state-task network (STN) process representation and unit-specific event-based continuous time formulation.

The rest of the chapter is organized as follows. In Section 3.2, the problem statement for scheduling of pipeless plants is presented followed by a discussion on STN representation. In Section 3.3, the proposed models are described. Finally, in Section 3.4 computational results are presented for two benchmark examples to illustrate the effectiveness of the proposed approaches.

3.2 PROBLEM STATEMENT

The scheduling problem addressed in this study can be stated as follows:

Given: (i) the design and layout of processing stations: design means that the number of processing stations and vessels are specified, and layout determines the transfer times between stations, (ii) the production recipe, that is, processing times for each task in suitable units, precedence of tasks, states, and units, (iii) capacity limits for processing stations and vessels, and (iv) the scheduling horizon and/or demands for final products.

Determine: (i) the optimal sequence of tasks taking place at each processing station, (ii) the start and end times of different tasks in each vessel and at each processing station, (iii) the amount of material processed and stored at each time in each vessel and at each processing station, and (iv) vessel utilization profiles.

The objective considered is minimizing the makespan for a specified demand of final products. The models presented do not take into account the routing of AGVs and it has been assumed that the conflict-free movement of vessels can be taken care of separately.

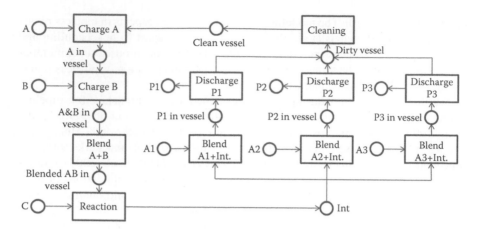

FIGURE 3.2 STN representation for a typical pipeless plant. (Adapted from *Comput Chem Eng*, 24, Huang, W., Chung, P.W.H., Scheduling of pipeless batch plants using constraint satisfaction techniques, 377–383, Copyright 2000, with permission from Elsevier.)

The first step in developing a scheduling model is to convert the process flow diagram into an STN representation. A typical STN for pipeless plants is shown in Figure 3.2 (Huang and Chung, 2000), where for instance, consider that there are three products to be produced using a specified layout of eight processing stations. All products must go through a specified sequence of operations: charging A, charging B, blending A and B, reaction, blending product-specific additives, discharging, and cleaning. Each operation is modeled as a task (represented by a rectangular block) and all raw materials, intermediates, and products are treated as separate states (represented by circles). A separate task must be defined for each operation suitable in multiple processing stations depending on the nature of the decision variables defined in the model. For instance, there are two blenders suitable for each blending task and two reactors suitable for reaction task. In addition, a separate task must be defined for each operation producing each product. For instance, task "charge A" is to be converted into three tasks suitable for three products. When the number of vessels available is less than the number of batches of all products to be produced, then vessels must be treated as separate states and the relevant tasks must be converted into as many tasks as the number of vessels. Otherwise, when there are more vessels available then vessels need not be considered in states and tasks.

3.3   MATHEMATICAL FORMULATION

In this section, we present an extension of the three-index unit-specific event-based model of Vooradi and Shaik (2012) for the scheduling of pipeless plants. Two scheduling models (M1 and M2) are presented which differ in the treatment of material balances, vessel balances, and production sequencing constraints. The first model (M1) is presented in Section 3.1, followed by the second model (M2) presented in Section 3.2.

3.3.1   MODEL M1[*]

3.3.1.1   Allocation Constraints

$$\sum_{i\in I_j} \sum_{\substack{n'\in N \\ n-\Delta n\le n'\le n}} \sum_{\substack{n''\in N \\ n\le n''\le n'+\Delta n}} w(i,n',n'') \le 1 \quad \forall j \in J, \, n \in N \tag{3.1}$$

This constraint states that at each processing station only one task can occur in only one vessel at each event. Here Δn is a parameter that allows tasks to occur over multiple events if necessary.

3.3.1.2   Capacity Constraints

$$B_i^{\min} w(i,n,n') \le b(i,n,n') \le B_i^{\max} w(i,n,n')$$
$$\forall i \in I, \, n,n' \in N, \, n \le n' \le n + \Delta n \tag{3.2}$$

This constraint states that the batch size to be produced for each task is within the available capacity limits of the vessels and processing stations.

3.3.1.3   Material Balances

$$ST(s,n) = ST(s,n-1) + \sum_{i\in I_s^p} \rho_{is} \sum_{\substack{n\in N \\ n-1-\Delta n\le n'\le n-1}} b(i,n',n-1) + \sum_{i\in I_s^c} \rho_{is} \sum_{\substack{n'\in N \\ n\le n'\le n+\Delta n}} b(i,n,n')$$

$$\forall s \in S, \, n \in N, \, n > 1$$

$$\tag{3.3a}$$

$$ST(s,n) = ST_o(s) + \sum_{i\in I_s^c} \rho_{is} \sum_{\substack{n'\in N \\ n\le n'\le n+\Delta n}} b(i,n,n')$$

$$\forall s \in S, \, n \in N, \, n = 1 \tag{3.3b}$$

The material balance equations are written for all states including vessels, and it takes into account production and consumption of each state at each event, which has to be as per the production sequence given in the STN, specified by parameter ρ_{is} which is positive for producing tasks and negative for consuming tasks.

3.3.1.4   Duration Constraints

$$T^f(i,n) = T^s(i,n) + (\alpha_i + \tau_i)w(i,n,n')$$
$$\forall i \in I, \, n,n' \in N, \, \Delta n = 0 \tag{3.4a}$$

---

[*] Adapted from *Comput Chem Eng*, 43, Vooradi, R., Shaik, M.A., Improved three index unit specific event based model for the short term scheduling of batch plants, 148–172, Copyright 2012, with permission from Elsevier.

$$T^f(i,n') \geq T^s(i,n) + (\alpha_i + \tau_i)w(i,n,n')$$
$$\forall i \in I, \, n,n' \in N, \, n \leq n' \leq n + \Delta n, \, \Delta n > 0 \tag{3.4b}$$

$$T^f(i,n') \leq T^s(i,n) + (\alpha_i + \tau_i)w(i,n,n') + M(1 - w(i,n,n'))$$
$$\forall i \in I, \, n,n' \in N, \, n \leq n' \leq n + \Delta n, \, \Delta n > 0 \tag{3.4c}$$

In duration constraints, the end time of each task is calculated based on processing and transfer times of each task at each processing station. For simplicity, the variable part of the processing time has been ignored.

3.3.1.5 Sequencing Constraints
Same Task in Same Unit

$$T^s(i,n+1) \geq T^f(i,n) \quad \forall i \in I, \, n \in N, \, n < N \tag{3.5a}$$

$$T^s(i,n+1) \leq T^f(i,n) + M\left(1 - \sum_{\substack{n' \in N \\ n - \Delta n \leq n' \leq n}} \sum_{\substack{n'' \in N \\ n \leq n'' \leq n' + \Delta n}} w(i,n',n'')\right)$$
$$\forall i \in I, \, n \in N, \, n < N, \, \Delta n > 0 \tag{3.5b}$$

Different Tasks in Same Unit

$$T^s(i,n+1) \geq T^f(i',n)$$
$$\forall i,i' \in I_j, \, i \neq i', \, j \in J_i, \, n < N \tag{3.5c}$$

The above sequencing constraints ensure that no overlapping of tasks occur at the same processing station and that a new task will start only after completion of a previous task.

Different Tasks in Different Units

$$T^s(i,n+1) \geq T^f(i',n) - M\left(1 - \sum_{\substack{n' \in N \\ n - \Delta n \leq n' \leq n}} w(i',n',n)\right)$$
$$\forall s,i,i',j,j',n \in N, \, n < N, \, s \in S, \, i \in I_j, \, i' \in I_{j'}, \, i \neq i', \, j \neq j', \, i \in I_s^c, \, i' \in I_s^p \tag{3.5d}$$

This constraint ensures that the tasks occur on a time axis as per the production recipe and sequence specified in the STN. This constraint is written for all intermediate states including the vessels, so that vessels can be reused after they are emptied and cleaned.

3.3.1.6   Tightening Constraint

$$\sum_{i \in I_j} \sum_{n \in N} \sum_{\substack{n' \in N \\ n \le n' \le n + \Delta n}} (\alpha_i + \tau_i) w(i,n,n') \le MS \quad \forall j \in J \tag{3.6}$$

This is a general tightening constraint which ensures that the total time spent at each processing station is within the makespan time.

3.3.1.7   Demand Constraint

$$ST(s,N) + \sum_{i \in I_s^p} \rho_{is} \sum_{n=N} \sum_{n - \Delta n \le n' \le n} b(i,n',n) \ge D_s \quad \forall s \in S^P \tag{3.7}$$

$$T^f(i,N) \le MS \quad \forall i \in I \tag{3.8}$$

The demand of final products to be produced by the end of the time horizon is specified in Equation 3.7 followed by the definition of makespan time in constraint (3.8).

3.3.1.8   Bounds on Variables

$$w(i,n,n') = 0; \; b(i,n,n') = 0 \quad \forall i \in I, \; \forall n' < n$$
$$b(i,n,n') \le B_i^{max} \qquad\qquad \forall i \in I, \; \forall n,n' \in N \tag{3.9a}$$

The general bounds on different variables are specified in Equation 3.9a.

$$ST_o(s) = ST_s^0 \quad \forall s \notin S^R; \; ST_o(s) = 1 \; \forall s \in S^V \tag{3.9b}$$

For all states except raw materials, the initial amount available, if any, is specified in Equation 3.9b followed by a specified number of each vessel type available (assumed to be unity here).

$$ST(s,N) = 0 \quad \forall s \in S^I, \; s \notin S^V; \; w(i,n,N) = 0, b(i,n,N) = 0 \quad \forall i \notin I_{s \in S^P}^P, \; n \in N \tag{3.9c}$$

No intermediate states are left out in a vessel by the end of the time horizon as specified in first part of Equation 3.9c. This will ensure that new batches of the production sequence cannot start unless they can be completed within the given time. Similarly, no tasks other than those producing the final products are allowed to finish as the last event as specified in second part of Equation 3.9c.

3.3.2   Model M2

In this model, the material balances are removed based on the premise that demand can be specified by simply fixing the number of batches of each product (B_s) that need to be produced. The batch size considerations can be implicitly taken through the size of the vessel used and hence, the variable $b(i,n,n')$ is eliminated. The variable $ST(s,n)$ is also eliminated for all states except vessels. Accordingly, the material balances in model M1 have been replaced by vessel balances and the following sequencing constraints. In addition to Equations 3.1, 3.4 through 3.6, 3.8, and 3.9, the following equations complete this model.

3.3.2.1   Vessel Balances

$$ST(s,n) = ST_o(s) + \sum_{i \in I_s^c} \sum_{\substack{n' \in N \\ n \leq n' \leq n + \Delta n}} \rho_{is} w(i,n,n') \quad \forall n = 1, \; s \in S^V \tag{3.10a}$$

$$ST(s,n) = ST(s,n-1) + \sum_{i \in I_s^p} \rho_{is} \sum_{\substack{n' \in N \\ n-1-\Delta n \leq n' \leq n-1}} w(i,n',n-1) + \sum_{i \in I_s^c} \rho_{is} \sum_{\substack{n' \in N \\ n \leq n' \leq n + \Delta n}} w(i,n,n')$$

$$\forall n > 1, s \in S^V \tag{3.10b}$$

The consumption and release of vessels are monitored at each event through the above vessel balances.

3.3.2.2   Sequencing Constraint

$$\sum_{\substack{n' \in N \\ n' \leq n}} \sum_{\substack{n'' \in N \\ n' \leq n'' \leq n' + \Delta n}} \sum_{i \in I_s^c} w(i,n',n'') \leq \sum_{\substack{n' \in N \\ n' \leq n}} \sum_{\substack{n'' \in N \\ n' \leq n'' \leq n' + \Delta n \\ n'' < n}} \sum_{i \in I_s^p} w(i,n',n'') \quad \forall n, s \in S^I, s \notin S^V \tag{3.11}$$

Since the material balances (Equation 3.3) are eliminated for all intermediate states, the production recipe for each product is specified now by the sequencing constraint (3.11) as per the STN.

3.3.2.3   Demand Constraint

$$\sum_{n \in N} \sum_{i \in I_s^p} \sum_{\substack{n' \in N \\ n \leq n' \leq n + \Delta n}} w(i,n,n') \geq B_s \quad \forall s \in S^P \tag{3.12}$$

The demand constraint is now specified in terms of the number of batches of each product to be produced.

3.4 COMPUTATIONAL RESULTS

Two benchmark examples are considered from literature (Bok and Park 1998) involving the scheduling of multiproduct and multipurpose pipeless plants. In Section 3.4.1, a benchmark problem involving a multiproduct pipeless plant has been solved. In Section 3.4.2, another problem involving a multipurpose and multiproduct pipeless plant has been solved. The objective function considered is minimizing the makespan in both cases. These examples are solved using GAMS 23.5/CPLEX 12.2 on a 3.33 GHz Intel Xeon processor with 32 GB RAM running on Linux operating system. The parameter Δn is treated as zero whenever it is not mentioned in the results.

EXAMPLE 3.1

This is a simple example from Bok and Park (1998) involving a multiproduct pipeless plant producing three products using the same processing sequence for each product. A maximum of three identical vessels are available for transporting material from one processing station to another. There are seven stages of processing and there are eight stations available for production as shown in Figure 3.1. We consider the case of a single batch of production ($B_s = 1$) for each product. The STN for this example is shown in Figure 3.2. The processing times for all units and the transfer times between stages are given in Tables 3.1 and 3.2 (Bok and Park, 1998).

TABLE 3.1
Processing Times (h) for Example 3.1

Processing Time (h)	U1	U2	U3	U4	U5	U6	U7	U8
P1	0.6	0.5	0.5	0.85	0.85	0.6	0.5	0.5
P2	0.5	0.5	0.7	0.75	0.75	0.5	0.5	0.5
P3	0.5	0.6	0.5	0.65	0.65	0.6	0.5	0.5

Source: Adapted with permission from Bok, J.K., Park, S. 1998. Continuous-time modelling for short-term scheduling of multipurpose pipeless plants. *Ind Eng Chem Res* 37: 3652–3659. Copyright 1998, American Chemical Society.

TABLE 3.2
Transfer Times (h) for Example 3.1

Transfer Time (h)	Charge A	Charge B	Blend A + B	Reaction	Blend Add. + Int.	Discharging	Cleaning
P1	0.05	0.06	0.05	0.06	0.07	0.05	0.1
P2	0.05	0.06	0.05	0.06	0.07	0.05	0.1
P3	0.05	0.06	0.05	0.06	0.07	0.05	0.1

Source: Adapted with permission from Bok, J.K., Park, S. 1998. Continuous-time modelling for short-term scheduling of multipurpose pipeless plants. *Ind Eng Chem Res* 37: 3652–3659. Copyright 1998, American Chemical Society.

TABLE 3.3
Computational Results for Example 3.1

No. of Vessels	Model	No. of Events	Makespan (h)	No. of Discrete Variables	No. of Continuous Variables	No. of Constraints	CPU Time (s)
3	M1	9	5.54	243	1075	2503	1.4
	M2	9	5.54	243	541	1855	0.9
	B&P[a]	5	5.54	840	123	1887	6.5
	FMGP[b]	–	5.54	66	43	162	–
2	M1	14	8.28	474	1967	5292	17.3
	M2	14	8.28	474	1037	4104	17.4
	B&P[a]	5	9.08[c]	840	123	1895	1.3
	FMGP[b]	–	8.28	66	43	174	–
1	M1	21	12.62	603	2572	6292	20.7
	M2	21	12.62	603	1282	4744	21.7
	B&P[a]	6	12.62	1008	139	2271	2.1
	FMGP[b]	–	–	–	–	–	–

[a] B&P: Bok and Park (1998).
[b] FMGP: Ferrer-Nadal et al. (2007).
[c] Suboptimal solution.

A comparison of results for both the models presented earlier (M1 and M2) is given in Table 3.3, where the results are compared with other literature models. It can be observed from the results that makespan increases with decrease in the number of vessels available, as expected; reassignment of vessels can only be done after a suitable batch of production is complete. It can be inferred from the results that model M2 performs better than model M1 as it requires fewer continuous variables and constraints, although the CPU times are comparable. The model of Ferrer-Nadal et al. (2007) performs the best in the first two cases for this example, owing to a smaller problem size, but it has limited applications for general multipurpose problems, owing to the use of precedence-based formulation. The results for the Bok and Park (1998) model given in Table 3.3 (referred to as B&P) are based on our reproducing of their model under the same hardware and software conditions. Although B&P requires fewer events compared to models M1 and M2, for the second case (two vessels available) it gives a suboptimal solution and is unable to find the optimal solution (makespan time of 8.28 h) even at a higher number of slots. The Gantt chart for the two-vessel case is shown in Figure 3.3.

EXAMPLE 3.2

This is a relatively complex example from Bok and Park (1998) involving a multipurpose and multiproduct plant producing six products with different processing sequence for each product. A maximum of six identical vessels are available for material transfer. There are seven processing steps and eight stations are available

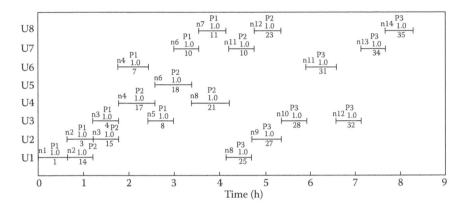

FIGURE 3.3 Gantt chart for the case of two vessels for Example 3.1.

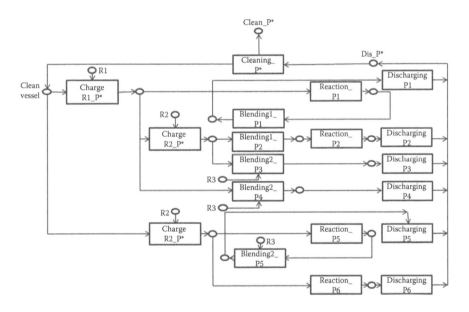

FIGURE 3.4 STN representation for Example 3.2.

for production. We consider the case of a single batch of production for each product. The STN for this example is shown in Figure 3.4. The relevant data is given in Tables 3.4 and 3.5 (Bok and Park, 1998).

A comparison of results for both models presented earlier (M1 and M2) is given in Table 3.6. It can be observed from the results that model M2 again performs better than model M1 as it requires fewer continuous variables and constraints. The CPU time required for model M2 is also less compared to model M1 in all cases, except for the single-vessel case. The Gantt chart for the two vessel case is shown in Figure 3.5.

TABLE 3.4
Processing Times (h) for Example 3.2

Processing Time (h)	U1	U2	U3	U4	U5	U6	U7	U8
P1	0.5	0.5	0.5	0.75	0.75	0.75	0.6	0.7
P2	0.6	0.4	0.6	0.65	0.55	0.55	0.5	0.5
P3	0.4	0.6	0.4	0.45	0.75	0.75	0.6	0.6
P4	0.3	0.6	0.5	0.75	0.65	0.65	0.7	0.5
P5	0.5	0.4	0.5	0.65	0.75	0.75	0.5	0.7
P6	0.5	0.6	0.6	0.75	0.65	0.65	0.5	0.5

Source: Adapted with permission from Bok, J.K., Park, S. 1998. Continuous-time modelling for short-term scheduling of multipurpose pipeless plants. *Ind Eng Chem Res* 37: 3652–3659. Copyright 1998, American Chemical Society.

TABLE 3.5
Transfer Times (h) for Example 3.2

Transfer Time (h)	Charge A	Charge B	Blending 1	Reaction	Blending 2		Discharging	Cleaning
					U3	U4		
P1	0.05	0.05	0.05	0.05	0.07	0.08	0.05	0.1
P2	0.05	0.05	0.05	0.05	0.07	0.08	0.05	0.1
P3	0.05	0.05	0.05	0.05	0.07	0.08	0.05	0.1
P4	0.05	0.05	0.05	0.05	0.07	0.08	0.05	0.1
P5	0.05	0.05	0.05	0.05	0.07	0.08	0.05	0.1
P6	0.05	0.05	0.05	0.05	0.07	0.08	0.05	0.1

Source: Adapted with permission from Bok, J.K., Park, S. 1998. Continuous-time modelling for short-term scheduling of multipurpose pipeless plants. *Ind Eng Chem Res* 37: 3652–3659. Copyright 1998, American Chemical Society.

TABLE 3.6
Computational Results for Example 3.2

No. of Vessels	Model	No. of Events	Makespan (h)	No. of Discrete Variables	No. of Continuous Variables	No. of Constraints	CPU Time (s)
6	M1	9	5.75	310	1333	3270	4.3
	M2	9		310	685	2474	5.4
5	M1	9	5.80	718	2650	17,571	222.1
	M2	9		718	1594	15,935	55.9
4	M1	10	5.95	690	2588	14,285	735.3
	M2	10		690	1521	12,695	259.2
3	M1	11	6.90	638	2452	10,917	103.3
	M2	11		638	1398	9421	27.9
2	M1	15	9.25	712	2815	9882	563.5
	M2	15		712	1531	8180	326.2
1	M1	29	17.75	1070	4422	12,127	3560.6
	M2	29		1070	2234	9491	6113.8

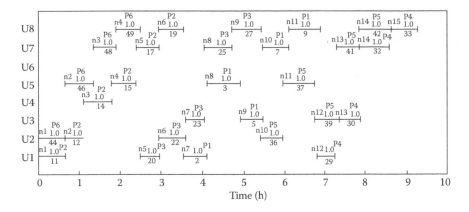

FIGURE 3.5   Gantt chart for the case of two vessels for Example 3.2.

3.5   CONCLUSION

In this work, two unit-specific event-based models are presented for scheduling pipe-less plants for a given design and layout. The proposed models are extensions of standard batch scheduling models using STN representation. By treating vessels as states, separate vessel balances and sequencing constraints are written. It has been observed that removal of material balances and batch size considerations, and replacing them with simple production sequencing constraints leads to a reduction in problem size and a better computational performance in most of the cases considered.

ACKNOWLEDGMENTS

The author gratefully acknowledges financial support received from the Department of Science and Technology (DST), India, under FAST TRACK scheme, Grant No. SR/FTP/ETA-0095/2011. The author would also like to acknowledge Mr. Pulkit Mathur (Masters Student from IIT Delhi) for help in implementing the case studies.

NOMENCLATURE

INDICES

i, i'	tasks
j, j'	units
n, n', n''	events
s	states

SETS

I	tasks
I_j	tasks which can be performed in unit j

I_s tasks which process state s and either produce or consume
I_s^p tasks which produce state s
I_s^c tasks which consume state s
J units
J_i units which are suitable for performing task i
N event points within the time horizon
S states
S^R states that are raw materials
S^I states that are intermediates
S^V states that are vessels
S^P states that are final products

PARAMETERS

α_i processing time of task i
Δn limit on the maximum number of events over which task i is allowed to continue
ρ_{is} proportion of state s produced ($\rho_{is} > 0$) or consumed ($\rho_{is} < 0$) by task i
τ_i transfer time of task i
B_i^{min} minimum capacity (batch size) of task i
B_i^{max} maximum capacity (batch size) of task i
B_s number of batches to be produced for products ($s \in S^P$)
M large positive number in big-M constraints
ST_s^0 initial amount of state s available

BINARY VARIABLES

$w(i,n,n')$ binary variable for assignment of task i that starts at event n and ends at event n'

POSITIVE VARIABLES

$b(i,n,n')$ amount of material undertaking task i that starts at event n and ends at event n'
$ST_o(s)$ initial amount of state ($s \in S^R$) that is required from external resources
$ST(s,n)$ excess amount of state s at event n
$T^s(i,n)$ time at which task i starts at event n
$T^f(i,n)$ time at which task i ends at event n

REFERENCES

Bok, J.K., Park, S. 1998. Continuous-time modelling for short-term scheduling of multipur-pose pipeless plants. *Ind Eng Chem Res* 37: 3652–3659.
Ferrer-Nadal, S., Mendez, C.A., Graells, M., Puigjaner, L. 2007. A novel continuous-time MILP approach for short term scheduling of multipurpose pipeless batch plants. In: V. Plesu and P. Agachi (Eds.), *17th European Symposium on Computer Aided Process Engineering (ESCAPE17).*

Floudas, C.A., Lin, X. 2004. Continuous-time versus discrete-time approaches for scheduling of chemical processes: A review. *Comp Chem Eng* 28: 2109–2129.

Gonzalez, R., Realff, M.J. 1998a. Operation of pipeless batch plants I. MILP schedules. *Comput Chem Eng* 22: 841–855.

Gonzalez, R., Realff, M.J. 1998b. Operation of pipeless batch plants II. Vessel dispatch rules. *Comput Chem Eng* 22: 857–866.

Huang, W., Chung, P.W.H. 2000. Scheduling of pipeless batch plants using constraint satisfaction techniques. *Comput Chem Eng* 24: 377–383.

Huang, W., Chung, P.W.H. 2005. Integrating routing and scheduling for pipeless plants in different layouts. *Comput Chem Eng* 29: 1069–1081.

Liu, R., McGreavy, C. 1996. A framework for operation strategy of pipeless batch plants. *Comput Chem Eng* 20: S1161–S1166.

Pantelides, C.C., Realff, M.J., Shah, N. 1995. Short-term scheduling of pipeless plants. *Chem Eng Res Des* 73: 431–444.

Patsiatzis, D.I., Xu, G., Papageorgiou, L.G. 2005. Layout aspects of pipeless batch plants. *Ind Eng Chem Res* 44: 5672–5679.

Piana, S., Engell, S. 2009. Constraint handling in the evolutionary optimization of pipeless chemical batch plants. In: *IEEE Congress on Evolutionary Computation*. pp. 2547–2553.

Piana, S., Engell, S. 2010. Hybrid evolutionary optimization of the operation of pipeless plants. *J Heuristics* 16: 311–336.

Realff, M.J., Shah, N., Pantelides, C.C. 1996. Simultaneous design, layout and scheduling of pipeless batch plants. *Comput Chem Eng* 20: 869–883.

Seid, R., Majozi, T. 2012. A robust mathematical formulation for multipurpose batch plants. *Chem Eng Sci* 68: 36–53.

Shaik, M.A., Floudas, C.A. 2008. Unit-specific event-based continuous-time approach for short-term scheduling of batch plants using RTN framework. *Comput Chem Eng* 32: 260–274.

Shaik, M.A., Floudas, C.A. 2009. Novel unified modeling approach for short-term scheduling. *Ind Eng Chem Res* 48: 2947–2964.

Shaik, M.A., Janak, S.L., Floudas, C.A. 2006. Continuous-time models for short-term scheduling of multipurpose batch plants: A comparative study. *Ind Eng Chem Res* 45: 6190–6209.

Vooradi, R., Shaik, M.A. 2012. Improved three index unit specific event based model for the short term scheduling of batch plants. *Comput Chem Eng* 43: 148–172.

Yoo, D.J., Lee I.-B., Jung, J.H. 2005. Design of pipeless chemical batch plants with queuing networks. *Ind Eng Chem Res* 44: 5630–5644.

Zeballos, J.L., Mendez, C.A. 2010. An integrated CP-based approach for scheduling of processing and transport units in pipeless plants. *Ind Eng Chem Res* 49: 1799–1811.

4 Evolution of Unit-Specific Event-Based Models in Batch Process Scheduling

Munawar A. Shaik

CONTENTS

4.1   INTRODUCTION

Scheduling is a decision making activity which involves efficient management of all resources such as time, equipment, storage, and materials; and optimal sequencing of different tasks over a given time horizon. In general there are specific objectives to be met such as maximizing profitability or productivity of a manufacturing facility minimizing total cost or makespan time. On the basis of the time horizon considered, process scheduling has been broadly classified into three types: (i) short-term scheduling (time horizon in days), (ii) medium-term scheduling (time horizon in weeks), and (iii) long-term scheduling (time horizon in months). In this work, the focus is on short-term scheduling of batch plants. Extensive reviews were written by Floudas and Lin (2004), Mendez et al. (2006), Shaik et al. (2006), Pitty and Karimi (2008), Sundaramoorthy and Maravelias (2011), Maravelias (2012), and Harjunkoski et al. (2014), who presented different developments and associated challenges in the scheduling of batch plants.

4.1.1   CLASSIFICATION OF PROCESS SCHEDULING MODELS

Different models proposed in the literature are classified into several types on the basis of time representation, handling of events, and process representation.

4.1.1.1   Time Representation

On the basis of time representation used, the mathematical models are classified into two main groups: discrete-time models and continuous-time models. Discrete-time models (Kondili et al. 1993, Pantelides 1994) divide the scheduling horizon into a finite number of discrete time intervals with known duration and provide a reference time grid for all operations. Therefore, formulating various constraints in a scheduling problem is simple and straightforward. In general, discrete-time models have two major disadvantages. First, the mathematical model and its computational efficiency is highly dependent on the number of time intervals postulated. Second, suboptimal or even infeasible schedules may be generated because of inflexibility in timing decisions. Owing to these inherent limitations of the discrete-time approaches, significant attention has been given to the development of continuous-time modeling approaches. As compared to discrete-time models, continuous-time models have the advantages of smaller problem sizes, reduction in number of variables, and flexibility in timing decisions. However, model complexity increases as we move from discrete- to continuous-time domain.

4.1.1.2   Handling of Events

Different types of event representations have been used to develop various mathematical formulations for scheduling problems. In the discrete-time formulation, global-time intervals are used. In continuous-time formulations, occurrences of events are handled using global-events, time slots, and unit-specific events. Global-event-based models use a set of events or time slots that are common across all tasks and units. Beginning and/or finishing times of tasks are linked to specific time points through binary variables (Maravelias and Grossmann 2003, Castro et al. 2004). Slot-based models divide the scheduling horizon into a number of predefined time periods with unknown duration. The starting and finishing activities of tasks are assigned to these slots. On the basis of the slots arrangement over the time horizon, these models are further classified into: process slots (or synchronous time grid) (Pinto and Grossmann 1995, Sundaramoorthy and Karimi 2005) and unit slots (or asynchronous time grid) (Susarla et al. 2010, Seid and Majozi 2012a). Precedence-based models are mostly used for sequential processes based on the concept of batch precedence (Mendez and Cerda 2003, Kopanos et al. 2010). These models are further classified into three types: unit-specific immediate precedence, immediate precedence, and general precedence.

Unit-specific event-based (USEB) models introduced the original concept of event points (Ierapetritou and Floudas 1998), which are heterogeneous locations of task occurrences along the time axis, allowing different tasks to start at different times for the same event point. Owing to this asynchronous location of event points and owing to tasks that can start and end at the same event point, these models require fewer events than global events and process-slot-based models (Janak et al. 2004, Shaik et al. 2006, Shaik and Floudas 2008, 2009) as shown in Figure 4.1 (Shaik et al. 2006).

Over the years researchers working on global-events/process-slots/uniform-time grid/synchronous-slot-based models (e.g., Maravelias and Grossmann 2003, Castro et al. 2004, Sundaramoorthy and Karimi 2005) have migrated into unit-specific events/nonuniform-time grid/multiple-time grid/unit-slot-based models (e.g., Castro and Grossmann 2005, Susarla et al. 2010), although using different synonyms, owing to the advantage of a compact problem size.

4.1.1.3   Process Representation

Based on the structure of process flowsheet used scheduling problems are classified into network and sequential represented processes. In sequential or multiproduct processes, different products follow the same processing sequence. Problems with less recipe similarities, and involving batch splitting and recycle are classified as network represented processes. State-task-network (STN) and resource-task-network (RTN) representations are popularly used for formulating mathematical models of scheduling problems, among other representations (recipe diagrams and

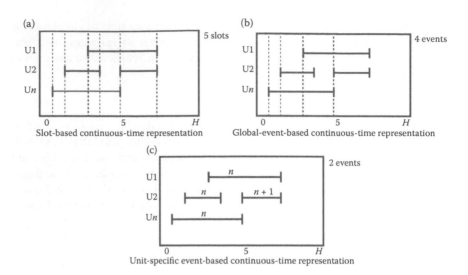

FIGURE 4.1 Schematic of different continuous-time representations. (Adapted with permission from Shaik, M.A., Janak, S.L., Floudas, C.A. 2006. Continuous-time models for short-term scheduling of multipurpose batch plants: A comparative study. *Ind Eng Chem Res* 45:6190–6209. Copyright 2006, American Chemical Society.)

state-sequence-networks). The STN representation (Kondili et al. 1993) of a process flow sheet has two types of distinctive nodes: a state node (circle) representing raw materials, intermediate states, or final products, and a task node (rectangle) representing different operations as shown in Figure 4.2a (Shaik et al. 2006). In an RTN representation (Pantelides 1994) processing equipment, storage tanks, and all resources are also explicitly shown, in addition to state and task nodes as shown in Figure 4.2b (Shaik and Floudas 2008). More importantly, the RTN representation offers unified treatment for different resources such as processing and storage units, material states, and utilities in contrast to the STN representation.

In a recipe diagram representation (Sundaramoorthy and Karimi 2005), nodes represent tasks, arcs represent various materials, and arc directions represent task precedence. In a state-sequence-network (SSN) representation (Majozi and Zhu 2001) only states are considered in network and state identity is maintained with respect to the production and consumption tasks and corresponding units.

4.1.2 IMPORTANT ISSUES RELATED TO SCHEDULING OF BATCH PLANTS

In general, batch plants are used to produce small amounts of diverse ranges of products using available resources. Therefore, batch process scheduling is extensively needed to use the shared resources effectively. Some of the important issues related to batch processes are handling of: (i) different storage policies, (ii) sequence-dependent and sequence-independent times, (iii) allowing tasks to occur over multiple events, (task splitting), (iv) utility resources, (v) unit-wait policies, (vi) material transfer, (vii) unconditional and conditional sequencing, among others.

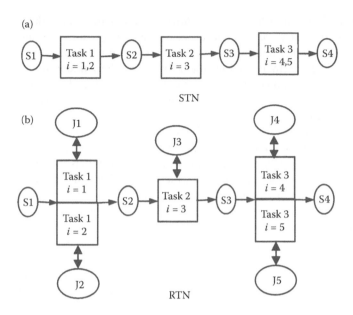

FIGURE 4.2 Schematic of (a) STN (Adapted with permission from Shaik, M.A., Janak, S.L., Floudas, C.A. 2006. Continuous-time models for short-term scheduling of multipurpose batch plants: A comparative study. *Ind Eng Chem Res* 45:6190–6209. Copyright 2006, American Chemical Society.) and (b) RTN process representations. (Adapted from *Comp Chem Eng*, 32, Shaik, M.A., Floudas, C.A. Unit-specific event-based continuous time approach for short-term scheduling of batch plants using RTN framework, 260–274, Copyright 2013, with permission from Elsevier.)

Storage policies are classified into many categories depending on whether the intermediate state or final product produced can be stored or not, or has to be consumed immediately, or has limited or unlimited storage capacity. In the dedicated finite intermediate storage (FIS) policy, the intermediate state has one or more storage tanks dedicated to each state. In the shared finite intermediate storage (SFIS) policy, each tank can store more than one state successively. In the no intermediate storage (NIS) policy, the intermediate state does not have additional storage capacity, but it can store within the same processing unit before or after processing. In the zero wait (ZW) policy, the intermediate state has to be consumed immediately as and when it is produced. In the unlimited intermediate storage (UIS) policy, the intermediate state has unlimited storage capacity. In addition to these policies, some states may also have storage time limitations.

Handling of *setup times* (or costs) is one of the most common complications in scheduling problems. The setup time can be generally defined as the time required to prepare the necessary resources (e.g., processing units, man power) to perform a specific task (e.g., job, operation). Scheduling problems involving setup times can be mainly categorized into sequence-dependent and sequence-independent setups. A *material transfer* represents the transfer of material from one batch unit to the other unit. In literature, most of the modeling approaches neglect transfer time between different batches. *Task splitting* refers to allowing tasks to occur over multiple events instead of a single event. A task occurring over multiple events typically starts at one event (where materials are consumed) and may continue over several events before finally ending at a later event (where materials are produced). In discrete-time and global-event-based models, tasks naturally occur over multiple events, whereas in USEB models this feature has been later incorporated to avoid suboptimal solutions.

Most of the batch processes require several *utility resources* to maintain the required operating conditions and these utilities are broadly classified into two types: (i) continuous resources which are available continuously (steam and cooling water [CW], etc.), (ii) discrete resources which are available in integer numbers (man power availability and processing units, etc.). *Unit-wait times* for FIS and NIS intermediate states in processing units are classified as (i) postprocessing unit-wait times and (ii) preprocessing unit-wait times. Postprocessing unit-wait time is the waiting time allowed for the state after production in that unit. Preprocessing unit-wait time refers to the waiting time allowed for a state before processing in that unit (perhaps waiting for production of another feed state). Preprocessing unit-wait times facilitate nonsimultaneous *material transfer* into processing units if multiple states are to be consumed by a task.

Explicit sequencing constraints between producing and consuming tasks are generally not required for global-event and process-slot-based models, whereas USEB models require explicit sequencing constraints. These constraints can be classified into two types: (i) unconditional sequencing, and (ii) conditional sequencing as discussed in Vooradi and Shaik (2013). In *unconditional-sequencing*-based models, different tasks in different units are always aligned without monitoring the actual material flows. In other words, a downstream consumption task is always aligned to occur after an upstream production task irrespective of whether the consuming task is actually receiving material from the producing task or not (it could receive material from inventory). In *conditional-sequencing* models, production and consumption tasks are aligned conditionally and only if the material produced by a given production task is actually used by a given consumption task, or if the material to be stored is not within the storage capacity.

The complexity, computational efficiency, and optimality of different batch-scheduling models strongly depend on how the above-mentioned different features are handled, the size of the problem considered, and size of the resulting model. Despite significant advancements in modeling and solution approaches in the past two decades, it is still a challenging task to solve larger scheduling problems with complex features. Hence, there is a global concern to develop innovative techniques to handle complex features of batch processes and to reduce the model size to enable their efficient solution in reasonable computational time.

The rest of the chapter is organized as follows. In Section 4.2, a brief literature review is presented accounting for different developments in the recent past based on USEB models. In Section 4.3, few mathematical models for handling some important features are summarized. Finally in Section 4.4, illustrative computational results are presented for selected benchmark examples from literature.

4.2 LITERATURE REVIEW

The primary goal of this review is to highlight different stages of the evolution of USEB models for short-term scheduling of batch processes based on the above-mentioned important features. Ierapetritou and Floudas (1998) proposed the first USEB model based on STN representation for short-term scheduling of multipurpose batch plants, which is applicable only for UIS policy (as shown in Shaik et al. 2006, who presented an improved version of this model applicable for UIS policy). For the FIS policy, the model of Ierapetritou and Floudas (1998) gives real-time storage violations (as shown in Shaik and Floudas 2008). The model of Ierapetritou and Floudas (1998) can be categorized as a two-index model (indices for tasks and events) since it does not allow task splitting, which is another important issue not addressed in their model. Unlike USEB models, global-event-based models proposed by Maravelias and Grossmann (2003) (STN based) and Castro et al. (2004) (RTN based) do not require explicit task splitting or time sequencing constraints because events across units are already aligned to a common time grid. Shaik et al. (2006) presented a comparison of different continuous-time approaches for UIS policy and for problems that do not involve any utility resources. They showed that USEB models outperform compared to process-slot-based and global-event-based models, based on results from selected benchmark examples drawn from literature.

Lin and Floudas (2001) extended the USEB model to account for design, synthesis, and scheduling. It is also a two-index model, but applicable for handling FIS by treating storage as a separate task. Janak et al. (2004) proposed an enhanced USEB model that used three different sets of two-index binary and continuous variables for assigning the start, end, and continuation of tasks to effectively handle task splitting, which is necessary for modeling utility resources. Their model can handle various storage policies (UIS, FIS, NIS, and ZW), and storage was treated as a separate task for handling FIS. The sequencing constraints used to handle utility resources were globally aligned in their model (as shown recently by Shaik and Vooradi 2013). Janak et al. (2004) reported better results compared to the global-event-based models (Maravelias and Grossmann 2003). Shaik and Floudas (2008) proposed the first RTN-based USEB formulation for batch scheduling. They proposed novel sequencing constraints for handling dedicated finite storage without the need for considering storage as a separate task (unlike Lin and Floudas 2001, Janak et al. 2004) and for handling sequence-dependent and sequence-independent changeovers. This model offered a first level unification of STN- and RTN-based USEB models for problems involving no utility resources. They have shown that the resulting constraints in STN- and RTN-based models are similar (except in the way the allocation constraints are written). This is also a two-index model that does not allow task splitting and is not applicable for problems involving utility resources.

Castro and Grossmann (2005) and Castro and Novais (2008) proposed multiple-time grid models for short-term scheduling of multistage multiproduct batch plants with UIS policy. Castro and Novais (2008) used explicit time grids for intermediate storage units, which result in fewer events compared to the two-index model of Shaik and Floudas (2008). However, their model is applicable only for UIS, whereas Shaik and Floudas (2008) model is general and applicable for FIS as well.

Janak and Floudas (2008) presented a USEB formulation by extending their earlier model (Janak et al. 2004) to consider partial task splitting (where only a few selected tasks are allowed to occur over multiple events). They could solve the challenging Westenberger–Kallrath benchmark problem (Kallrath 2002), which involves complex features such as variable fractions of production and different storage policies. For large problem instances, their model could not close the integrality gap. They initially solved the problem without considering storage tasks and if there were any storage violations then the storage constraints were considered for the states violating the storage capacity.

Shaik and Floudas (2009) proposed a generic unified model for short-term scheduling of batch plants using three-index binary and continuous variables, where the three indices correspond to task, starting event, and ending event. It was shown that task splitting is a necessary feature to find optimal solutions to batch-scheduling problems involving no utility resources as well. Their unified model could handle both problems involved with and without utility resources effectively, unlike the model of Janak et al. (2004) that was originally proposed to handle problems with utility resources. In their comparative study, they demonstrated that USEB models perform better than the slot-based and global-event-based models and require a minimum number of events to find optimal solutions for both UIS and FIS policies and for problems involving utility resources as well. The generality of the Shaik and Floudas (2009) model (relative to the two-index RTN model of Shaik and Floudas 2008) was further demonstrated by Li et al. (2010) who presented additional complex scenarios.

Li and Floudas (2010) extended the model of Shaik and Floudas (2009) by considering the postprocessing unit-wait policy. Recently, Vooradi and Shaik (2012a) extended this model and proposed improved allocation, duration, and sequencing constraints through effective use of the three-index variables based on the concept of the active task leading to fewer constraints and big-M constraints. Owing to these improvements, they could effectively solve the Westenberger–Kallrath challenging benchmark problem. In all the above USEB models, including Shaik and Floudas (2008, 2009), nonsimultaneous material transfers are not allowed and they assume no waiting of material in a unit except at the last event, unlike in the models of Li and Floudas (2010) and Vooradi and Shaik (2012a) where the postprocessing unit-wait policy is modeled. Susarla et al. (2010) presented a unit-slot-based MILP model that can handle nonsimultaneous material transfers, and various unit-wait and storage policies. Similar to USEB models, this formulation requires a minimum number of slots/events and has better computational efficiency compared to single-time grid approaches. However, all the above USEB models inadvertently assumed *unconditional sequencing* of producing and consuming tasks, and also they used *unconditional alignment* of producing and consuming tasks irrespective of whether there is enough storage capacity or not for storing the amount produced at a previous event (as shown recently by Seid and Majozi 2012a).

Seid and Majozi (2012a) proposed a USEB model, an improvement over their earlier model based on SSN representation (Majozi and Zhu 2001), that can handle additional features such as nonsimultaneous material transfers, different unit-wait policies, and conditional sequencing where production and consumption tasks are aligned conditionally. They modeled these features without the need for explicitly allowing tasks to occur over multiple events, and demonstrated that the above features have the potential to reduce the number of events required for solving batch-scheduling problems. Their model cannot handle utility or discrete resources. Li and Floudas (2010) and Seid and Majozi (2012b) presented approaches to estimate the number of events required for USEB models. Recently, Seid and Majozi (2013) extended their model for solving design, synthesis, and scheduling problems simultaneously.

Vooradi and Shaik (2013) proposed a rigorous scheduling model based on four-index binary and continuous variables that overcomes the various limitations of earlier USEB models in terms of handling nonsimultaneous material transfers, and comprehensive sequencing constraints for handling different storage policies, unit-wait policies, and utility resources. They identified some limitations of the model of Seid and Majozi (2012a) that: (i) they assumed *partial conditional sequencing*, that is, they align all production tasks with all consumption tasks even if a single consumption task uses material from a production task; (ii) a similar assumption for the alignment of producing and consuming tasks of dedicated FIS states that when the material to be stored is not within its storage capacity, then all producing and all consuming tasks are aligned irrespective of whether such alignment is required for all tasks or not; and (iii) that their model may lead to real-time storage violations. To overcome these limitations, Vooradi and Shaik (2013) proposed *rigorous conditional sequencing* where production and consumption tasks are aligned conditionally by accurately monitoring the material flow from each production task to each consumption task. They extended this concept to effectively handle different storage policies, unit-wait policies, and utility resources, resulting in a further reduction in the number of events required through rigorous alignment compared to partial alignment (as done by Seid and Majozi 2012a).

All batch-scheduling models in the literature traditionally assumed that a consumption task typically starts at a later event relative to a production task. Recently, Vooradi and Shaik (2012b) extended the USEB models to allow production and consumption tasks to occur at the same event, for the first time. So far, this concept was only used in scheduling continuous plants. They demonstrated that this feature leads to further reduction in the number of events required for solving batch-scheduling problems that do not have recycle states and utility resources (Vooradi 2013).

For problems involving utility resources, recently, Shaik and Vooradi (2013) presented a unification of the STN- and RTN-based USEB models through a novel resource balance that allows unit-specific alignment for sequencing of utility-related tasks (unlike the global alignment used by Janak et al. 2004, Shaik and Floudas 2009). This feature resulted in a reduction in the number of events required to find optimal solutions for USEB models.

A summary of different features and the capabilities of selected recent USEB models is given in Table 4.1 (Vooradi and Shaik 2013) to enable direct comparison. From the detailed literature survey presented earlier, it is evident that the USEB (or unit-slot or multiple-time grid) models offer a promising tool, among others, for solving batch-scheduling problems. Although there have been several approaches presented in the literature leading to a reduction in the number of events and the

TABLE 4.1
Summary of Comparison of USEB Models

Issues/Models	Susarla et al. (2010)	Seid and Majozi (2012a)	Vooradi and Shaik (2012a)	Shaik and Vooradi (2013)	Vooradi and Shaik (2013)
Continuous-time representation	Unit slots	USEB	USEB	USEB	USEB
Sequencing of production and consumption tasks	Unconditional (always aligned)	Partial conditional	Unconditional (always aligned)	Unconditional (always aligned)	Rigorous conditional
Unit-wait policy	Can handle post and preprocessing unit-wait policies	• Can handle post and preprocessing unit-wait policies • Sequencing constraints between unit wait and processing tasks are missing	Can handle postprocessing unit-wait policy	Can handle postprocessing unit-wait policy	Can handle post and preprocessing unit-wait policy
Tasks allowed to occur over multiple events	Modeled using 0–1 continuous variables, but without using Δn	Not modeled explicitly	Modeled using three-index variables and used Δn to control number of events over which a task can continue	Modeled using three-index variables and used Δn to control number of events over which a task can continue	Modeled using three-index variables and used Δn to control number of events over which a task can continue
Utility resources	Not modeled	Not modeled	Can be handled based on global alignment and unconditional sequencing of utility-related tasks	Can be handled based on unit-specific alignment and unconditional sequencing of utility-related tasks	Can be handled based on unit-specific alignment and rigorous conditional sequencing of utility-related tasks
ZW/NIS	Modeled using unconditional sequencing	• Not modeled explicitly (but can be handled) • If modeled, it will be similar to unconditional sequencing	Modeled using unconditional sequencing	Modeled using unconditional sequencing	Modeled using rigorous conditional sequencing

Source: Adapted with permission from Vooradi, R., Shaik, M.A. 2013. Rigorous unit-specific event-based model for short-term scheduling of batch plants with conditional sequencing and unit-wait times. *Ind Eng Chem Res* 52:12950–12972. Copyright 2013, American Chemical Society.

resulting problem size, the computational complexity of the resulting models is still an important issue especially when dealing with large-scale or industrial-scale scheduling problems. Despite the significant advances in computing power and computational software used for solving optimization problems in the recent times, it is still difficult to find optimal or good feasible solutions in industrial applications (Harjunkoski et al. 2014).

4.3 SELECTED MATHEMATICAL MODELS

In this section, we present selected recent USEB models and describe how the above-mentioned important features have been incorporated over the years. The constraints used in different models are summarized below based on a common nomenclature adapted across different models to avoid confusion.

4.3.1 MODEL OF SHAIK AND FLOUDAS (2009)[*]

Three-index binary and continuous variables were defined for effectively handling task splitting and parameter Δn was used to control the problem size.

4.3.1.1 Allocation Constraints

$$\sum_{i \in I_j} \sum_{\substack{n' \in N \\ n \leq n' \leq n + \Delta n}} w(i,n,n') \leq 1 \quad \forall j \in J, \, n \in N \tag{4.1a}$$

$$\sum_{i \in I_j} \sum_{\substack{n \in N \\ n' - \Delta n \leq n \leq n'}} w(i,n,n') \leq 1 \quad \forall j \in J, \, n' \in N, \, \Delta n > 0 \tag{4.1b}$$

$$\sum_{\substack{i' \in I_j \\ i \neq i'}} \sum_{\substack{n' \in N \\ n - \Delta n \leq n' \leq n}} w(i',n',n) \leq 1 - \sum_{\substack{n' \in N \\ n \leq n' \leq n + \Delta n}} w(i,n,n') \quad \forall i \in I_j, \, j \in J, \, n \in N, \, \Delta n > 0 \tag{4.1c}$$

$$\sum_{\substack{n' \in N \\ n \leq n' \leq n + \Delta n}} w(i,n,n') \leq 1 - \sum_{\substack{n' \in N \\ n' < n}} \sum_{\substack{n'' \in N \\ n' \leq n'' \leq n' + \Delta n}} \sum_{j \in J_i} \sum_{i' \in I_j} w(i',n',n'')$$

$$+ \sum_{\substack{n'' \in N}} \sum_{\substack{n' \in N, n' < n \\ n'' \leq n' \leq n'' + \Delta n}} \sum_{j \in J_i} \sum_{i' \in I_j} w(i',n'',n') \quad \forall i \in I, \, n \in N, \, n > 1, \, \Delta n > 0 \tag{4.1d}$$

[*] Adapted with permission from Shaik, M.A., Floudas, C.A. 2009. Novel unified modeling approach for short-term scheduling. *Ind Eng Chem Res* 48:2947–2964. Copyright 2009, American Chemical Society.

$$\sum_{\substack{n'\in N \\ n-\Delta n \leq n' \leq n}} w(i,n',n) \leq \sum_{\substack{n'\in N \\ n'\leq n}} \sum_{\substack{n''\in N \\ n'\leq n''\leq n'+\Delta n}} w(i,n',n'') - \sum_{\substack{n''\in N}} \sum_{\substack{n'\in N, n'<n \\ n''\leq n'\leq n''+\Delta n}} w(i,n'',n')$$

$$\forall i \in I,\ n \in N,\ \Delta n > 0 \tag{4.1e}$$

Five different allocations constraints were defined to allow tasks to occur over multiple events in order to avoid overlapping with other tasks.

4.3.1.2 Capacity Constraints

$$B_i^{min} w(i,n,n') \leq b(i,n,n') \leq B_i^{max} w(i,n,n') \quad \forall i \in I, n,n' \in N, n \leq n' \leq n + \Delta n \tag{4.2}$$

4.3.1.3 Material Balances

$$E(s,n) = E(s,n-1) + \sum_{i\in I_s^p} \rho_{is} \sum_{\substack{n'\in N \\ n-1-\Delta n \leq n' \leq n-1}} b(i,n',n-1) + \sum_{i\in I_s^c} \rho_{is} \sum_{\substack{n'\in N \\ n\leq n'\leq n+\Delta n}} b(i,n,n')$$

$$\forall s \in S,\ n \in N,\ n > 1 \tag{4.3a}$$

$$E(s,n) = E_s^0 + \sum_{i\in I_s^c} \rho_{is} \sum_{\substack{n'\in N \\ n\leq n'\leq n+\Delta n}} b(i,n,n') \quad \forall s \in S,\ n = 1 \tag{4.3b}$$

These are standard material balances written event-wise and owing to the unit-specific nature of events; these balances are to be accompanied by proper sequencing constraints for different tasks in different units so that the production and consumption tasks are aligned in real time.

4.3.1.4 Duration Constraints

$$T^f(i,n) = T^s(i,n) + \gamma_i w(i,n,n) + \delta_i b(i,n,n) \quad \forall i \in I, n \in N, \Delta n = 0 \tag{4.4a}$$

$$T^f(i,n) \geq T^s(i,n) \quad \forall i \in I,\ n \in N,\ \Delta n > 0 \tag{4.4b}$$

$$T^f(i,n') \geq T^s(i,n) + \gamma_i w(i,n,n') + \delta_i b(i,n,n') - M(1 - w(i,n,n'))$$
$$\forall i \in I, n, n' \in N,\ n \leq n' \leq n + \Delta n,\ \Delta n > 0 \tag{4.4c}$$

$$T^f(i,n') \leq T^s(i,n) + \gamma_i w(i,n,n') + \delta_i b(i,n,n') + M(1 - w(i,n,n'))$$
$$\forall i \in I, n,n' \in N,\ n \leq n' \leq n + \Delta n,\ \Delta n > 0 \tag{4.4d}$$

Since tasks are allowed to start and end at the same event or at a later event, the duration of a task is calculated appropriately.

4.3.1.5   Time Sequencing Constraints

Same Task in the Same Unit:

$$T^s(i, n+1) \geq T^f(i,n) \quad \forall i \in I, \, n \in N, \, n < N \tag{4.5a}$$

$$T^s(i,n+1) \leq T^f(i,n) + M \left[1 - \left(\sum_{\substack{n' \in N \\ n' \leq n}} \sum_{\substack{n'' \in N \\ n' \leq n'' \leq n' + \Delta n}} w(i,n',n'') - \sum_{n'' \in N} \sum_{\substack{n' \in N, n' < n \\ n'' \leq n' \leq n'' + \Delta n}} w(i,n'',n') \right) \right]$$

$$+ M \sum_{\substack{n' \in N \\ n - \Delta n \leq n' \leq n}} w(i,n',n) \quad \forall i \in I, \, n \in N, \, n < N, \, \Delta n > 0 \tag{4.5b}$$

Different Tasks in the Same Unit:

$$T^s(i, n+1) \geq T^f(i', n) \quad \forall i, \, i' \in I_j, \, i \neq i', \, j \in J, \, n \in N, \, n < N \tag{4.5c}$$

Different Tasks in Different Units:

$$T^s(i, n+1) \geq T^f(i', n) - M \left(1 - \sum_{\substack{n' \in N \\ n - \Delta n \leq n' \leq n}} w(i', n', n) \right)$$

$$\forall s \in S, i \in I^c_s, i' \in I^p_s, i \in I_j, i' \in I_{j'}, i \neq i', j, j' \in J, j \neq j', n \in N, n < N \tag{4.5d}$$

Owing to the unit-specific nature of events, the starting and ending times of different tasks must be properly aligned using the above sequencing constraints to avoid overlapping of tasks in real time and to enforce the production-sequencing order. These production-sequencing constraints are one of the unique features of USEB models. Equation 4.5d enforces unconditional sequencing because it always aligns an active production task with a consumption task even though the material produced by the production task is not used by the consumption task.

4.3.1.6   Tightening Constraints

$$\sum_{i \in I_j} \sum_{n \in N} \sum_{\substack{n' \in N \\ n \leq n' \leq n + \Delta n}} \left(\gamma_i w(i, n, n') + \delta_i b(i, n, n') \right) \leq H \quad \forall j \in J \tag{4.6}$$

Depending on the nature of the objective function, either H or MS (makespan) is used in the tightening constraint.

4.3.1.7   Resource Constraints

Utility Resource Balance:

$$
\sum_{i \in I_u} \left[\gamma_{iu} \left(\sum_{\substack{n' \in N \\ n' \leq n}} \sum_{\substack{n'' \in N \\ n' \leq n'' \leq n' + \Delta n}} w(i,n',n'') - \sum_{\substack{n'' \in N \\ n'' \leq n' \leq n'' + \Delta n}} \sum_{\substack{n' \in N, n' < n}} w(i,n'',n') \right) \right]
$$

$$
+ \sum_{i \in I_u} \left[\delta_{iu} \left(\sum_{\substack{n' \in N \\ n' \leq n}} \sum_{\substack{n'' \in N \\ n' \leq n'' \leq n' + \Delta n}} b(i,n',n'') - \sum_{\substack{n' \in N \\ n'' \leq n' \leq n'' + \Delta n}} \sum_{\substack{n'' \in N, n' < n}} b(i,n'',n') \right) \right] \leq E_u^0 \quad \forall u \in U, \ n \in N
$$

(4.7a)

Sequencing of Utility-Related Tasks:

$$
T_{ut}^s(u, n+1) \geq T_{ut}^s(u,n) \quad \forall u \in U, \ n \in N, \ n < N \tag{4.7b}
$$

$$
T_{ut}^s(u,n) \geq T^s(i,n) - M \left[1 - \left(\sum_{\substack{n' \in N \\ n' \leq n}} \sum_{\substack{n'' \in N \\ n' \leq n'' \leq n' + \Delta n}} w(i,n',n'') - \sum_{\substack{n'' \in N \\ n'' \leq n' \leq n'' + \Delta n}} \sum_{\substack{n' \in N, n' < n}} w(i,n'',n') \right) \right]
$$

$$
\forall u \in U, \ i \in I_u, \ n \in N \tag{4.7c}
$$

$$
T_{ut}^s(u,n) \leq T^s(i,n) + M \left[1 - \left(\sum_{\substack{n' \in N \\ n' \leq n}} \sum_{\substack{n'' \in N \\ n' \leq n'' \leq n' + \Delta n}} w(i,n',n'') - \sum_{\substack{n'' \in N \\ n'' \leq n' \leq n'' + \Delta n}} \sum_{\substack{n' \in N, n' < n}} w(i,n'',n') \right) \right]
$$

$$
\forall u \in U, \ i \in I_u, \ n \in N \tag{4.7d}
$$

$$
T^f(i, n-1) \geq T_{ut}^s(u,n) - M \left[1 - \left(\sum_{\substack{n' \in N \\ n' \leq n-1}} \sum_{\substack{n'' \in N \\ n' \leq n'' \leq n' + \Delta n}} w(i,n',n'') - \sum_{\substack{n'' \in N \\ n'' \leq n' \leq n'' + \Delta n}} \sum_{\substack{n' \in N, n' < n-1}} w(i,n'',n') \right) \right]
$$

$$
-M \sum_{\substack{n' \in N \\ n-1-\Delta n \leq n' \leq n-1}} w(i,n',n-1) \quad \forall u \in U, \ i \in I_u, \ n \in N, \ n > 1 \tag{4.7e}
$$

$$
T^f(i, n-1) \leq T_{ut}^s(u,n) + M \left[1 - \left(\sum_{\substack{n' \in N \\ n' \leq n-1}} \sum_{\substack{n'' \in N \\ n' \leq n'' \leq n' + \Delta n}} w(i,n',n'') - \sum_{\substack{n'' \in N \\ n'' \leq n' \leq n'' + \Delta n}} \sum_{\substack{n' \in N, n' < n-1}} w(i,n'',n') \right) \right]
$$

$$
\forall u \in U, \ i \in I_u, \ n \in N, \ n > 1 \tag{4.7f}
$$

To effectively account for utility consumption at each event, the start and end times of utility consumption were globally aligned to a common reference time grid.

4.3.1.8   Different Storage Policies

$$T^s(i, n+1) \leq T^f(i', n) + M \left(2 - \sum_{\substack{n' \in N \\ n - \Delta n \leq n' \leq n}} w(i', n', n) - \sum_{\substack{n' \in N \\ n+1 \leq n' \leq n+1+\Delta n}} w(i, n+1, n') \right)$$

$$\forall s \in S^{ZW}, S^{NIS}, S^{DFIS}, i \in I_s^c, i' \in I_s^p, i \in I_j, i' \in I_{j'}, i \neq i', j, j' \in J, j \neq j', n \in N, n < N$$

$$(4.8a)$$

$$T^f(i', n) \geq T^s(i, n) - M \left(1 - \sum_{\substack{n' \in N \\ n - \Delta n \leq n' \leq n}} w(i', n', n) \right)$$

$$\forall s \in S^{DFIS}, i \in I_s^c, i' \in I_s^p, i \in I_j, i' \in I_{j'}, i \neq i', j, j' \in J, j \neq j', n \in N \qquad (4.8b)$$

$$E(s, n) = 0, \quad \forall s \in S^{ZW}, \ S^{NIS}, \ \forall n \in N \qquad (4.8c)$$

$$E(s, n) \leq E_s^{max}, \quad \forall s \in S^{DFIS}, \ \forall n \in N \qquad (4.8d)$$

$$\sum_{n=N} \left(E(s, n) + \sum_{i \in I_s^p} \rho_{is} \sum_{\substack{n' \in N \\ n - \Delta n \leq n' \leq n}} b(i, n', n) \right) = 0 \quad \forall s \in S^{ZW} \qquad (4.8e)$$

For handling dedicated FIS policies and to avoid real-time storage violations, Equations 4.5a, 4.8a, and 4.8b were used without considering storage as a separate task and, for flexible FIS storage, must be treated as a separate task.

4.3.1.9   Objective Function

Maximization of Profit:

$$\text{Max Profit} = \sum_{s \in S^P} P_s \sum_{n=N} \left(E(s, n) + \sum_{i \in I_s^p} \rho_{is} \sum_{\substack{n' \in N \\ n - \Delta n \leq n' \leq n}} b(i, n', n) \right) \qquad (4.9)$$

$$\text{Min } MS \qquad (4.10a)$$

$$E(s,N) + \sum_{n=N}\sum_{i \in I_s^p} \rho_{is} \sum_{\substack{n' \in N \\ n-\Delta n \leq n' \leq n}} b(i,n',n) \geq D_s \quad \forall s \in S^f \tag{4.10b}$$

$$T^f(i,N) \leq MS \quad \forall i \in I \tag{4.10c}$$

4.3.2 MODEL OF VOORADI AND SHAIK (2012a)[*]

The model of Shaik and Floudas (2009) was improved by Vooradi and Shaik (2012a) by taking advantage of the three-index variables and effectively incorporating the concept of the active task leading to fewer constraints and big-M terms based on improvements in allocation, duration, and sequencing constraints.

4.3.2.1 Allocation Constraints

$$\sum_{i \in I_j}\sum_{\substack{n' \in N \\ n-\Delta n \leq n' \leq n}} \sum_{\substack{n'' \in N \\ n \leq n'' \leq n' + \Delta n}} w(i,n',n'') \leq 1 \quad \forall j \in J,\, n \in N \tag{4.11}$$

A single allocation constraint (4.11) was proposed replacing the five different allocation constraints Equations 4.1a through 4.1e based on the concept of the active task.

4.3.2.2 Duration Constraints

Without Unit-Wait Times:

$$T^f(i,n') \geq T^s(i,n) + \gamma_i w(i,n,n') + \delta_i b(i,n,n')$$
$$\forall i \in I,\, n,\, n' \in N,\, n \leq n' \leq n + \Delta n,\, \Delta n > 0 \tag{4.12a}$$

In the model of Shaik and Floudas (2009), the unit-wait policy is not allowed except at the last event and, accordingly, the duration constraints to be used here are Equations 4.4a, 4.12a, and 4.4d.

With Unit-Wait Times:

$$T^f(i,n) \geq T^s(i,n) + \gamma_i w(i,n,n) + \delta_i b(i,n,n) \quad \forall i \in I,\, n \in N,\, \Delta n = 0 \tag{4.12b}$$

$$T^f(i,n) \leq T^s(i,n) + \gamma_i w(i,n,n) + \delta_i b(i,n,n) + UW_i w(i,n,n)$$
$$\forall i \in I,\, n \in N,\, \Delta n = 0 \tag{4.12c}$$

---

[*] Adapted from *Comput Chem Eng*, 43, Vooradi, R., Shaik, M.A., Improved three index unit specific event based model for the short term scheduling of batch plants, 148–172, Copyright 2012, with permission from Elsevier.

$$T^f(i,n') \leq T^s(i,n) + \gamma_i w(i,n,n') + \delta_i b(i,n,n') + UW_i w(i,n,n''')$$
$$+ M\big(1 - w(i,n,n')\big) \quad \forall i \in I, n, n' \in N, n \leq n' \leq n + \Delta n, \Delta n > 0 \quad (4.12d)$$

To handle different postprocessing unit-wait policies, the modified duration constraints 4.12a through 4.12d should be used.

4.3.2.3   Time Sequencing Constraints

$$T^s(i,n+1) \leq T^f(i,n) + M\left[1 - \sum_{i \in I_j} \sum_{\substack{n' \in N \\ n-\Delta n < n' \leq n}} \sum_{\substack{n'' \in N \\ n < n'' \leq n' + \Delta n}} w(i,n',n'')\right],$$

$$\forall i \in I, \, n \in N, \, n < N, \, \Delta n > 0 \quad (4.13)$$

To handle the same task in the same unit, Equations 4.5a and 4.13 are used based on an active task concept, and there is no change in the constraint for different tasks in the same unit. For different tasks in different units, they proposed an alternate constraint by removing the big-M term.

4.3.2.4   Resource Constraints

Utility Resource Balance:

$$\sum_{i \in I_u}\left[\gamma_{iu} \sum_{\substack{n' \in N \\ n-\Delta n \leq n' \leq n}} \sum_{\substack{n'' \in N \\ n \leq n'' \leq n' + \Delta n}} \sum w(i,n',n'')\right] + \sum_{i \in I_u}\left[\delta_{iu} \sum_{\substack{n' \in N \\ n-\Delta n \leq n' \leq n}} \sum_{\substack{n'' \in N \\ n \leq n'' \leq n' + \Delta n}} b(i,n',n'')\right]$$

$$\leq E_u^0, \quad \forall u \in U, n \in N \quad (4.14a)$$

Sequencing of Utility-Related Tasks:

$$T_{ut}^s(u,n) \geq T^s(i,n) - M\left[1 - \sum_{\substack{n' \in N \\ n-\Delta n \leq n' \leq n}} \sum_{\substack{n'' \in N \\ n \leq n'' \leq n' + \Delta n}} w(i,n',n'')\right]$$

$$\forall u \in U, \, i \in I_u, \, n \in N \quad (4.14b)$$

$$T_{ut}^s(u,n) \leq T^s(i,n) + M\left[1 - \sum_{\substack{n' \in N \\ n-\Delta n \leq n' \leq n}} \sum_{\substack{n'' \in N \\ n \leq n'' \leq n' + \Delta n}} w(i,n',n'')\right]$$

$$\forall u \in U, \, i \in I_u, \, n \in N \quad (4.14c)$$

$$T^f(i,n-1) \leq T_{ut}^s(u,n) + M\left[1 - \sum_{\substack{n' \in N \\ n-1-\Delta n \leq n' \leq n-1}} \sum_{\substack{n'' \in N \\ n-1 \leq n'' \leq n'+\Delta n}} w(i,n',n'')\right]$$

$$\forall u \in U, i \in I_u, n \in N, n > 1 \tag{4.14d}$$

$$T^f(i,n-1) \geq T_{ut}^s(u,n) - M\left[1 - \sum_{\substack{n' \in N \\ n-\Delta n \leq n' \leq n-1}} \sum_{\substack{n'' \in N \\ n \leq n'' \leq n'+\Delta n}} w(i,n',n'')\right]$$

$$\forall u \in U, i \in I_u, n \in N, n > 1 \tag{4.14e}$$

On the basis of the active task concept, the resource balances and sequencing constraints are modified and Equations 4.7b, 4.14a, 4.14b through 4.14e are to be used.

4.3.3   MODEL OF VOORADI AND SHAIK (2012b)

Vooradi and Shaik (2012b) proposed a USEB model where production and consumption tasks corresponding to a given state are allowed to occur at the same event point, unlike conventional batch-scheduling models in literature where consumption tasks typically start at the next event relative to production tasks (Vooradi 2013).

4.3.3.1   Material Balances

$$E(s,n) = E(s,n-1) + \sum_{i \in I_s^p} \rho_{is} \sum_{\substack{n' \in N \\ n-\Delta n \leq n' \leq n}} b(i,n',n) + \sum_{i \in I_s^c} \rho_{is} \sum_{\substack{n' \in N \\ n \leq n' \leq n+\Delta n}} b(i,n,n')$$

$$\forall s \in S, n \in N, n > 1 \tag{4.15a}$$

$$E(s,n) = E_s^0 + \sum_{i \in I_s^p} \rho_{is} \sum_{\substack{n' \in N \\ n-\Delta n \leq n' \leq n}} b(i,n',n) + \sum_{i \in I_s^c} \rho_{is} \sum_{\substack{n' \in N \\ n \leq n' \leq n+\Delta n}} b(i,n,n')$$

$$\forall s \in S, n \in N, n = 1 \tag{4.15b}$$

The standard material balances, Equations 4.3a and 4.3b, are replaced with Equations 4.15a and 4.15b, where production and consumption tasks occur at the same event n as presented in Vooradi (2013). Accordingly, the sequencing constraint equation 4.5d for different tasks in different units is also modified as shown in the following equation:

$$T^s(i,n) \geq T^f(i',n) - M \left(1 - \sum_{\substack{n' \in N \\ n-\Delta n \leq n' \leq n}} w(i',n',n) \right)$$

$$\forall s \in S, i, i', j, j' \in J, n \in N, i \in I_j, i' \in I_{j'}, i \neq i', j \neq j', i \in I_s^c, i' \in I_s^p \quad (4.16)$$

One limitation of this model is that it is applicable only for states that do not involve any recycle streams. The normal material balances Equations 4.3a and 4.3b and the sequencing constraints in Equation 4.5d are used for recycle states.

4.3.4   MODEL OF SHAIK AND VOORADI (2013)[*]

In the above USEB models, there is no unified treatment for handling the utility or discrete resources because of the heterogeneous location of events. Usually, global alignment is enforced for different tasks that use the same utility, whereas for material states unit-specific alignment is used. Shaik and Vooradi (2013) proposed a unified framework for the handling of material states and utility resources by incorporating a novel resource balance, which led to the unification of STN- and RTN-based USEB models in the presence of utility resources.

4.3.4.1   Unified Resource Balances

$$E(r,n) = E(r,n-1) + \sum_{i \in I_r^p} \left(\gamma_{ir} \sum_{\substack{n' \in N \\ n-1-\Delta n_i \leq n' \leq n-1}} w(i,n',n-1) + \delta_{ir} \sum_{\substack{n' \in N \\ n-1-\Delta n_i \leq n' \leq n-}} b(i,n',n-1) \right)$$

$$- \sum_{i \in I_r^c} \left(\gamma_{ir} \sum_{\substack{n' \in N \\ n \leq n' \leq n+\Delta n_i}} w(i,n,n') + \delta_{ir} \sum_{\substack{n' \in N \\ n \leq n' \leq n+\Delta n_i}} b(i,n,n') \right) \forall r \in R, n \in N, n > 1$$

(4.17a)

$$E(r,n) = E_r^0 - \sum_{i \in I_r^c} \left(\gamma_{ir} \sum_{\substack{n' \in N \\ n \leq n' \leq n+\Delta n_i}} w(i,n,n') + \delta_{ir} \sum_{\substack{n' \in N \\ n \leq n' \leq n+\Delta n_i}} b(i,n,n') \right)$$

$$\forall r \in R, \ n \in N, \ n = 1 \quad (4.17b)$$

Instead of writing separate balances for material and utility resources (Equations 4.3a, 4.3b, and 4.14a), a common balance was used in Equations 4.17a and 4.17b. Here the parameters γ_{ir} and δ_{ir} have different meanings for different resources, as explained in Shaik and Vooradi (2013). Similarly, for unit-specific alignment of utility-related tasks, Equation 4.17c was used instead of Equations 4.7b, 4.14b through 4.14e.

$$T^s(i,n+1) \geq T^f(i',n)$$

$$\forall u \in U, i, i' \in I_u, i \neq i', j, j' \in J, j \neq j', i \in I_j, i' \in I_{j'}, n \in N, n < N \qquad (4.17c)$$

In addition, they presented modified sequencing constraints for handling sequence-dependent and sequence-independent changeovers. They also presented simplified handling of multiple orders using the concept of active task, and improved sequencing constraints for flexible storage cases.

4.3.5 MODEL OF VOORADI AND SHAIK (2013)[*]

Vooradi and Shaik (2013) proposed a rigorous unit-specific event-based model that allows for nonsimultaneous material transfer and rigorous conditional sequencing of each production and consumption task, only if the material produced by a given production task is used by a given consumption task. Their model can effectively handle scheduling problems with different storage, unit-wait policies, and utility resources based on four-index binary and continuous variables.

4.3.5.1 Sequencing Constraints: Based on Whether There Is Enough Material in Storage or Not

$$-\sum_{i' \in I_s^c} \rho_{i's} \sum_{\substack{n' \in N \\ n \leq n' \leq n+\Delta n}} b(i',n,n') \leq E(s,n-1) + \sum_{i' \in I_s^c} \sum_{i \in I_s^p} b1(i,i',s,n)$$

$$s \in S^I, \ s \notin S^{ZW}, \ s \notin S^{NIS}, \ n \in N, \ n > 1 \qquad (4.18a)$$

$$\rho_{is} \sum_{\substack{n' \in N \\ n-1-\Delta n \leq n' \leq n-1}} b(i,n',n-1) \geq \sum_{i' \in I_s^c} b1(i,i',s,n)$$

$$\forall s \in S^I, \ s \notin S^{ZW}, s \notin S^{NIS}, \ i \in I_s^p, \ n \in N, \ n > 1 \qquad (4.18b)$$

$$-\rho_{i's} \sum_{\substack{n' \in N \\ n \leq n' \leq n+\Delta n}} b(i',n,n') \geq \sum_{i \in I_s^p} b1(i,i',s,n)$$

$$\forall s \in S^I, \ s \notin S^{ZW}, \ s \notin S^{NIS}, \ i' \in I_s^c, \ n \in N, \ n > 1 \qquad (4.18c)$$

---

[*] Adapted with permission from Vooradi, R., Shaik, M.A. 2013. Rigorous unit-specific event-based model for short-term scheduling of batch plants with conditional sequencing and unit-wait times. *Ind Eng Chem Res* 52:12950–12972. Copyright 2013, American Chemical Society.

$$b1(i,i',s,n) \leq z(i,i',s,n)\rho_{is}B_i^{max}$$

$$\forall s \in S^I, s \notin S^{NIS}, i \in I_s^p, i' \in I_s^c, j, j' \in J, j \neq j', i \in I_j, i' \in I_{j'}, n \in N, n > 1$$

$$(4.18d)$$

Here, four-index variables $b1$ and z are activated in Equations 4.18a through 4.18d, if a consumption task receives material from a production task, accordingly the unconditional sequencing constraints are relaxed in Equations 4.19a and 4.19b.

$$T^s(i',n) \geq T^f(i,n-1) - M\left(1 - z(i,i',s,n)\right)$$

$$\forall s \in S^I, i \in I_s^p, i' \in I_s^c, j, j' \in J, j \neq j', i \in I_j, i' \in I_{j'}, n \in N, n > 1 \quad (4.19a)$$

$$T^s(i',n) \geq T^f(i,n-2) - M\left(1 - \sum_{\substack{n' \in N \\ n-2-\Delta n \leq n' \leq n-2}} w(i,n',n-2)\right)$$

$$\forall s \in S^I, s \notin S^{ZW}, s \notin S^{NIS}, i \in I_s^p, i' \in I_s^c, j, j' \in J, j \neq j', i \in I_j, i' \in I_{j'}, n \in N, n > 1$$

$$(4.19b)$$

4.3.5.2 Sequencing Constraints: Based on Whether There Is Enough Storage Capacity or Not (FIS Storage Policy)

$$\sum_{i \in I_s^p} \rho_{is} \sum_{\substack{n' \in N \\ n-1-\Delta n \leq n' \leq n-1}} b(i,n',n-1) + E(s,n-1) \leq E_s^{max} + \sum_{i \in I_s^p}\sum_{i' \in I_s^c} b2(i,i',s,n)$$

$$\forall s \in S^{FIS}, n \in N, n > 1 \tag{4.20a}$$

$$\rho_{is} \sum_{\substack{n' \in N \\ n-1-\Delta n \leq n' \leq n-1}} b(i,n',n-1) \geq \sum_{i' \in I_s^c} b2(i,i',s,n) \quad \forall s \in S^{fis}, i \in I_s^p, n \in N, n > 1$$

$$(4.20b)$$

$$-\rho_{i's} \sum_{\substack{n' \in N \\ n \leq n' \leq n+\Delta n}} b(i',n,n') \geq \sum_{i \in I_s^p} b2(i,i',s,n) \quad \forall s \in S^{FIS}, i' \in I_s^c, n \in N, n > 1$$

$$(4.20c)$$

$$b2(i,i',s,n) \leq \rho_{is}B_i^{max}\left(x(i,i',s,n) + v(i,i',s,n)\right)$$

$$\forall s \in S^{FIS}, i \in I_s^p, i' \in I_s^c, j, j' \in J, j \neq j', i \in I_j, i' \in I_{j'}, n \in N, n > 1 \quad (4.20d)$$

$$x(i,i',s,n) + v(i,i',s,n) \leq 1$$

$$\forall s \in S^{FIS}, i \in I_s^p, i' \in I_s^c, j, j' \in J, j \neq j', i \in I_j, i' \in I_{j'}, n \in N \quad (4.20e)$$

Here, four-index variable $b2$ is activated only if the total amount of material available for consumption at an event exceeds the storage capacity and, accordingly, either x or v is activated depending on whether the material is consumed immediately or it can wait in the unit based on the following sequencing constraints:

$$T^s(i',n) \leq T^f(i,n-1) + M\left(1 - x(i,i',s,n)\right)$$

$$\forall s \in S^{FIS}, i \in I_s^p, i' \in I_s^c, j, j' \in J, j \neq j', i \in I_j, i' \in I_{j'}, n \in N, n > 1 \quad (4.21a)$$

$$T^f(i'',n-1) \leq T^f(i,n-1) + M\left(1 - v(i,i',s,n)\right)$$

$$\forall s \in \left(S^{FIS} \cup S^{NIS}\right), i \in I_s^p, i' \in I_s^c, j, j' \in J, j \neq j', i \in I_j, i' \in I_{j'}, i'' \in I_{j'}, n \in N, n > 1$$

$$(4.21b)$$

4.3.5.3 Sequencing Constraints: Based on Whether There Is Enough Excess Utility Available or Not

$$\sum_{i \in I_u} \left(\gamma_{iu} \sum_{\substack{n' \in N \\ n \leq n' \leq n + \Delta n}} w(i,n,n') + \delta_{iu} \sum_{\substack{n' \in N \\ n \leq n' \leq n + \Delta n}} b(i,n,n') \right) \leq E(u,n-1)$$

$$+ \sum_{i' \in I_u} \sum_{i \in I_u} b3(i,i',s,n) \quad \forall u \in U, n \in N, n > 1 \quad (4.22a)$$

$$\gamma_{iu} \sum_{\substack{n' \in N \\ n-1-\Delta n \leq n' \leq n-1}} w(i,n',n-1) + \delta_{iu} \sum_{\substack{n' \in N \\ n-1-\Delta n \leq n' \leq n-1}} b(i,n',n-1) \geq \sum_{i' \in I_u} b3(i,i',s,n)$$

$$\forall u \in U, i \in I_u, n \in N, n > 1 \quad (4.22b)$$

$$\gamma_{i'u} \sum_{\substack{n' \in N \\ n \leq n' \leq n + \Delta n}} w(i',n,n') + \delta_{i'u} \sum_{\substack{n' \in N \\ n \leq n' \leq n + \Delta n}} b(i',n,n') \geq \sum_{i \in I_u} b3(i,i',s,n)$$

$$\forall u \in U, i' \in I_u, n \in N, n > 1 \quad (4.22c)$$

$$b3(i,i',s,n) \leq y(i,i',s,n) \, E_u^0$$

$$\forall u \in U, \ i, i' \in I_u, j, j' \in J, j \neq j', i \in I_j, i' \in I_{j'}, n \in N, n > 1 \quad (4.22d)$$

Here, four-index variables $b3$ and y are activated only if the excess amount of a utility available at a previous event is not enough for the consumption tasks at event n, and they are used to align the utility-related tasks in the following sequencing constraints:

$$T^s(i',n) \geq T^f(i,n-1) - M\left(1 - y(i,i',s,n)\right)$$
$$\forall u \in U, i,i' \in I_u, j, j' \in J, j \neq j', i \in I_j, i' \in I_{j'}, n \in N, n > 1 \quad (4.23a)$$

$$T^s(i',n) \geq T^f(i,n-2) - M\left(1 - \sum_{\substack{n' \in N \\ n-2-\Delta n \leq n' \leq n-2}} w(i,n',n-2)\right)$$
$$\forall u \in U, i,i' \in I_u, j, j' \in J, j \neq j', i \in I_j, i' \in I_{j'}, n \in N, n > 1 \quad (4.23b)$$

Similarly, their model extended the rigorous conditional sequencing concept to effectively handle ZW and NIS policies.

4.4   COMPUTATIONAL RESULTS

To illustrate the computational performance and to enable comparison among the different USEB models summarized in Table 4.1, selected benchmark examples drawn from batch-scheduling literature are considered. Standard performance metrics and practices used in batch-scheduling literature (Sundaramoorthy and Karimi 2005, Shaik et al. 2006) are adapted for comparison of different models. The metrics used for comparison are: optimality, the best possible solution at root node (relaxed MILP or RMIP), number of events and/or Δn required (if any), problem size in terms of the number of binary, continuous variables, and constraints, number of nodes required, CPU time, and integrality gap. All the examples are solved under similar software and hardware conditions to enable comparison of CPU time required to find the optimal solution. For each example reported in the results, the problems are solved using a higher number of events and/or Δn (if any) to ascertain optimality. Whenever results are directly shown as reported in the original paper for some models, some metrics such as CPU time and nodes are not given in the results, because no direct comparison is possible in these cases. Benchmark problems involving selected important features such as with and without utility resources, UIS and FIS storage policies, unit-wait policies, and two different objective functions: maximization of profit for different time horizons, and minimization of makespan for specified demands are considered.

For examples involving no utility resources, the computational results are given in Section 4.1, followed by the results for problems with utility resources given in Section 4.2. Results to challenge the Westenberger–Kallrath problem are given in Section 4.3. All the examples are solved using GAMS 23.5/CPLEX 12.2 on 2.66 GHz Intel Core 2 Duo processor with 3 GB RAM running on a Linux operating system.

By default, consider Δn as zero and the relative gap as zero when they are not mentioned in all the subsequent results.

4.4.1 Examples with No Utility Resources

EXAMPLE 4.1

This is a simple example for the STN shown in Figure 4.2a (Shaik et al. 2006), and the relevant data is taken from Shaik and Floudas (2009). The production recipe has one raw material producing two intermediates and one final product. The raw material is processed in three sequential tasks, where the first task is suitable in two units (J1 and J2), the second task is suitable in one unit (J3), and the third task is suitable in two units (J4 and J5). Since the first and third tasks are suitable in two units, two different tasks are defined in the STN. The initial stock level for all intermediates is assumed to be zero. The price of product S4 is $5/µm.

For the objective of maximizing profit, this example is solved over a time horizon of 10 h using UIS and FIS policies. The computational results are shown in Table 4.2. For UIS, it can be observed that the model of Vooradi and Shaik (2012b), referred as V&S3, requires only three events compared to other models that require five events. For the same number of events the model of Vooradi and Shaik (2013), referred as V&S2, results in more binary, continuous variables, and constraints because of the four-index variables used for rigorous-conditional alignment of production and consumption tasks.

For the FIS policy, states S2 and S3 are assumed to have a dedicated finite storage capacity of 200 and 250 µm, respectively. The computational results are additionally compared with the literature models of Susarla et al. (2010), referred as SLK2; and Seid and Majozi (2012a), referred as S&M. The results for SLK2 and S&M models are directly reported from their papers and hence, CPU times for these two models are not directly compared here owing to a difference in hardware. For SLK model, N event indicates $N - 1$ slots. The models of Shaik and Floudas (2009), referred as S&F; Vooradi and Shaik (2012a), referred as V&S1; and V&S3 require one extra event for handling FIS compared to UIS. The model of Vooradi and Shaik (2013), referred as V&S2, requires the same number of events ($n = 5$) for both UIS and FIS cases. The model of V&S3 requires the least number of events ($n = 4$).

For the objective of the minimization of makespan, Example 4.1 is solved for the FIS policy with a demand of product S4 specified as 2000 µm. $M = 50$ h is used in big-M constraints. The computational results are shown in Table 4.3. The model of V&S2 has the best computational performance for this case as it gives the optimal solution (27.881 h) using 14 events requiring 3537.85 s to solve zero integrality gap. All other models are either unable to close the integrality gap in the specified CPU time limit (40,000 s) or give suboptimal solution. So, it can be observed from these results that even for a simple benchmark example most of the models are finding it difficult to close the integrality gap for the objective of minimization of makespan, which is more difficult to solve compared to the objective of maximization of profit.

EXAMPLE 4.2

This is a popular example that has been solved by many authors in the literature. The production recipe has five processing stages: heating, reactions 1, 2, and 3, and separation, for producing two products as shown in the STN representation

TABLE 4.2

Computational Results for Examples 4.1 and 4.2 for Maximization of Profit

Model	Events	RMILP ($)	MILP ($)	CPU Time(s)	Nodes	Binary Variables	Continuous Variables	Constraints
			Example 4.1 ($H = 10$) with UIS					
S&F	5	3000.00	2628.18	0.02	16	15	87	117
V&S1	5	2914.84	2628.18	0.04	0	15	87	117
V&S2	5	3000.00	2628.18	0.05	56	29	101	201
V&S3	3	2914.84	2628.18	0.02	0	15	59	85
			Example 4.2 ($H = 10$) with UIS					
S&F($\Delta n = 1$)	6	2730.66	1962.69	4.17	9336	60	214	680
V&S1($\Delta n = 1$)	6	2727.11	1962.69	4.52	7044	60	214	505
V&S2	5	2436.69	1962.69	0.16	281	58	209	436
V&S3($\Delta n = 1$)	6	2749.79	1962.69	8.74	19,908	76	230	527

(*Continued*)

TABLE 4.2 (Continued)
Computational Results for Examples 4.1 and 4.2 for Maximization of Profit

Model	Events	RMILP ($)	MILP ($)	CPU Time(s)	Nodes	Binary Variables	Continuous Variables	Constraints
			Example 4.1 ($H = 10$) with FIS					
SLK2	7	4000.00	2628.19	–	2764	60	359	491
S&M	6	4000.00	2628.19	–	6215	60	205	368
S&F	6	3973.70	2628.19	0.16	303	20	106	137
V&S1	6	3361.12	2628.19	0.11	175	20	106	185
V&S2	5	3000.00	2628.19	0.07	63	57	118	311
V&S3	4	3361.12	2628.19	0.13	217	20	78	137
			Example 4.2 ($H = 10$) with FIS					
SLK2	7	2730.70	1962.69	–	19,043	72	449	645
S&M	6	2730.70	1962.69	–	23,837	88	383	779
S&F($\Delta n = 1$)	6	2730.66	1962.69	6.10	10,628	60	214	790
V&S1($\Delta n = 1$)	7	2775.41	1962.69	109.20	222,524	76	255	724
V&S2	5	2436.69	1962.69	0.23	254	122	265	768
V&S3($\Delta n = 1$)	7	2787.01	1962.69	154.21	268,300	92	271	746

Source: S&F: Shaik and Floudas (2009); V&S1: Vooradi and Shaik (2012a); V&S2: Vooradi and Shaik (2013); V&S3: Vooradi and Shaik (2012b); SLK2: Susarla et al. (2010); S&M: Seid and Majozi (2012a).

TABLE 4.3
Computational Results for Example 4.1 for Minimization of Makespan

Model	Events	RMILP (h)	MILP (h)	CPU Time (s)	Nodes	Binary Variables	Continuous Variables	Constraints	Relative Gap (%)
				Example 4.1 ($D_4 = 2000\ \mu m$) with FIS					
SLK2	17	24.2	28.772[b]	–	328,879	160	901	1330	–
S&M	13	24.2	27.98[b]	–	4,804,688	130	444	796	–
S&F	20	24.236	28.032[b]	40,000[a]	29,941,133	90	372	698	7.25
V&S1	20	24.492	27.881	40,000[a]	84,604,531	90	372	694	3.10
V&S2	14	24.236	27.881	3537.85	5,722,414	210	361	1009	0.00
V&S3	18	24.492	27.881	40,000[a]	87,031,983	90	344	646	3.98

Source: S&F: Shaik and Floudas (2009); V&S1: Vooradi and Shaik (2012a); V&S2: Vooradi and Shaik (2013); V&S3: Vooradi and Shaik (2012b); SLK2: Susarla et al. (2010); S&M: Seid and Majozi (2012a).

[a] Resource limit reached.
[b] Suboptimal.

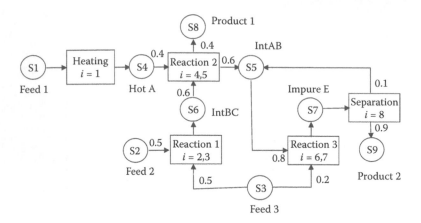

FIGURE 4.3 STN representation for Example 4.2. (Adapted with permission from Shaik, M.A., Janak, S.L., Floudas, C.A. 2006. Continuous-time models for short-term scheduling of multipurpose batch plants: A comparative study. *Ind Eng Chem Res* 45:6190–6209. Copyright 2006, American Chemical Society.)

given in Figure 4.3 (Shaik et al. 2006). The relevant data is taken from Shaik and Floudas (2009). Since each of the reaction tasks can take place in two different reactors, each reaction is represented by two separate tasks. The initial stock level for all intermediates is assumed to be zero. The price of products 1 and 2 is $10/μm.

For the objective of maximizing profit, this example is solved over a time horizon of 10 h using UIS and FIS policies. For FIS policy, states S4, S5, S6, and S7 are assumed to have a dedicated finite storage capacity of 100, 200, 150, and 200 μm, respectively. The computational results are shown in Table 4.2, where it can be observed that the model of V&S2 performs best for both UIS and FIS as it requires the least number of events ($n = 5$ and $\Delta n = 0$) and solves faster owing to rigorous conditional sequencing. For FIS, V&S1 and V&S3 require one event higher compared to the UIS case owing to the removal of the big-M term in constraint for different tasks in different units as discussed in Vooradi and Shaik (2012a). If Equation 4.5d is used, then these models also require six events ($n = 6$ and $\Delta n = 1$) similar to S&F. Since this problem involves a recycle stream (for state S5), for all states except S5, production and consumption tasks are allowed to occur at the same event in the model of V&S3. However, unlike Example 4.1, here the model of V&S3 does not give a further reduction in the number of events, owing to the presence of recycle streams which is a limitation of V&S3.

4.4.2 Examples with Utility Resources

EXAMPLE 4.3

This is an example from Maravelias and Grossmann (2003) that has been solved by many authors (Janak et al. 2004, Shaik and Floudas 2009). The problem involves four tasks in three units processing seven states as shown in the STN

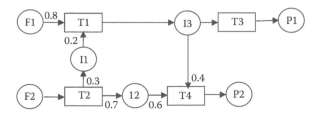

FIGURE 4.4   STN representation for Example 4.3. (Adapted with permission from Shaik, M.A., Janak, S.L., Floudas, C.A. 2006. Continuous-time models for short-term scheduling of multipurpose batch plants: A comparative study. *Ind Eng Chem Res* 45:6190–6209. Copyright 2006, American Chemical Society.)

of Figure 4.4 (Shaik et al. 2006). Two reactors of type I (R1 and R2) and one reactor of type II (R3) are available, and four reactions are suitable in these reactors. Reactions T1 and T2 require a type I reactor, whereas reactions T3 and T4 require a type II reactor. In addition, the heat required for endothermic reactions T1 and T3 is provided by steam (HS) available in limited amounts. Reactions T2 and T4 are exothermic, and the required CW is also available in limited amounts. Each reactor allows variable batch sizes, where the minimum batch size is half the capacity of the reactor. The processing times and the utility requirements include a fixed term and a variable term that is proportional to the batch size. For raw materials and final products, unlimited storage is available (say 1000 kg), whereas for intermediates, finite storage is available. On the basis of the resource availability, two different cases are considered: in the first case (Example 4.3a) the availability of both HS and CW is 40 kg/min and in the second case (Example 4.3b) it is 30 kg/min. Also, two different objective functions—maximization of profit and minimization of makespan—are considered. For problems involving utility resources, the models of SLK, S&M, and V&S3 are not directly applicable. The model of Shaik and Vooradi (2013) is additionally included for comparison, where S&V1 refers to unified STN model (M1) and S&V2 refers to unified RTN model (M3) from their paper.

For the objective of maximizing profit, both cases are solved over a time horizon of 8 h. The model statistics and computational results are reported in Table 4.4. The V&S2 model requires fewer events and $\Delta n = 0$ compared to all other models, and it has a comparable CPU time, despite its large problem size owing to four-index variables.

Similar observations can be made for the objective of minimization of makespan, where the above two cases are solved for a fixed demand of 100 kg of P1 and 80 kg of P2. The optimal solution is 8.5 h in the first case (Example 4.3a) and 9.025 h in the second case (Example 4.3b). The computational results are reported in Table 4.5. In constraints involving big-M terms, a common value of $M = 10$ was used. For problem instance 3b, all the models are able to find optimal solution using $\Delta n = 0$.

EXAMPLE 4.4

This is an example from Maravelias and Grossmann (2003) which has been solved by many authors (Janak et al. 2004, Shaik and Floudas 2009). The

TABLE 4.4

Computational Results for Examples 4.3 and 4.4 for Maximization of Profit with Resources

Model	Events	RMILP ($)	MILP ($)	CPU Time (s)	Nodes	Binary Variables	Continuous Variables	Constraints
				Example 4.3a ($H = 8$)				
S&F ($\Delta n = 1$)	6	10,713.78	5904	1.47	1708	42	167	767
V&S1 ($\Delta n = 1$)	6	10,713.81	5904	1.48	1149	42	167	639
V&S2	6	10,713.81	5904	1.55	1935	182	339	974
S&V1 ($\Delta n = 1$)	6	10,769.78	5904	1.24	1340	42	167	558
S&V2 ($\Delta n = 1$)	6	10,769.78	5904	0.50	661	42	167	562
				Example 4.3b ($H = 8$)				
S&F ($\Delta n = 1$)	6	8291.55	5227.77	1.28	1378	42	167	767
V&S1 ($\Delta n = 1$)	6	8107.30	5227.77	0.56	715	42	167	639
V&S2	5	6213.28	5227.77	0.09	54	140	274	776
S&V1 ($\Delta n = 1$)	5	8797.93	5227.77	0.09	715	30	136	444
S&V2 ($\Delta n = 1$)	5	8797.93	5227.77	0.08	47	30	151	448

Example 4.4a (H = 12)

S&F (Δn = 2)	9	21,394.73	13,000	497.97	300,644	126	421	1840
V&S1 (Δn = 2)	9	21,394.73	13,000	287.29	187,058	126	421	1535
V&S2	7	18,639.64	13,000	1.39	1078	242	579	1557
S&V1 (Δn = 1)	9	21,394.73	13,000	167.61	189,867	94	392	1210
S&V2 (Δn = 1)	9	21,394.73	13,000	168.74	174,149	94	446	1225

Example 4.4b (H = 14)

S&F (Δn = 3)	9	24,960.52	16,350	269.24	85,669	148	443	1982
V&S1 (Δn = 3)	9	24,960.52	16,350	223.90	170,263	148	443	1677
V&S2	7	19,030.98	16,350	0.36	574	242	579	1557
S&V1 (Δn = 2)	9	24,960.52	16,350	106.10	125,288	126	424	1382
S&V2 (Δn = 2)	9	24,960.52	16,350	89.05	108,522	126	478	1397

Source: S&F: Shaik and Floudas (2009); V&S1: Vooradi and Shaik (2012a); V&S2: Vooradi and Shaik (2013); S&V1: Shaik and Vooradi (2013); S&V2: Shaik and Vooradi (2013) STN model M1; S&V2: Shaik and Vooradi (2013) RTN model M3.

TABLE 4.5

Computational Results for Example 4.3 for Minimization of Makespan with Resources

Model	Events	RMILP (h)	MILP (h)	CPU Time (s)	Nodes	Binary Variables	Continuous Variables	Constraints
				Example 4.3a ($D_{p1} = 100$ kg, $D_{p2} = 80$ kg)				
S&F ($\Delta n = 1$)	7	5.077	8.5	2.87	1835	54	200	929
V&S1 ($\Delta n = 1$)	7	5.077	8.5	2.59	1691	54	200	774
V&S2	6	5.077	8.5	0.36	106	182	339	981
S&V1 ($\Delta n = 1$)	7	5.077	8.5	3.14	3475	54	200	679
S&V2 ($\Delta n = 1$)	7	5.077	8.5	3.36	2787	54	223	683
				Example 4.3b ($D_{p1} = 100$ kg, $D_{p2} = 80$ kg)				
S&F	6	5.077	9.025	0.19	39	24	149	492
V&S1	6	5.336	9.025	0.22	49	24	149	484
V&S2	6	5.077	9.025	0.28	162	182	339	981
S&V1	6	5.336	9.025	0.10	27	24	149	403
S&V2	6	5.336	9.025	0.13	46	24	169	407

Source: S&F: Shaik and Floudas (2009); V&S1: Vooradi and Shaik (2012a); V&S2: Vooradi and Shaik (2013); S&V1: Shaik and Vooradi (2013) STN model M1; S&V2: Shaik and Vooradi (2013) RTN model M3.

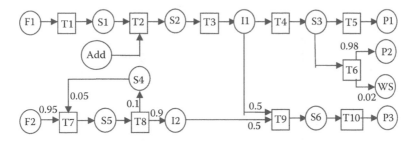

FIGURE 4.5   STN representation for Example 4.4. (Adapted with permission from Shaik, M.A., Janak, S.L., Floudas, C.A. 2006. Continuous-time models for short-term scheduling of multipurpose batch plants: A comparative study. *Ind Eng Chem Res* 45:6190–6209. Copyright 2006, American Chemical Society.)

problem involves 10 tasks in 6 units processing 14 states as shown in the STN representation of Figure 4.5 (Shaik et al. 2006). Raw materials F1 and F2, intermediates I1 and I2, final products P1–P3 and WS states have UIS capacity, while states S3 and S4 have FIS, states S2 and S6 have NIS, and states S1 and S5 have ZW policies. Three different renewable utilities are available for this process: with tasks T2, T7, T9, and T10 requiring CW; tasks T1, T3, T5, and T8 requiring low-pressure steam (LPS); and tasks T4 and T6 requiring high-pressure steam (HPS). The maximum availabilities of CW, LPS, and HPS are 25, 40, and 20 kg/min, respectively.

For the objective of maximizing profit, this example is solved for two different time horizons (a: $H = 12$ h and b: $H = 14$ h). The model statistics and computational results for these cases are reported in Table 4.4. In both problem instances, V&S2 model performs best as it requires $\Delta n = 0$, two events lesser, and solves faster compared to other models.

To illustrate the effect of the unit-wait policy, Example 4.4 is solved again by considering an unlimited unit-wait policy for FIS and NIS states and the results are compared in Table 4.6. The results of other models (Castro et al. 2004, Janak et al. 2004) are reported from their original paper and hence, CPU time is not directly compared, owing to differences in hardware and software. By comparing the results of S&V1 and S&V2 from Tables 4.4 and 4.6, it can be seen that owing to simply allowing a unit-wait policy there is a reduction in the number of events and Δn.

4.5   WESTENBERGER–KALLRATH BENCHMARK EXAMPLE

In this section, we present a solution to the challenging Westenberger–Kallrath benchmark scheduling problem (Blomer and Gunther 2000, Kallrath 2002), which has been solved by many authors (Janak and Floudas 2008, Vooradi and Shaik 2012a,b). This benchmark problem has several complex features including variable fractions of products produced from a task with multiple product streams, different storage policies for intermediates, multiple tasks suitable in a single machine, and multiple machines for a single task, and task-specific capacity ranges for machines

TABLE 4.6

Computational Results for Example 4.4 for Maximization of Profit with Unit-Wait Policy

Model	Events	RMILP ($)	MILP ($)	CPU Time (s)	Nodes	Binary Variables	Continuous Variables	Constraints
			Example 4.4a (H = 2)					
Castro et al. (2004) (Δt = 2)	9	21,587	13,000	–	2525	150	523	628
Janak et al. (2004)	8	19,000	13,000	–	222	110	1077	3318
S&V1 (Δn = 1)	8	21,394	13,000	6.82	7167	74	339	1053
S&V2 (Δn = 1)	8	21,394	13,000	8.41	8371	74	387	1068
			Example 4.4b (H = 14)					
Castro et al. (2004)(Δt = 2)	8	19,189	16,350	–	45	130	459	549
Janak et al. (2004)	8	19,000	16,350	–	2869	110	1077	3354
S&V1 (Δn = 1)	7	19,054	16,350	0.29	241	54	286	896
S&V2 (Δn = 1)	7	19,207	16,350	0.23	202	54	328	911

Source: S&V1: Shaik and Vooradi (2013) STN model M1; S&V2: Shaik and Vooradi (2013) RTN model M3.

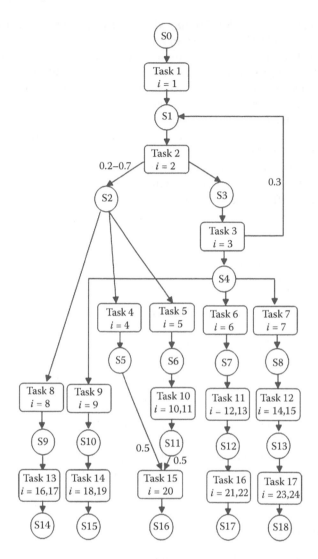

FIGURE 4.6 STN representation for Westenberger–Kallrath problem. (Reprinted from *Comp Chem Eng*, 43, Vooradi, R., Shaik, M.A. Improved three index unit specific event based model for the short term scheduling of batch plants, 148–172, Copyright 2012, with permission from Elsevier.)

and so on. The production problem consists of a network of 17 processing tasks, 19 states (S0–S18), and 9 production units. A task that can take place in more than one unit is considered as a separate task. The STN representation of the production process is shown in Figure 4.6 (Blomer and Gunther 2000, Vooradi and Shaik 2012a). The states S2 and S3 have variable fractions of the amount produced by task 2 as shown in Figure 4.6. All intermediate states are subject to some restriction on material storage. States S1–S4, S6–S8, S11, and S13 are subject to FIS, whereas states S5, S9, S10, and S12 are subject to ZW.

The intermediate states S1 and S3 are involved in the recycle loop; hence, for the model of V&S3 conventional material balances are used to handle the material flows, and for all other states the material transfers are modeled at the same event. In order to handle the variable fractions of the states S2 and S3 being produced by task 2, standard material balances must be modified to allow the fraction of states produced by task 2 to vary between the specified lower and upper bounds (for details refer to Vooradi and Shaik 2012a,b).

The problem is solved for 14 different demand instances based on the objective of minimizing the makespan. Each demand instance has up to five final products (S14–S18) with a different demand structure for each problem instance. The first set (P1–P10) has a total demand of 60, and the second set (P11–P14) has a total demand of 90. The first set (P1–P10) has simple demand structure in which exactly three products need to be produced. In these instances, preprocessing is done to exclude those tasks that are associated with the products that do not have a demand. Problem instances P11–P14 have slightly larger demand structure where we have demands for all the final products and no tasks are effectively excluded during preprocessing. The model statistics and computational results for all the problem instances are reported in Table 4.7.

Problems with higher demands of products are generally found to be more difficult to solve, because they require more tasks and events. The converged optimal solution is found in all instances except for P9 and P11 for which the best makespan is reported within the specified CPU time limit of 40,000 s. Although it is a relatively larger problem with recycle streams, in all problem instances V&S3 model still gives a reduction of one to two events to find the optimal solution compared to the V&S1 model, except for problems P11 and P14 where it is unable to close the integrality gap and gives suboptimal solution owing to CPU time limit. When the number of events are the same, then V&S1 model performs better than V&S3 model and gives optimal converged solution.

4.6 CONCLUSION

Unit-specific event (or unit-slot or nonuniform time grid)-based continuous-time representation has become a promising tool, among others, for solving batch-scheduling problems. In view of this, a detailed and critical review of batch-scheduling models based on USEB formulations is presented in this chapter along with illustrative case studies, including analysis on how these models have evolved over time in terms of efficiently dealing with different aspects of scheduling problems in batch plants. There have been several innovative approaches presented recently in the literature leading to reducing the number of events and the resulting problem size. The computational complexity of the resulting models is still an important issue that needs to be addressed in the future, particularly when dealing with large-scale or industrial-scale scheduling problems.

TABLE 4.7
Computational Results for Westenberger–Kallrath Problem

Problem Instance	Model	Events	RMILP (h)	Make Span (h)	CPU Time (s)	Binary Variables	Continuous Variables	Constraints
P1	V&S1	5	24	28	0.07	46	239	560
	V&S3	4	24	28	0.08	54	208	525
P2	V&S1	7	24	28	0.45	73	335	811
	V&S3	5	24	28	0.66	68	255	653
P3	V&S1	7	24	28	0.33	73	342	837
	V&S3	5	24	28	0.39	68	260	664
P4	V&S1	6	18.667	27	0.61	65	326	813
	V&S3	5	18.667	27	2.44	78	310	794
P5	V&S1	6	18.667	26	0.94	65	332	835
	V&S3	5	18.667	26	2.01	78	300	790
P6	V&S1 ($\Delta n = 1$)	8	16	30	106.13	188	534	1782
	V&S3 ($\Delta n = 1$)	6	16	30	528.63	173	433	1429
P7	V&S1	7	18.667	28	4.05	81	385	972
	V&S3	6	18.667	28	12.96	94	354	941
P8	V&S1	7	18.667	28	4.91	82	392	999
	V&S3	5	18.667	28	2.5	78	300	790
P9	V&S1 ($\Delta n = 1$)	11	17.797	33	40,000[a,b]	305	813	2734

(Continued)

TABLE 4.7 (Continued)
Computational Results for Westenberger–Kallrath Problem

Problem Instance	Model	Events	RMILP (h)	Make Span (h)	CPU Time (s)	Binary Variables	Continuous Variables	Constraints
	V&S3 ($\Delta n = 1$)	9	17.797	33	40,000[a,c]	286	702	2352
P10	V&S1	9	26.667	35	88.15	124	568	1572
	V&S3	7	26.667	35	48.79	124	469	1328
P11	V&S1	12	30.667	40	40,000[a,d]	245	1003	2935
	V&S3	12	30.667	41	40,000[a,e]	286	1044	3091
P12	V&S1	8	30	34	69.66	148	655	1878
	V&S3	7	30	34	70.87	166	609	1771
P13	V&S1	10	28	36	408.53	196	829	2406
	V&S3	8	28	36	321.33	190	696	2035
P14	V&S1	10	24	36	7092.17	196	829	2406
	V&S3	10	24	38	40,000[a,f]	238	870	2563

Source: V&S1: Vooradi and Shaik (2012a); V&S3: Vooradi and Shaik (2012b).

[a] Resource limit reached.
[b] Relative gap: 3.30%.
[c] Relative gap: 9.09%.
[d] Relative gap: 4.52%.
[e] Relative gap: 21.9%.
[f] Relative gap: 17.89%.

ACKNOWLEDGMENTS

The author would like to acknowledge Mr Ramsagar Vooradi (PhD student from IIT Delhi) for help in implementing the case studies.

NOMENCLATURE

INDICES

i, i'	tasks
j, j'	units or equipment resources
n, n', n''	events
s	states or material resources
u	utility resources
r	all resources

SETS

I	tasks
I_j	tasks which can be performed in unit j
I_s^p	tasks which produce state s
I_s^c	tasks which consume state s
I_u	tasks which consume utility u
J	units or equipment resources
N	total event points postulated in the time horizon
S	states or material resources
S^R	states that are raw materials
S^P	states that are final products
S^I	states that are intermediates
$S^{DFIS}, S^{ZW}, S^{NIS}$	intermediate states with dedicated FIS, ZW, and NIS cases, respectively
U	utility resources

PARAMETERS

δ_i	coefficient of variable term of processing time of task i (in duration constraints)
δ_{iu}	coefficient of variable term of consumption of utility u by task i
Δn	limit on the maximum number of events over which a task is allowed to continue
γ_i	coefficient of constant term of processing time of task i (in duration constraints)
γ_{iu}	coefficient of constant term of consumption of utility u by task i
ρ_{is}	fractions of state s produced ($\rho_{is} \geq 0$) or consumed ($\rho_{is} \leq 0$) by task i
B_i^{min}	minimum batch size of task i
B_i^{max}	maximum batch size of task i
D_s	demand for state s

E_s^0 initial amount available for state s

E_u^0 maximum availability of utility u

E_s^{max} maximum storage capacity for state s

H short-term scheduling horizon

M large positive number in Big-M constraints

P_s price of state s

UW_i task-specific maximum unit-wait time

BINARY VARIABLES

$v(i, i', s, n)$ binary variable for state s which takes a value of 1 if the material produced by task i at event $n - 1$ can wait in consumption unit before consumed by task i' at event n

$w(i, n, n')$ binary variable for task i that starts at event n and ends at event n'

$x(i, i', s, n)$ binary variable for state s which takes a value of 1 if the material produced by task i at event $n - 1$ is consumed immediately by task i' at event n

$y(i, i', s, n)$ binary variable for utility u which takes a value of 1 if the utility released by task i at event $n - 1$ is used by task i' at event n

$z(i, i', s, n)$ binary variable for state s which takes a value of 1 if the material produced by task i at event $n - 1$ is used by task i' at event n

POSITIVE VARIABLES

$b(i, n, n')$ batch size of task i that starts at event n and ends at event n'

$b1(i, i', s, n)$ amount of state s consumed by task i' at event n from the total amount produced by task i at event $n - 1$

$b2(i, i', s, n)$ excess amount of FIS state s produced by task i at event $n - 1$, beyond the maximum storage capacity, which is to be either immediately consumed or can wait in consumption unit for consumption by task i' at event n

$b3(i, i', s, n)$ amount of utility used by task i' at event n from the total amount released by task i

E_s^0 initial amount of state s required from external resources

$E(s,n)$ excess amount of state s that needs to be stored at event n

$E(u, n)$ excess amount of utility u available at event n

$T^s(i, n)$ start time of task i at event n

$T^f(i, n)$ end time of task i at event n

$T_{ut}^s(u,n)$ start time at which there is a change in the consumption of utility u at event n

REFERENCES

Blomer, F., Gunther, H.-O. 2000. LP-based heuristics for scheduling chemical batch processes. *Int J Prod Res* 38:1029–1051.

Castro, P.M., Barbosa-Povoa, A.P., Matos, H.A., Novais, A.Q. 2004. Simple continuous-time formulation for short-term scheduling of batch and continuous processes. *Ind Eng Chem Res* 43:105–118.

Castro, P.M., Grossmann, I.E. 2005. New continuous-time MILP model for the short-term scheduling of multistage batch plants. *Ind Eng Chem Res* 44:9175–9190.

Castro, P.M., Novais, A.Q. 2008. Short-term scheduling of multistage batch plants with unlimited intermediate storage. *Ind Eng Chem Res* 47:6126–6139.

Floudas, C.A., Lin, X. 2004. Continuous-time versus discrete-time approaches for scheduling of chemical processes: A review. *Comput Chem Eng* 28:2109–2129.

Harjunkoski, I., Maravelias, C.T., Bongers, P., Castro, P.M., Engell, S., Grossmann, I.E., Hooker, J., Mendez, C., Sand, G., Wassick, J. 2014. Scope for industrial applications of production scheduling models and solution methods. *Comput Chem Eng* 62:161–193.

Ierapetritou, M.G., Floudas, C.A. 1998. Effective continuous-time formulation for short-term scheduling: 1. Multipurpose batch processes. *Ind Eng Chem Res* 37:4341–4359.

Janak, S.L., Floudas, C.A. 2008. Improving unit-specific event based continuous-time approaches for batch processes: Integrality gap and task splitting. *Comput Chem Eng* 32:913–955.

Janak, S.L., Lin, X., Floudas, C.A. 2004. Enhanced continuous-time unit-specific event-based formulation for short-term scheduling of multipurpose batch processes: Resource constraints and mixed storage policies. *Ind Eng Chem Res* 43:2516–2533.

Kallrath, J. 2002. Planning and scheduling in the process industry. *OR Spect* 24:219–250.

Kondili, E., Pantelides, C.C., Sargent, R.W.H. 1993. A general algorithm for short-term scheduling of batch operations. Part 1. MILP formulation. *Comput Chem Eng* 17:211–227.

Kopanos, G.M., Mendez, C.A., Puigjaner, L. 2010. MIP-based decomposition strategies for large-scale scheduling problems in multiproduct multistage batch plants: A benchmark scheduling problem of the pharmaceutical industry. *Eur J Oper Res* 207:644–655.

Lin, X., Floudas, C.A. 2001. Design, synthesis and scheduling of multipurpose batch plants via an effective continuous-time formulation. *Comput Chem Eng* 25:665–674.

Li, J., Floudas, C.A. 2010. Optimal event point determination for short-term scheduling of multipurpose batch plants via unit-specific event-based continuous-time approaches. *Ind Eng Chem Res* 49:7446–7469.

Li, J., Susarla, N., Karimi, I.A., Shaik, M.A., Floudas, C.A. 2010. An analysis of some unit-specific event-based models for the short-term scheduling of non continuous processes. *Ind Eng Chem Res* 49:633–647.

Majozi, T., Zhu, X.X. 2001. A novel continuous-time MILP formulation for multipurpose batch plants. 1. Short-term scheduling. *Ind Eng Chem Res* 40:5935–5949.

Maravelias, C. T. 2012. A General framework and modeling approach classification for chemical production scheduling. *AIChE J* 58:1812–1828.

Maravelias, C.T., Grossmann, I.E. 2003. New general continuous-time state-task network formulation for short-term scheduling of multipurpose batch plants. *Ind Eng Chem Res* 42:3056–3074.

Mendez, C.A., Cerda, J. 2003. An MILP continuous-time framework for short-term scheduling of multipurpose batch processes under different operation strategies. *Optim Eng* 4:7–22.

Mendez, C.A., Cerda, J., Grossmann, I.E., Harjunkoski, I., Fahl, M. 2006. State-of-the-art review of optimization methods for short-term scheduling of batch processes. *Comput Chem Eng* 30:913–946.

Pantelides, C.C. (1994). Unified frameworks for optimal process planning and scheduling. In: D.W.T. Rippin, J.C. Hale, and J. Davis (Eds.), *Proceedings of the Second International Conference on Foundations of Computer-Aided Process Operations*. New York: Cache Publications, pp. 253–274.

Pinto, J.M., Grossmann, I.E. 1995. A continuous time mixed integer linear programming model for short term scheduling of multistage batch plants. *Ind Eng Chem Res* 34:3037–3051.

Pitty, S.S., Karimi, I.A. 2008. Novel MILP models for scheduling permutation flowshops. *Chem Prod Process Model* 3:1–46.

Seid, R., Majozi, T. 2012a. A robust mathematical formulation for multipurpose batch plants. *Chem Eng Sci* 68:36–53.

Seid, R., Majozi, T. 2012b. A novel technique for prediction of time points for scheduling of multipurpose batch plants. *Chem Eng Sci* 68:54–71.

Seid, R., Majozi, T. 2013. Design and synthesis of multipurpose batch plants using a robust scheduling platform. *Ind Eng Chem Res.* 52:16301–16313.

Shaik, M.A., Floudas, C.A. 2008. Unit-specific event-based continuous time approach for short-term scheduling of batch plants using RTN framework. *Comput Chem Eng* 32:260–274.

Shaik, M.A., Floudas, C.A. 2009. Novel unified modeling approach for short-term scheduling. *Ind Eng Chem Res* 48:2947–2964.

Shaik, M.A., Janak, S.L., Floudas, C.A. 2006. Continuous-time models for short-term scheduling of multipurpose batch plants: A comparative study. *Ind Eng Chem Res* 45:6190–6209.

Shaik, M.A., Vooradi, R. 2013. Unification of STN and RTN based models for short-term scheduling of batch plants with shared resources. *Chem Eng Sci* 98:104–124.

Sundaramoorthy, A., Karimi, I.A. 2005. A simpler better slot-based continuous-time formulation for short-term scheduling in multipurpose batch plants. *Chem Eng Sci* 60:2679–2702.

Sundaramoorthy, A., Maravelias, C.T. 2011. Computational study of network-based mixed-integer programming approaches for chemical production scheduling. *Ind Eng Chem Res* 50:5023–5040.

Susarla, N., Li, J., Karimi, I.A. 2010. A novel approach for short-term scheduling of multipurpose batch plants using unit-slots. *AIChE J* 56:1859–1879.

Vooradi, R. 2013. Unit specific event based models for short-term scheduling of batch plants. PhD thesis. Indian Institute of Technology (IIT) Delhi.

Vooradi, R., Shaik, M.A. 2012a. Improved three index unit specific event based model for the short term scheduling of batch plants. *Comput Chem Eng* 43:148–172.

Vooradi, R., Shaik, M.A. 2012b. Advanced unit-specific event based modeling approach for short-term scheduling of batch plants. In *AIChE Annual Meeting*. USA: Pittsburgh.

Vooradi, R., Shaik, M.A. 2013. Rigorous unit-specific event-based model for short-term scheduling of batch plants with conditional sequencing and unit-wait times. *Ind Eng Chem Res* 52:12950–12972.

5 Planning and Scheduling in the Biopharmaceutical Industry
An Overview

Miguel Vieira, Tânia Pinto-Varela, and Ana Paula Barbosa-Póvoa

CONTENTS

5.1 INTRODUCTION

The pharmaceutical industry is aiming to develop improved and efficient manufacturing processes to address current market challenges. Stronger competition among drug manufacturers and increasing government regulations over prescriptions have endured particular focus on production costs. Pharmaceutical substances are still commonly acknowledged as the backbone of modern medicinal therapy. This fact is driving research and development (R&D) processes toward more effective drug treatments for the wide variety of human diseases, with interesting developments in the biotechnology sector. Biopharmaceuticals have been gathering particular attention in this context due to its efficiency in the treatment of complex health diseases, including cancer, autoimmune and inflammatory disorders, organ transplant rejection, and many more are in clinical trials. However, the complexity of the production process deals with high manufacturing costs to assure its mass distribution. And unlike many areas of the process industry, the development of decision tools to support the production management of these biochemical processes have been fairly fostered.

This work aims to provide an insight over process management optimization in the pharmaceutical industry, first with the characterization of the main problems of the sector with a review of current research topics. Particular relevance is given to planning and scheduling problems in biopharmaceutical processes, featuring a review of literature to date. Second, current research work is classified according to the roadmap defined in the review of Méndez et al. (2006) for optimization models applied to batch scheduling problems. On the basis of the analysis performed, future research opportunities and potential areas are identified and discussed.

5.2   PHARMACEUTICAL INDUSTRY OVERVIEW

The management of a pharmaceutical process comprises the scheduling and allocation of resources across a broad number of competing activities of different drug projects, typically over a multiyear time frame and subject to a set of uncertainty factors. The process life cycle involves complex drug development steps, as outlined in Figure 5.1, roughly spanning from 7 to 12 years—from the moment a new discovery is promoted to the status of a lead molecule.

With the discovery of a lead chemical/biological compound, the initial step of first human dose (FHD) trials encompasses the planning activities for administering the drug to a selected group of volunteers, including several studies of pharmaco-kinetic properties. Phase I is the first clinical trial with the drug administered to a small number of healthy volunteers, to establish safe dosages and gather information on absorption, distribution, metabolic effects, excretion, and toxicity of the compound.

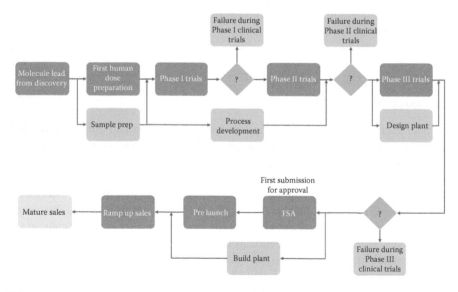

FIGURE 5.1   Pharmaceutical drug development pipeline. (Adapted from Varma, V. A. et al. 2008. *Computers & Chemical Engineering* 32(4–5): 1000–1015.)

In Phase II trials are conducted with subjects who have the targeted disease, and designed to obtain evidence on safety and preliminary data efficacy. Phase III represents a large-scale clinical trial on subjects with the disease, confirming their efficacy, toxicology, and uncovering potential side effects. While the number of subjects in phase II trials can total in the hundreds, it can rise to thousands during phase III (Varma et al. 2008). In these initial drug development phases, if the results fail to treat the disease symptoms or if it has inferior effectiveness compared to competitive products, the component is terminated and returned to the research phase. If the drug successfully surpasses the trials, the following step is the First Submission for Approval where all previous evidence of safety and efficacy are compiled and submitted to the regulatory authorities for market approval. Simultaneously, marketing strategy decisions and production process developments can be taken. Prelaunch activities represent the final preparation stage, with the approval of the regulatory agency and defining the global commercial strategy, leading to the mass distribution of the product. At the operation stage, most pharmaceutical products involve primary active ingredients (AI) production (often multistage chemical synthesis or bioprocesses) and secondary (formulation) production.

The characteristics of the life cycle of a pharmaceutical product differ from other sectors of the chemical industry, as highlighted in Laínez et al. (2012): (1) a significant investment is required and low success rate of product discovery or clinical development is observed, owing to the confined understanding of the different diseases therapeutics; (2) major costs associated with the clinical trials phase, with an extended timeline and uncertainty of high rate of failure; (3) heavy regulatory burden over its entire cycle time; (4) a limited product shelf life, since it contains active and sensitive molecular entities that can degrade with time; (5) the complex structure of the product supply chain, requiring the coordination of multiple manufacturing sites of AI, drug products or packaging, including contract manufacturers, wholesalers, and retailers; (6) the supply chain often implies a longer cycle time in comparison with consumer products, up to 6–9 months, with the AI production representing the rate-limiting step (characterized by low manufacturing velocities and numerous quality assurance activities); and (7) the ferocious generic competition after the end of its 20-year patent life. These factors constrain the optimal design and operation of these industries, posing significant challenges for enterprise-wide decision making.

Shah (2004) identified five key players of the pharmaceutical industry: (1) large R&D-based multinationals; (2) large generic manufacturers operating in the international market; (3) local companies based in their country producing either generic or branded products; (4) contract manufacturers with their own portfolio; and (5) biotechnology companies with a small portfolio and manufacturing capacity. The first group is undoubtedly the most extensively reviewed considering the complexity of its supply chain, but the emerging biotechnology industrial sector also offers relevant operational problems owing to the specific nature of biochemical processes. The wide R&D multinationals, typically based on chemical synthesis and separation stages of AI, are not only economically the most important but also the most vulnerable to global regulatory effects. On the other end, biopharmaceutical companies are linked to a scalable production of complex biological components, with special emphasis on fermentation and purification stages. However, the market for biopharmaceuticals is expanding because of the exceptional effectiveness of its genetically engineered therapeutics, providing significant interest to the medical industry.

To deal with these process characteristics, enterprise-wide optimization (EWO) has become a common decision-making approach in current pharmaceutical industries, supporting the optimization of supply, manufacturing, and distribution activities to reduce costs, inventories, and environmental impact (Grossmann 2012). Main operation activities include planning, scheduling and control, aiming to maximize profit, and responsiveness across the various company sectors. The increased availability of enterprise information and modern computing technology has driven the research for mathematical models and simulators applied to optimization frameworks. However, the latest reviews of current research works show that it was still not able to address decision-making spanning the entire product life cycle, once a series of cloistered applications for the different stages have been developed.

The inherent complexity of the drug development and manufacturing processes has become an interesting management problem to study, in particular, because of the planning and scheduling problems that arise during its long life cycle. The most relevant aspects will be detailed in the following section.

5.3 DRUG DEVELOPMENT PROCESS MANAGEMENT

The R&D of new therapeutics is struggling with the industrial constraints of the pharmaceutical sector. Flexible multiproduct facilities have been adapted to respond to changing customer demand and increase plant utilization. However, the greater complexity of these processes has led the research toward improved scheduling and planning frameworks. Three broad problem categories can arise in the pharmaceutical industry: (1) product development pipeline management, (2) capacity planning, and (3) supply chain management.

The first category—product development pipeline management—covers multilevel decisions impacting a number of tasks and associated milestones: the selection of the most promising portfolio of drug, the prioritization of assigned tasks, or the respective assignment of resources. These decisions are taken considering budget limitations, resource constrains, and technical or regulatory uncertainties. Three main areas where the most relevant work has been reported were identified: (i) on the early testing of AI toxicity and stability; (ii) the portfolio selection and execution; and (iii) the management of a clinical trial supply chain. Regarding these topics, Varma et al. (2008) proposed a framework called SIM-OPT as an integrated resource management tool with the goal of maximizing the portfolio expected net present value (ENPV), controlling risk, and reducing cycle times. The framework includes the combination of a stochastic simulation of the process work flow with a mixed integer linear programming (MILP) model that is able to generate optimal resource allocation and schedules. Likewise, Colvin and Maravelias (2010) discussed methods for the solution of multistage stochastic programming and developed a new branch and cut algorithm for the resource-constrained scheduling of clinical trials in the pharmaceutical pipeline.

Concerning the second category—capacity planning—while strategic decisions over the product portfolio are maintained for longer time periods, tactical or midterm planning relate to the magnitudes of material and cash flows across manufacturing and distribution networks. These problems often address the capacity planning, with the determination of the optimal sourcing, manufacturing, and distribution network. For example, in the pharmaceutical industry, it is relevant to consider the capacity expansion/reductions of the existing network to accommodate the outcomes of the clinical trials. On the basis of the work from Papageorgiou et al. (2001), Gatica et al. (2003) have extended an MILP approach to the problem of capacity planning under clinical trial uncertainty, with four accountable scenarios influencing demand: high, target, low success, and failure. More recently, Tsang et al. (2007a,b) presented a two-stage stochastic program for the capacity planning of vaccine manufacturing, and further extended the study to different risk metrics (expected downside risk, opportunity value, value-at-risk, and conditional value-at-risk).

The third category—supply chain management—addresses global supply chain optimization formulations, a current practice in most business areas to increase profit margins. Several authors addressed the pharmaceutical industry supply chain issues: Papageorgiou et al. (2001) proposed a deterministic model to manage the portfolio of a company and the production stage is formulated in detail; Levis and Papageorgiou (2004) later extended the same work, accounting for uncertainty in a demand forecast based on the results of the clinical trials of each product; Amaro and Barbosa-Póvoa (2008) considered the integration of planning and scheduling on a real pharmaceutical supply chain with the existence of reverse flows, and later the same authors (Amaro and Barbosa-Póvoa 2009) studied the supply chain planning and different partnership structures under uncertainty. More recently, Sousa et al. (2011) reviewed the research topic claiming that, so far, most of the problems refer to relatively small systems (reduced number of sites and products), and that the long strategic planning of large pharmaceutical companies has not yet been addressed. In their work, they propose a dynamic allocation and planning optimization decomposition MILP models for the global supply chain, from production stages at primary (AI production) and secondary sites to distribution markets.

The aggregation of supply chain decision methods encompasses all the operational and financial aspects of manufacturing and delivering an innovative product to the patients. The primary manufacturing deals with the campaign planning and inventory management of AI to satisfy the demand of the secondary manufacturing; these secondary sites are often geographically scattered, where AI are added to excipient materials with further processing and packaging to the final product. The customer-facing end is considered a "pull" process (driven by orders), but since the primary manufacturing stage has long cycle times it is difficult to ensure end-to-end responsiveness. This means that primary production is effectively a "push" process, driven by medium- and long-term forecasts, and demanding relatively large stocks of AI to ensure good service levels (Shah 2004). This wide process complexity resumes the requirement of computer-aided decision-support tools, generally grounded on optimization model frameworks. The optimization of planning and scheduling operations, alongside with portfolio management, has become one of the most interesting areas of study in this industry. Although the scheduling level consists of models and methods to determine the assignment, sequencing, and timing of use of shared resources, the planning level sets manufacturing targets and makes resource allocations (Stephanopoulos and Reklaitis 2011). Planning and scheduling decisions should be jointly taken, integrating in one model the production plan (strategic and tactical planning) and a daily operation schedule with the assignment of task and resources (Mockus et al. 2002). However, the model representation of complex process constraints is one of the many challenges that have driven significant research topics to overcome optimization problems.

5.3.1 PLANNING AND SCHEDULING OPTIMIZATION RESEARCH

Over the last years, the general problem of planning and scheduling operations has been gathering extensive research in the process industry, with relevance in modeling and optimization techniques for short-term batch scheduling, as remarked in Méndez et al. (2006). Barbosa-Póvoa (2007) later reviewed the design/retrofit-design optimization on multiproduct and multipurpose batch plants, for which the models are often linked with scheduling problems to address production flexibility. More recently, Harjunkoski et al. (2014) reviewed scheduling methodologies developed for process industries and discussed their characteristics and challenges focused on industrial problems. Mathematical programming is clearly noticed in the majority of the model formulations, but heuristic approaches have also been explored to overcome computational limitations. It is acknowledged that some of the major issues in the development of mathematical modeling tools arise when dealing with nonlinearities, multiscale temporal and spatial integration, uncertainty parameters, commercial software versus tailored-made methods, and feasible solutions for industrial-sized problems.

Following the research reviewed by Méndez et al. (2006), several authors have been tackling some of these modeling challenges. Presented here are some recent achievements: Liu et al. (2009) developed a compact MILP formulation for the medium-term planning of single-stage multiproduct continuous plants with parallel processors that are subject to sequence-dependent changeovers. On the basis of a classic Travelling Salesman Problem (TSP), formulation considering a hybrid discrete/continuous time representation, Terrazas-Moreno and Grossmann (2011) proposed two decomposition methods to solve multisite multimarket simultaneous planning and scheduling problems using a bi-level decomposition or a hybrid method where the planning level is decomposed through spatial Lagrangean relaxation. Castro et al. (2011) presented a new algorithm for large-scale scheduling problems, incorporating an MILP continuous formulation and a discrete-event simulation model, and addressing the complexity of multistage batch plant procedures featuring single unit per stage, zero-wait storage policies, and a single transportation device for moving lots between stages. Li et al. (2012) developed, based on previous work by the same authors, a deterministic operational planning model for large-scale multiproduct continuous plants and a robust optimization for demand due date parameter uncertainty. Marchetti et al. (2012) suggested two new MILP monolithic approaches combining lot sizing and scheduling of multistage batch facilities, handling multiple customer orders, variable processing times, and sequence-dependent changeover times. Lastly, Velez et al. (2013) developed a constraint propagation algorithm for the calculation of parameters that are used to generate valid inequalities, which greatly enhances the computational performance of a mixed integer programming (MIP) scheduling model.

As noted, the progress of optimization model formulations has thrived from traditional MIP models to incorporate hybrid heuristic/metaheuristic methods (e.g., genetic algorithms, Tabu search or simulated annealing), stochastic programming, or decomposition techniques. Regarding the pharmaceutical industry, since large R&D companies are mainly grounded in chemical-based synthesis processes, some research over these specific problems of planning and scheduling management have not yet been considered. However, it is verified that literature specifically addressing the pharmaceutical topic has been increasingly explored within the wide extent of chemical processes. Roe et al. (2005) proposed a hybrid constraint logic programming (CLP) and MILP algorithm that decomposes a scheduling problem applied to pharmaceutical production, first solving the aggregate planning using the MILP model and then sequencing the problem using CLP techniques. Kopanos et al. (2010) developed an iterative two-step MIP-based solution strategy (a constructive step followed by an improvement step) for the resolution of large-scale scheduling problems applied to multiproduct multistage pharmaceutical batch plants. Stefansson et al. (2011) addressed a large real-world scheduling problem from a pharmaceutical company and compared the results of both discrete and continuous time formulations. Lastly, Moniz et al. (2013) proposed an MILP discrete time formulation studying some production constraints, such as sequence-dependent changeovers, temporary storage in processing units, lots blending, and materials traceability applied to a real chemical–pharmaceutical industry.

However, the increasing importance of the biopharmaceutical sector and the nature of the biochemical manufacturing steps have arisen specific optimization problems. In particular, the high operating costs of downstream process bottlenecks have endured exponential research interest in this area. Considering the importance of planning and scheduling optimization explored in other areas of process industry, the following sections discuss the biopharmaceutical manufacturing process along with the analysis of the respective planning and scheduling process issues.

5.4 BIOPHARMACEUTICAL MANUFACTURING PROCESSES

Biopharmaceutical drugs refer to medicinal biomolecules with pharmacological activity used for therapeutic or *in vivo* diagnostic purposes. Since the introduction of recombinant human insulin in the 1980s, such molecules are produced by means of genetically engineered biological organisms other than direct extraction from native sources (Walsh 2003). These AI differ from conventional chemical drugs by its complex macromolecular structure. The ability to genetically manipulate (by recombinant DNA or hybridoma technology) highly effective therapeutics to specific diseases has represented a breakthrough in the pharmaceutical industry (Walsh 2010). The main drugs are proteins, including monoclonal antibodies (mAb), nucleic acids comprised of DNA, RNA or antisense oligonucleotides, and live vaccines, where virulence is reduced by the process of attenuation (Ryu et al. 2012). However, the safety and efficacy profiles of the biopharmaceutical drugs are highly dependent on their complex manufacturing process.

As previously mentioned, biopharmaceutical manufacturing is also subject to high pressures to increase flexibility and contain costs. This is due to the long term of process development, the high risk of clinical failure and the requirements to enhance capacity utilization, as well as the greater variability of biopharmaceuticals in comparison to the chemical drugs. Simaria et al. (2012) identified that these facilities will tend to adopt a smaller scale with multiple bioreactors appropriately scheduled, able to reduce the capital cost while optimizing the number of production batches to match uncertain demand. Decision-support tools with advanced optimization formulations are also essential to develop consistent frameworks for the optimal design and operation of biopharmaceutical facilities.

The process to bring a biopharmaceutical product to market can take on average 7.7 years, and roughly 85% of the drugs entering clinical trials will not reach the market (Rajapakse et al. 2005; Cuatrecasas 2006). With such a wide development horizon and owing to the highly uncertainty of the drug's success, these companies typically develop a pipeline of compounds under trial. Besides the risk of failure during the phases of the clinical testing, operating multiproduct facilities poses several challenges to the planning and scheduling of the available resources. According to Lakhdar et al. (2005), one of the challenges is the considerable time and cost associated with campaign changeover owing to equipment setup and cleaning, two critical process stages subject to strict regulatory control.

The biopharmaceutical process is mainly composed of two complex manufacturing steps referred to as upstream and downstream processing. Upstream processing includes all tasks associated with the culture and maintenance of cells, and downstream processing is related to the chemical and physical separations necessary for the isolation and purifications of the drug product, as shown in Figure 5.2 (Sekhon 2010).

In upstream batch fermentation, cells derived from the culture bank progressively inoculate larger media volumes till the bioreactor, followed by downstream purification processes composed by a series of filtration/chromatography steps, with

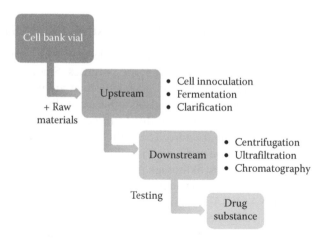

FIGURE 5.2 Outline of biopharmaceutical manufacturing.

intermediate filtration and viral clearance operations. These plants typically operate on a long campaign basis, running for an uninterrupted series of batch sequences in order to minimize changeovers and cross-product contamination. Some facilities have been adopting continuous/semicontinuous process steps owing to the improved control over operational conditions and equipment reliability, using smaller bioreactors with excess volumetric productivity (titer). For example, perfusion processes can be advantageous when dealing with less-stable products since the continuous harvesting decreases the bioreactor residence time.

As an example, the case of mAb production commonly establishes three chromatography operations for purification, after the batch culture harvesting where a majority of processes use mammalian cells. Simaria et al. (2012) classified these chromatography steps as critical steps in an mAb purification particularly owing to the use of expensive affinity matrices (namely, Protein A) as well as the large amounts of buffer reagents required. With increasing upstream productivities to carry the utmost market potential of these drugs, purification yields can represent one of the most relevant bottlenecks. Farid (2007) discussed some of the production challenges such as the equipment alternative for disposable bioreactors with proved economic advantages, but limitations in the scale capacity. Increasing cell culture titers is driving the research for novel approaches in downstream processes to comply with smaller purification costs. Intermediate product storage shelf life, biological variability of fermentation titers, regulation policies, and uncertain product demand are also considered relevant factors in the planning and scheduling of biopharmaceutical facilities.

5.4.1 BIOPROCESSES OPTIMIZATION

Biopharmaceutical processes often require a wider timeline than chemical-based pharmaceuticals, which can hamper the optimization modeling. The bioprocess automation has been enhancing applications to control and monitor the manufacturing parameters, as reviewed by Junker and Wang (2006), with relevant achievements in applications of process analytical technology (PAT) for quality and performance attributes. Progressively, with the vast potential of possible solutions and methods development, bioprocess management has been gathering significant interest in computer-aided tools that will be able to incorporate process and business needs in the manufacturing decision making.

The reviewed research area covers from the overall portfolio management of biopharmaceutical drugs development to the design of optimization models for specific steps of the manufacturing process. On the one hand, Rajapakse et al. (2005) developed a prototype decision-making application based on simulation tools to assist in the management of the R&D portfolio by accessing both the drug development activities and resources flows. Farid et al. (2007) presented the *SimBiopharma* software tool, to evaluate manufacturing alternatives in terms of cost, time, yield, resource utilization, and risk in modeling a biopharmaceutical manufacture. Considering a hybrid simulation-optimization approach, Brunet et al. (2012) also addressed the design of upstream/downstream units in a single-product process with a mixed integer dynamic optimization. On the other hand, Montagna et al. (2000)

and Asenjo et al. (2000) proposed mixed integer nonlinear programming (MINLP) formulations for the design structure and process variables optimization in biotechnological multiproduct batch plants. Vasquez-Alvarez and Pinto (2004) and Simaria et al. (2012) proposed optimization models for the downstream purification stages of a manufacturing process, the former by an MILP formulation and the latter using a meta-heuristic model addressing different levels of decision such as facility, product sequence, and unit operation. Liu et al. (2013) suggested the optimization of the chromatography column-sizing design in the antibody purification processes using MINLP by minimizing the total cost.

Regarding the planning/scheduling optimization of resource allocations and assignments applied to biopharmaceutical processes, research topics remain fairly unnoticed. Understandably, the overall hierarchy of process operations, in some decision levels, can be quite similar to mature chemical processes, but cannot address the specific constraints and complexity of large-scale biologic processes. It is noticed that a relatively small number of research papers addressing biochemical processes modeling, such as encompassing the process performance as well as capital investment, operating costs, resource utilization, or uncertainties. Lim et al. (2004) reviewed that custom bioprocesses planning methods include spreadsheet-based evaluators often with static and limited numerical solving techniques (supported by enterprise resource planning (ERP) or manufacturing resource planning (MRP) systems) and most of the literatures only discuss general bioprocess simulation software by their creators (e.g., Miller et al. 2010a,b). The same authors recommend that commercial simulators to also be used for modeling industrial bioprocesses, but do not support extensive dynamic customization. More recently, discrete-event simulation techniques have gained popularity for modeling the logistic operations. Even though mathematical formulation frameworks are still gathering the most attention, exploring time representation and process constraints to accurately represent the production development, as well as provide a feasible solution to these large-scale problems.

Focusing on planning and scheduling optimization models reviewed for biopharmaceutical processes, two research topics are highlighted and summarized in Table 5.1: (i) scheduling and planning modeling and (ii) modeling under uncertainty of operational parameters.

5.4.1.1 Planning and Scheduling Modeling

Samsatli and Shah (1996a,b) referred that the majority of work published in this area is focused on individual unit operations, neglecting the process integration. These authors presented one of the first frameworks for the design and optimization of biochemical processes: the problem is decomposed into two stages, the first to determine the processing conditions of all the unit operations using dynamic optimization; the second for detailed scheduling, using the solution obtained from the first stage to determine the sequence of operations of the whole process, as well as the design of the intermediate storage. Formulated as a discrete time MILP, this model is based on the state task network (STN) formulation proposed by Kondili et al. (1993) for a cyclic schedule within a 48–68 h time horizon. Iribarren et al. (2004) also presented another example of short-term scheduling together with process development. Addressing longer-term planning, Lakhdar et al. (2005) proposed a medium-term

TABLE 5.1
Resume of Biopharmaceutical Literature Planning and Scheduling Optimization

References	Model Characteristics	Type of Model/Solution Methods
Samsatli and Shah (1996a,b)	• Two stages optimization: first stage, processing rates and conditions of unit operations/equipment capacities through dynamic optimization (gProms); second stage, scheduling and design adjustments of intermediate storage.	*Second stage* • MILP model (STN framework) • Discrete time representation • Cyclic schedule (48 h/68 h) • Max. operating profit
Lakhdar et al. (2005)	• Medium-term planning and scheduling (1–2 years); • Determines campaigns durations and sequence campaigns, production quantities, inventories and product sales; • Considers storage constraints.	• MILP model • Max. operating profit
Lakhdar et al. (2006)	• Medium-term planning (1–3 years) considering uncertainty in the fermentation titers; • Considers storage constrains; • Results compared within deterministic model (Lakhdar et al. 2005), a two-stage programming model accompanied by an iterative construction algorithm, and a proposed CCP model.	• MILP model derived using CCP • Max. operating profit • Multiscenario stochastic programming
Lakhdar et al. (2007)	• Long-term planning, first solved as single objective problem (max. profit) and capacity analysis was conducted; then extended to allow multiple objectives through goal programming; • Determines campaigns durations and sequence campaigns, production quantities, inventories and product sales; • Storage and risk constraints.	• MILP model • Max. manufacturing profits • Min. total adverse deviations to targets: cost, service level, and capacity utilization (multiple objectives)
Lakhdar and Papageorgiou (2008)	• Medium-term planning under uncertain fermentation titers; • Storage constraints; • Propose future extension to multistage framework to allow uncertainty to be revealed gradually at any time period; further, proposed decomposition and approximation solution methods.	• MILP model (two stage) • Iterative algorithm for large-scale problem • Max. operating profit
Miller et al. (2010a,b)	• Core mathematical programming solver designed around a uniform discretization model and customized outer layer to address biologics process behavior; • VirtECS Scheduler Software; • Intermediate material storage consideration.	• MILP model (RTN framework) • Discrete time representation • Monte Carlo stochastic parameters

(Continued)

TABLE 5.1 (*Continued*)
Resume of Biopharmaceutical Literature Planning and Scheduling Optimization

References	Model Characteristics	Type of Model/Solution Methods
Gicquel et al. (2012)	• Hybrid flow shop scheduling problem; • Zero intermediate capacity and limited waiting time between processing; • Suggest heuristic solution as future work.	• MILP model • Discrete time representation • Min. total weighted tardiness
Kabra et al. (2013)	• Unit-specific event-based continuous-time representation; • Multiperiod scheduling of multistage multiproduct process; • Based on STN representation.	• MILP model • Continuous time representation • Max. profit

horizon of 1–2 years, using a discrete time MILP model, to determine the optimal production and cost-effective sequence of manufacturing tasks. More recently, Kabra et al. (2013) compared this later model with a newly developed continuous-time multiperiod scheduling of a multistage, multiproduct process based on an STN framework, verifying improved results.

Considering a wider time horizon, Lakhdar et al. (2007) proposed a multiobjective long-term planning horizon (14 years), first solved as a single objective problem to maximize profit and analyze the required capacity, and then extended to allow for multiple objectives (cost, service level, and capacity utilization) using goal programming.

5.4.1.2 Modeling under Uncertainty of Operational Parameters

Accounting for uncertain conditions in bioprocess manufacturing—batch yield titers, contamination rates and campaign length—is considered crucial to keep optimal operational performance. Sahinidis (2004) reviewed the topics of optimization under uncertainty, including two-stage programming, probabilistic (chance) programming, fuzzy programming, and dynamic stochastic programming. Typically, the most common approach is two-stage stochastic programming formulation, whereby strategic decisions are made first (here and now), while operational decisions are made in a second stage (wait and see). This approach, however, generates an extensive number of possible scenarios outcomes, which normally depend on heuristics, aggregation, and/or decomposition solver procedures. With application to biopharmaceutical processes, Farid et al. (2005) analyzed the impact of various uncertainty parameters on both operational and financial outputs, subject to fluctuating product demands, fermentation titers, and market success. Lakhdar et al. (2006) considered uncertainty in the fermentation titers over a medium-term planning, comparing a proposed chance-constrained programming (CCP) model with a deterministic model and a two-stage programming model accompanied by an iterative construction algorithm. Lakhdar and Papageorgiou (2008) updated the work presented in 2005, proposing an efficient iterative algorithm to solve large-scale problems considering uncertainty in fermentation titers.

5.4.2   BIOPHARMACEUTICAL OPTIMIZATION MODELS CLASSIFICATION

Owing to the wide diversity of scheduling problem formulations of processes, Méndez et al. (2006) proposed a roadmap for optimizing model classification focused on four main aspects: time representation, material balances, event representation, and objective function:

- Time representation constitutes the most relevant aspect of optimization modeling: two common approaches are discrete and continuous time formulations, whether considering the scheduling events to take place in predetermined time intervals or to occur at any moment of the time horizon.
- Regarding material balances, two categories are considered: the first category refers to monolithic approaches, dealing simultaneously with the optimal set of batches, the allocation and sequencing of manufacturing resources, and processing time of tasks. These models employ the representation methodologies, STN, and resource task network (RTN), developed by Kondili et al. (1993) and Pantelides (1994), respectively. The second category comprises models assuming that the number of batches of each size is known in advance. These solution algorithms decompose the problem into two stages, batching and schedule, to be applied to larger problems but restricted to sequential recipes.
- Event representation identifies the different concepts that arrange events of the schedule over time, guaranteeing that the maximum capacity of resources is never exceeded. The author classifies five different types, which can be seen in Figure 5.3. Event representations are usually oriented toward the solution of either arbitrary network processes or sequential batch processes. For discrete time formulations, the definition of global time intervals (common fixed time grid) is the option for general networks and sequential processes. In contrast, continuous time formulations involve

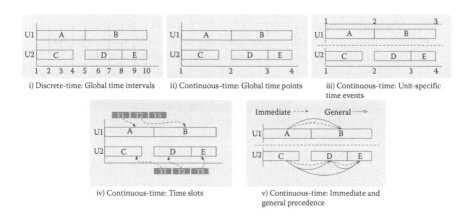

i) Discrete-time: Global time intervals   ii) Continuous-time: Global time points   iii) Continuous-time: Unit-specific time events

iv) Continuous-time: Time slots   v) Continuous-time: Immediate and general precedence

FIGURE 5.3   Time representations for scheduling problems. (Adapted from Méndez, C. A. et al. 2006. *Computers & Chemical Engineering* 30(6–7): 913–946.)

extensive alternative event representations, which are focused on different types of processes: for general network processes global time points and unit-specific time events can be used, whereas in the case of sequential processes, the alternatives involve the use of time slots and different batch precedence-based approaches. The global time point representation corresponds to a generalization of global time intervals where the timing of time intervals is treated as an additional model variable.

- The objective function represents the selected criteria used to define an optimal solution, with direct impact on the model computational performance. According to the authors, six were selected: makespan, earliness, tardiness, profit, inventory, and cost.

On the basis of these criteria, the previous reviewed papers are classified, as shown in Table 5.2.

5.4.3 DISCUSSION AND DIRECTIONS FOR FUTURE WORK

Considering Table 5.2, where the reviewed biopharmaceutical literature is classified according to the parameter roadmap defined by Méndez et al. (2006) for scheduling models, it is possible to analyze and define a set of guidelines for future work:

- It is acknowledged that only a small number of references were identified addressing the specific nature of planning and scheduling problems in biopharmaceutical processes. This verifies the claim from different authors that the specific area of biopharmaceutical optimization has been sparsely researched.
- Mainly discrete time representation and batch processes addressed, only until recently have new continuous formulations using network representations (e.g., STN) been explored. It is verified that general multipurpose model networks have been gathering significant research interest owing to its holistic formulation to approach process problems. While discrete time representation offers a rather simplified approach, continuous time formulations are able to explore more complex time modeling systems.
- Most MILP models' objective functions are focused on profit optimization.

As already discussed, with the research in the biopharmaceutical topic in its infancy, future research topics should explore the different model formulations and the results should be compared. Within the area of scheduling operations, recent developments have focused the research on the formulation of general networks (STN and RTN) integrating diverse operational events (e.g., variable/fixed batch size, batch/continuous plants, storage/transfer policies, shelf-life handling, lot blending/traceability, energy integration, changeovers, discrete/continuous time representation) and multiple objective criteria. These general models can provide a more simplified formulation, but often produce a larger model combinatory complexity with an increased number of variables when compared to specific designed batch models. Concerning time representation, alternative event representations can be

TABLE 5.2

Classification of Reviewed Literature according to Méndez et al. (2006) Roadmap

	Time Representation		Material Balances		Type of Model/Solution Methods					Objective Function					
	Discrete	Continuous	Network Batch	Lots/Batch	Global Time Intervals	Global Time Points	Unit-Specific Time Events	Time Slots	Precedence Based	Makespan	Earliness	Tardiness	Profit	Inventory	Cost
Samsatli and Shah (1996a,b)	X		X		X										X
Lakhdar et al. (2005)	X			X	X										X
Lakhdar et al. (2006)	X			X	X										X
Lakhdar et al. (2007)	X			X	X										X
Lakhdar and Papageorgiou (2008)	X			X	X										X
Miller et al. (2010 a,b)	X		X		X										
Gicquel et al. (2012)	X			X	X							X			
Kabra et al. (2013)		X	X				X								X

explored as suitable, as well as multiperiod or cyclic formulations. The generalization of the existent models to describe the specific characteristics of the biopharmaceutical problems should also be explored, fostering real-case study applications
addressing multiobjective criteria such as cost, service level, and capacity utilization.

The integration of uncertainty factors to provide a more accurate representation of the model to real operation, as proposed in the recent works of Farid et al.
(2005) and Lakhdar and Papageorgiou (2008), which used the Monte Carlo simulation technique to imitate the randomness inherent in manufacturing subjects to
technical and market uncertainties (e.g., fermentation titers), is also an important
aspect to develop.

Although the main problem often falls in the feasibility of the execution of
such complex mathematical models, the trend is toward determining of the best
possible feasible solution, accepting the limitations of optimality when demanding a high computational effort (Castro and Novais 2009). It is also acknowledged
that an optimal result could be forfeited during its implementation, owing to the
limited capability to model the dynamic nature of real process conditions. For
that reason, hybrid heuristic methods, decomposition/aggregation techniques, and
improved optimization-based techniques are getting widely embedded in mathematical models. These model features have been gathering significant advantages
in real industrial cases, once the solution responsiveness can be reached in less
time and sent promptly to the plant floor (Kopanos et al. 2010).

5.5   CONCLUSIONS

This work has presented an overview of the biopharmaceutical industrial sector, sus-
taining the importance of optimal planning and scheduling for efficient management
of these facilities. We have detailed the complex production process, from drug development to manufacturing, and characterized the main assigned problems toward cost
savings and regulatory policies of the sector. A review of current research topics was
performed, with evidence of a short explored scope when considering planning and
scheduling optimization applied to biopharmaceutical processes.

The research for improved planning and scheduling systems for the management
of industrial processes can be mathematically challenging, as claimed by Applequist
et al. (1997). Even though significant progress has been made in batch scheduling
for the resolution of large-scale industrial problems, it is still an unresolved issue in
the pharmaceutical area. Since no single approach exists that fits all problems, the
selection of the right modeling and solution method is the key to surpass the main
challenges in computational complexity. This chapter hopes to stimulate further
research as it identifies a fundamental gap in the application of scheduling methods
to biopharmaceutical processes and its specific operation problems.

ACKNOWLEDGMENTS

The authors would like to acknowledge the financial support of Fundação para a
Ciência e Tecnologia under the grant SFRH/BD/51594/2011.

REFERENCES

Amaro, A. C. S. and A. P. F. D. Barbosa-Póvoa. 2008. Planning and scheduling of industrial supply chains with reverse flows: A real pharmaceutical case study. *Computers & Chemical Engineering* 32(11): 2606–2625.

Amaro, A. C. S. and A. P. F. D. Barbosa-Póvoa. 2009. A continuous-time approach to supply chain planning: Managing product portfolios subject to market uncertainty under different partnership structures. *Computer Aided Chemical Engineering* 26: 973–978

Applequist, G., O. Samikoglu, J. Pekny, and G. Reklaitis. 1997. Issues in the use, design and evolution of process scheduling and planning systems. *ISA Transactions* 36(2): 81–121.

Asenjo, J. A., J. M. Montagna, A. R. Vecchietti, O. A. Iribarren, and J. M. Pinto. 2000. Strategies for the simultaneous optimization of the structure and the process variables of a protein production plant. *Computers & Chemical Engineering* 24(9): 2277–2290.

Barbosa-Póvoa, A. P. 2007. A critical review on the design and retrofit of batch plants. *Computers & Chemical Engineering* 31(7): 833–855.

Brunet, R., G. Guillen-Gosalbez, J. R. Perez-Correa, J. A. Caballero, and L. Jimenez. 2012. Hybrid simulation-optimization based approach for the optimal design of single-product biotechnological processes. *Computers & Chemical Engineering* 37: 125–135.

Castro, P. M. and A. Q. Novais. 2009. Scheduling multistage batch plants with sequence-dependent changeovers. *AIChE Journal* 55(8): 2122–2137.

Castro, P. M., A. M. Aguirre, L. J. Zeballos, and C. A. Mendez. 2011. Hybrid mathematical programming discrete-event simulation approach for large-scale scheduling problems. *Industrial & Engineering Chemistry Research* 50(18): 10665–10680.

Cuatrecasas, P. 2006. Drug discovery in jeopardy. *The Journal of Clinical Investigation* 116(11): 2837–2842.

Colvin, M. and C. T. Maravelias. 2010. Modeling methods and a branch and cut algorithm for pharmaceutical clinical trial planning using stochastic programming. *European Journal of Operational Research* 203(1): 205–215.

Farid, S. S. 2007. Process economics of industrial monoclonal antibody manufacture. *Journal of Chromatography B* 848(1): 8–18.

Farid, S. S., J. Washbrook, and N. J. Titchener-Hooker. 2005. Decision-support tool for assessing biomanufacturing strategies under uncertainty: Stainless steel versus disposable equipment for clinical trial material preparation. *Biotechnology Progress* 21(2): 486–497.

Farid, S. S., J. Washbrook, and N. J. Titchener-Hooker. 2007. Modelling biopharmaceutical manufacture: Design and implementation of SimBiopharma. *Computers & Chemical Engineering* 31(9): 1141–1158.

Gatica, G., L. G. Papageorgiou, and N. Shah. 2003. Capacity planning under uncertainty for the pharmaceutical industry. *Chemical Engineering Research & Design* 81(A6): 665–678.

Gicquel, C., L. Hege, M. Minoux, and W. van Canneyt. 2012. A discrete time exact solution approach for a complex hybrid flow-shop scheduling problem with limited-wait constraints. *Computers & Operations Research* 39(3): 629–636.

Grossmann, I. E. 2012. Advances in mathematical programming models for enterprise-wide optimization. *Computers & Chemical Engineering* 47: 2–18.

Harjunkoski, I., C. Maravelias, P. Bongers, P. Castro, S. Engell, I. Grossmann et al. 2014. Scope for industrial applications of production scheduling models and solution methods. *Computers & Chemical Engineering* 62: 161–193.

Iribarren, O. A., J. M. Montagna, A. R. Vecchietti, B. Andrews, J. A. Asenjo, and J. M. Pinto. 2004. Optimal process synthesis for the production of multiple recombinant proteins. *Biotechnology Progress* 20(4): 1032–1043.

Junker, B. H. and H. Y. Wang. 2006. Bioprocess monitoring and computer control: Key roots of the current PAT initiative. *Biotechnology and Bioengineering* 95(2): 226–261.

Kabra, S., M. A. Shaik, and A. S. Rathore. 2013. Multi-period scheduling of a multi-stage multi-product bio-pharmaceutical process. *Computers & Chemical Engineering* 57:95–103.

Kondili, E., C. C. Pantelides, and R. W. H. Sargent. 1993. A general algorithm for short-term scheduling of batch-operations.1. Milp formulation. *Computers & Chemical Engineering* 17(2): 211–227.

Kopanos, G. M., C. A. Mendez, and L. Puigjaner. 2010. MIP-based decomposition strategies for large-scale scheduling problems in multiproduct multistage batch plants: A benchmark scheduling problem of the pharmaceutical industry. *European Journal of Operational Research* 207(2): 644–655.

Laínez, J. M., E. Schaefer, and G. V. Reklaitis. 2012. Challenges and opportunities in enterprise-wide optimization in the pharmaceutical industry. *Computers & Chemical Engineering* 47: 19–28.

Lakhdar, K., J. Savery, L. G. Papageorgiou, and S. S. Farid. 2007. Multiobjective long-term planning of biopharmaceutical manufacturing facilities. *Biotechnology Progress* 23(6): 1383–1393.

Lakhdar, K. and L. G. Papageorgiou. 2008. An iterative mixed integer optimisation approach for medium term planning of biopharmaceutical manufacture under uncertainty. *Chemical Engineering Research & Design* 86(A3): 259–267.

Lakhdar, K., S. S. Farid, N. J. Titchener-Hooker, and L. G. Papageorgiou 2006. Medium term planning of biopharmaceutical manufacture with uncertain fermentation titers. *Biotechnology Progress* 22(6): 1630–1636.

Lakhdar, K., Y. Zhou, J. Savery, N. J. Titchener-Hooker, and L. G. Papageorgiou. 2005. Medium term planning of biopharmaceutical manufacture using mathematical programming. *Biotechnology Progress* 21(5): 1478–1489.

Levis, A. A. and L. G. Papageorgiou 2004. A hierarchical solution approach for multi-site capacity planning under uncertainty in the pharmaceutical industry. *Computers & Chemical Engineering* 28(5): 707–725.

Li, J., P. M. Verderame, and C. A. Floudas. 2012. Operational planning of large-scale continuous processes: Deterministic planning model and robust optimization for demand amount and due date uncertainty. *Industrial & Engineering Chemistry Research* 51(11): 4347–4362.

Lim, A. C., Y. H. Zhou, J. Washbrook, N. J. Titchener-Hooker, and S. Farid. 2004. A decisional-support tool to model the impact of regulatory compliance activities in the biomanufacturing industry. *Computers & Chemical Engineering* 28(5): 727–735.

Liu, S., A. S. Simaria, S. S. Farid, and Papageorgiou, L. G. 2013. Designing cost-effective biopharmaceutical facilities using mixed-integer optimization. *Biotechnology Progress* 29(6), 1472–1483.

Liu, S., J. M. Pinto, and L. G. Papageorgiou. 2009. MILP-based approaches for medium-term planning of single-stage continuous multiproduct plants with parallel units. *Computational Management Science* 7(4): 407–435.

Marchetti, P. A., C. A. Mendez, and J. Cerda. 2012. Simultaneous lot sizing and scheduling of multistage batch processes handling multiple orders per product. *Industrial & Engineering Chemistry Research* 51(16): 5762–5780.

Méndez, C. A., J. Cerdá, I. E. Grossmann, I. Harjunkoski, and M. Fahl. 2006. State-of-the-art review of optimization methods for short-term scheduling of batch processes. *Computers & Chemical Engineering* 30(6–7): 913–946.

Miller, D. L., D. Schertz, C. Stevens, and J. F. Pekny. 2010a. Mathematical programming for the design and analysis of a biologics facility. *BioPharm International* 23(2): 26–38.

Miller, D. L., D. Schertz, C. Stevens, and J. F. Pekny. 2010b. Mathematical programming for the design and analysis of a biologics facility—Part 2. *BioPharm International* 23(3): 40–52.

Mockus, L., J. M. Vinson, and K. Luo. 2002. The integration of production plan and operating schedule in a pharmaceutical pilot plant. *Computers & Chemical Engineering* 26(4–5): 697–702.

Moniz, S., A. P. Barbosa-Povoa, and J. P. de Sousa. 2013. New general discrete-time scheduling model for multipurpose batch plants. *Industrial & Engineering Chemistry Research* 52(48): 17206–17220.

Montagna, J. M., A. R. Vecchietti, O. A. Iribarren, J. M. Pinto, and J. A. Asenjo. 2000. Optimal design of protein production plants with time and size factor process models. *Biotechnology Progress* 16(2): 228–237.

Pantelides, C. C. 1994. Unified frameworks for optimal process planning and scheduling. *Proceedings on the Second Conference on Foundations of Computer Aided Operations.* New York: Cache Publications.

Papageorgiou, L. G., G. E. Rotstein, and N. Shah. 2001. Strategic supply chain optimization for the pharmaceutical industries. *Industrial & Engineering Chemistry Research* 40(1): 275–286.

Rajapakse, A., N. J. Titchener-Hooker, and S. S. Farid. 2005. Modelling of the biopharmaceutical drug development pathway and portfolio management. *Computers & Chemical Engineering* 29(6): 1357–1368.

Roe, B., L. G. Papageorgiou, and N. Shah. 2005. A hybrid MILP/CLP algorithm for multipurpose batch process scheduling. *Computers & Chemical Engineering* 29(6): 1277–1291.

Ryu, J. K., H. S. Kim, and D. H. Nam. 2012. Current status and perspectives of biopharmaceutical drugs. *Biotechnology and Bioprocess Engineering* 17(5): 900–911.

Sahinidis, N. V. 2004. Optimization under uncertainty: state-of-the-art and opportunities. *Computers & Chemical Engineering* 28(6–7): 971–983.

Samsatli, N. J. and N. Shah. 1996a. An optimization based design procedure for biochemical processes. Part I: Preliminary design and operation. *Food and Bioproducts Processing* 74(4): 221–231.

Samsatli, N. J. and N. Shah 1996b. An optimization based design procedure for biochemical processes. Part II: Detailed scheduling. *Food and Bioproducts Processing* 74(4): 232–242.

Sekhon, B. S. 2010. Biopharmaceuticals: An overview. *Thailand Journal of Pharmaceutical Science* 34: 1–19.

Shah, N. 2004. Pharmaceutical supply chains: Key issues and strategies for optimisation. *Computers & Chemical Engineering* 28(6–7): 929–941.

Simaria, A. S., R. Turner, and S. S. Farid. 2012. A multi-level meta-heuristic algorithm for the optimisation of antibody purification processes. *Biochemical Engineering Journal* 69: 144–154.

Sousa, R. T., S. S. Liu, L. G. Papageorgiou, and N. Shah. 2011. Global supply chain planning for pharmaceuticals. *Chemical Engineering Research & Design* 89(11A): 2396–2409.

Stefansson, H., S. Sigmarsdottir, P. Jensson, and N. Shah. 2011. Discrete and continuous time representations and mathematical models for large production scheduling problems: A case study from the pharmaceutical industry. *European Journal of Operational Research* 215(2): 383–392.

Stephanopoulos, G. and G. V. Reklaitis. 2011. Process systems engineering: From Solvay to modern bio- and nanotechnology. A history of development, successes and prospects for the future. *Chemical Engineering Science* 66(19): 4272–4306.

Terrazas-Moreno, S. and I. E. Grossmann. 2011. A multiscale decomposition method for the optimal planning and scheduling of multi-site continuous multiproduct plants. *Chemical Engineering Science* 66(19): 4307–4318.

Tsang, K. H., N. J. Samsatli, and N. Shah. 2007a. Capacity investment planning for multiple vaccines under uncertainty: 1: Capacity planning. *Food and Bioproducts Processing* 85(2): 120–128.

Tsang, K. H., N. J. Samsatli, and N. Shah. 2007b. Capacity investment planning for multiple vaccines under uncertainty: 2: Financial risk analysis. *Food and Bioproducts Processing* 85(2): 129–140.

Varma, V. A., J. F. Pekny, G. E. Blau, and G. V. Reklaitis. 2008. A framework for addressing stochastic and combinatorial aspects of scheduling and resource allocation in pharmaceutical R&D pipelines. *Computers & Chemical Engineering* 32(4–5): 1000–1015.

Vasquez-Alvarez, E. and J. M. Pinto. 2004. Efficient MILP formulations for the optimal synthesis of chromatographic protein purification processes. *Journal of Biotechnology* 110(3): 295–311.

Velez, S., A. Sundaramoorthy, and C. T. Maravelias. 2013. Valid inequalities based on demand propagation for chemical production scheduling MIP models. *AIChE Journal* 59(3): 872–887.

Walsh, G. 2003. *Biopharmaceuticals: Biochemistry and Biotechnology*, 2nd Edition. John Wiley & Sons, Chichester, UK.

Walsh, G. 2010. Biopharmaceutical benchmark 2010. *Nature Biotechnology* 28(9): 917–924.

Part II

Design and Synthesis

6 Design and Synthesis of Multipurpose Batch Plants

Esmael Reshid Seid, Jui-Yuan Lee,
and Thokozani Majozi

CONTENTS

6.1   INTRODUCTION

The last two decades have been characterized by a significant increase in the attention to batch processes in general and in particular to multipurpose batch operations. This is mainly due to the fact that batch processes are inherently flexible, which renders them an ideal choice in volatile or unstable market conditions. These plants easily adapt to changes in product specifications and operating conditions. They are also suitable for producing products of different recipes within the same facility. Despite this advantage, optimum design, synthesis, and scheduling of multipurpose batch plants remain poorly understood. This chapter presents a method that addresses design, synthesis, and scheduling in a holistic manner. As explained in detail in Chapter 2, the scheduling platform has a significant impact on the computational performance of the overall model. Given that optimum design and synthesis of batch plants is dependent on optimum scheduling, it is evident that the scheduling model of choice is of utmost importance in design and synthesis. Consequently, in this chapter, the scheduling model presented in Chapter 2 is adopted due to its proven computational efficiencies. It is hereby extended to incorporate design and synthesis. Furthermore, computational studies are presented to illustrate the effectiveness of the proposed model. A comparison with earlier formulations demonstrates that better computational times and objective functions are obtained through this formulation.

6.2   PROBLEM STATEMENT

In a nutshell, the problem that is of essence in this chapter can be summarized as follows.

Given:

- The product recipes (STNs) describing the production of one or more products over a single campaign structure
- The plant flowsheet with all possible equipment units to be installed and the involved connectivity
- The equipment units' suitability to perform the process/storage tasks
- The connections' suitability to transfer materials
- The operating and capital cost data involved in the plant
- The time horizon of planning
- The production requirements over the time horizon

Determine:

- The optimal plant configuration (i.e., number and type of equipment and the optimal design size of the equipment).
- A process schedule that allows the selected resources to achieve the required production (i.e., the starting and finishing times of all tasks, storage policies, batch sizes, amounts transferred, and allocation of tasks to equipment), so as to optimize the economic performance of the plant, measured in terms of the capital expenditure and the operating costs and revenues.

6.3   MATHEMATICAL FORMULATION

The research trend in the past two decades in the formulation of scheduling models for multipurpose batch plants focused on improving the problem complexity by reducing the event points required and a better description of sequencing constraints for processing tasks. If the number of event points required in the scheduling model is reduced, then the number of variables, constraints, and the search space is reduced considerably since the variables and constraints are defined at each event point. The model by Seid and Majozi (2012) results in a reduction of event points required, as a result gives better performance in terms of objective value and CPU time required when compared to previous literature models. This model is extended and used as a platform for the design and synthesis problem, since it has been proven that the solution performance is dependent on the scheduling framework used (Barbosa-Povoa 2007, Pinto et al. 2008). The design and synthesis model contains the following constraints.

6.3.1   EQUIPMENT EXISTENCE CONSTRAINTS

Constraint (6.1) states that in order to execute a task in the unit, the unit must be selected first.

$$\sum_{s_{in,j} \in S_{in,j}^*} y\left(s_{in,j}, p\right) \leq e(j), \quad \forall j \in J, \ p \in P \tag{6.1}$$

6.3.2   UNIT SIZE CONSTRAINTS

Constraint (6.2) implies that if the unit is selected, the design capacity should be between the minimum and maximum design capacity.

$$V_j^l e(j) \leq ss(j) \leq V_j^u e(j), \quad \forall j \in J \tag{6.2}$$

6.3.3   CAPACITY CONSTRAINTS

Constraint (6.3) implies that the total amount of all the states consumed at time point p is limited by the capacity of the unit, which consumes the states and represents the lower and upper bounds in the capacity of a given unit that processes the effective state.

$$\gamma_{s_{in,j}}^L ss(j) - V_j^L \left(1 - y\left(s_{in,j}, p\right)\right) \leq mu(s_{in,j}, p) \leq \gamma_{s_{in,j}}^U ss(j),$$
$$\forall \ p \in P, \ j \in J, \ s_{in,j} \in S_{in,J} \tag{6.3}$$

Constraints (6.4) and (6.5) ensure the amount of material stored at any time point p is limited by the capacity of the storage.

$$q_s(s,p) \le s(v), \quad \forall\, s \in S,\ p \in P,\ v \in V \tag{6.4}$$

$$V_v^L eu(v) \le s(v) \le V_v^U eu(v), \quad \forall\, s \in S,\ p \in P,\ v \in V \tag{6.5}$$

6.3.4 MATERIAL BALANCE FOR STORAGE

Constraint (6.6) states that the amount of material stored at each time point p is the amount stored at the previous time point adjusted by some amount, resulting from the difference between state s produced by tasks at the previous time point $(p-1)$ and used by tasks at the current time point p. This constraint is used for a state other than a product, since the latter is not consumed, but only produced within the process.

$$q_s(s,p) = q_s(s,p-1) - \sum_{j \in J_s^{sc}} muu(s,j,p) + \sum_{j \in J_s^{sp}} muu(s,j,p-1),$$

$$\forall\, p \in P,\ p \ge 1,\ s \in S \tag{6.6}$$

Constraint (6.7) is used for the material balance around storage at the first time point.

$$q_s(s,p) = QO(s) - \sum_{j \in J_s^{sc}} muu(s,j,p), \quad \forall\, p \in P,\ p = 1,\ s \in S \tag{6.7}$$

Constraint (6.8) states that the amount of product stored at time point p is the amount stored at the previous time point and the amount of product produced at time point p.

$$q_s(s^p,p) = q_s(s^p,p-1) + \sum_{s_{in,j} \in s_{in,J}^p} \rho_{s_{in,j}}^{sp} mu(s_{in,j},p), \quad \forall\, p \in P,\ s^p \in S^p \tag{6.8}$$

6.3.5 MATERIAL BALANCE AROUND THE PROCESSING UNIT

Constraint (6.9) is used to cater for the material processed in the unit and equals the amount of material directly coming from the unit producing it and from the storage.

$$\rho\left(s_{in,j}^{sc}\right) mu(s_{in,j},p) = muu(s,j,p) + \sum_{j' \in j_s^{sp}} mux(s,j,j',p),$$

$$\forall\, p \in P, j, j' \in J, \quad s_{in,j} \in S_{in,J},\quad s \in S \tag{6.9}$$

Constraint (6.10) is used to define the amount of material produced at time point p, which is sent to storage at the same time point p and to units that consume it at time point $(p+1)$.

$$\rho\left(s_{in,j}^{sp}\right) mu(s_{in,j},p) = muu(s,j,p) + \sum_{j' \in j_s^{sc}} mux(s,j',j,p+1),$$

$$\forall \, p \in P, \; j,j' \in J, \; s_{in,j} \in S_{in,J}, \; s \in S \qquad (6.10)$$

6.3.6   EXISTENCE CONSTRAINTS FOR PIPING

Constraints (6.11) and (6.12) are applicable to ensure whether pipe connections exist between the processing units and storage, as well as to ensure the amount of material transferred through the connection is limited by the design capacity of the connection. The material transfer time is assumed fixed.

$$V_{j,v}^L z(j,v) \le pip(j,v) \le V_{j,v}^U z(j,v), \quad \forall \, j \in J, \, v \in V \qquad (6.11)$$

$$muu(s,j,p) \le pip(j,v), \quad \forall \, j \in J, \quad v \in V, \; p \in P, \; s \in S \qquad (6.12)$$

Constraints (6.13) and (6.14) are similar to constraints (6.11) and (6.12) and are used when the connection is between processing units.

$$V_{j,j'}^L w(j,j') \le pipj(j,j') \le V_{j,j'}^U w(j,j'), \quad \forall \, j \in J \qquad (6.13)$$

$$mux(s,j,j',p) \le pipj(j,j'), \quad \forall \, j,j' \in J, \; p \in P, \; s \in S \qquad (6.14)$$

6.3.7   DURATION CONSTRAINTS (BATCH TIME AS A FUNCTION OF BATCH SIZE)

Constraint (6.15) describes the duration constraint modeled as a function of batch size where the processing time is a linear function of the batch size. For zero-wait (ZW), only the equality sign is used.

$$t_p(s_{in,j},p) \ge t_u(s_{in,j},p) + \tau(s_{in,j}) y(s_{in,j},p) + \beta(s_{in,j}) mu(s_{in,j},p),$$

$$\forall \, j \in J, \; p \in P, \; s_{in,j} \in S_{in,J} \qquad (6.15)$$

6.3.8   SEQUENCE CONSTRAINTS

The following two subsections address the proper allocation of tasks in a given unit to ensure the starting time of a new task is later than the finishing time of the previous task.

6.3.8.1   Same Task in Same Unit

Constraint (6.16) states that a state can only be used in a unit, at any time point, after all the previous tasks are complete.

$$t_u(s_{in,j},p) \ge t_p(s_{in,j},p-1), \quad \forall \, j \in J, \; p \in P, \; s_{in,j} \in S_{in,j}^* \qquad (6.16)$$

6.3.8.2 Different Tasks in Same Unit

Constraint (6.17) states that a task can only start in the unit after the completion of all the previous tasks that can be performed in the unit.

$$t_u(s_{in,j},p) \geq t_p\left(s'_{in,j},p-1\right), \quad \forall j \in J, \ p \in P,$$
$$s_{in,j} \neq s'_{in,j}, \ s_{in,j}, s'_{in,j}, \in S^*_{in,j} \tag{6.17}$$

If the state is consumed and produced in the same unit, where the produced state is unstable then, in addition to constraint (6.17), constraints (6.18) and (6.19) are used.

$$t_p\left(s_{in,j}^{usp},p-1\right) \geq t_u\left(s_{in,j}^{usc},p\right) - H\left(1 - y\left(s_{in,j}^{usp},p-1\right)\right),$$
$$\forall j \in J, \ p \in P, \ s_{in,j}^{usc} \in S_{in,j}^{usc}, \ s_{in,j}^{usp} \in S_{in,j}^{usp} \tag{6.18}$$

$$t_p\left(s_{in,j}^{usp},p\right) \geq t_p\left(s_{in,j}^{usp},p-1\right), \quad \forall j \in J, \ p \in P, \ s_{in,j}^{usp} \in S_{in,j}^{usp} \tag{6.19}$$

It should be noted that in this particular situation $t_u\left(s_{in,j}^{usc},p\right) = t_u(s_{in,j},p)$ and $t_p\left(s_{in,j}^{usc},p\right) = t_p(s_{in,j},p)$. Consequently, constraints (6.17) and (6.18) enforce $t_u\left(s_{in,j}^{usc},p\right) = t_p\left(s_{\sin,j}^{usc},p-1\right)$.

6.3.9 Sequence Constraints for Different Tasks in Different Units

These constraints state that for different tasks that consume and produce the same state, the starting time of the consuming task at time point p must be later than the finishing time of any task at the previous time point $(p-1)$ provided that the state is used.

If an intermediate state s is produced from one unit:

Constraints (6.20) and (6.21) work together in the following manner:

$$\rho\left(s_{in,j}^{sp}\right)mu(s_{in,j},p-1) \leq q_s(s,p) + V_j^U t(j,p),$$
$$\forall j \in J, \ p \in P, \ s_{in,j} \in S_{in,J}^{sp} \tag{6.20}$$

$$t_u(s_{in,j'},p) \geq t_p(s_{in,j},p-1) - H\left(\left(2 - y(s_{in,j},p-1) - t(j,p)\right)\right)$$
$$\forall j \in J, \ p \in P, \ s_{in,j} \in S_{in,J}^{sp}, \ s_{in,j'} \in S_{in,J}^{sc} \tag{6.21}$$

Constraint (6.20) states that if state s is produced from unit j at time point $(p-1)$, but is not consumed at time point p by another unit j', that is, $t(j,p) = 0$, then the amount produced cannot exceed allowed storage, that is, $q_s(s,p)$. On the other hand, if state s produced from unit j at time point $(p-1)$ is used by another unit j' then the amount of state s stored at time point p, that is, $q_s(s,p)$, is less than the amount of state s produced at time point $(p-1)$. The outcome is that the binary variable $t(j,p)$

becomes 1 in order for constraint (6.20) to hold. Constraint (6.21) states that the starting time of a task consuming state s at time point p must be later than the finishing time of a task that produces state s at the previous time point $(p-1)$, provided that state s is used. Otherwise, the sequence constraint is relaxed.

If an intermediate state is produced from more than one unit:

Constraint (6.22) states that the amount of state s used at time point p can either come from storage or from other units that produce the same state, depending on the binary variable $t(j,p)$. If the binary variable $t(j,p)$ is 0, which means that state s produced from unit j at time point $(p-1)$ is not used at time point p, then constraint (6.21) is relaxed. If $t(j,p)$ is 1, state s produced from unit j at time point $(p-1)$ is used, as a result constraint (6.21) holds. Although constraint (6.22) is nonlinear it can be linearized exactly using the Glover transformation (Glover 1975).

$$\sum_{s_{in,j} \in S_{in,J}^{sc}} \rho_{s_{in,j}}^{sc} mu(s_{in,j},p) \le qs(s,p-1) + \sum_{s_{in,j} \in S_{in,J}^{sp}} \rho_{s_{in,j}}^{sp} mu(s_{in,j},p-1)\, t(j,p)$$

$$\forall\, j \in J,\ p \in P \tag{6.22}$$

Sequence constraints for completion of previous tasks:

Constraint (6.23) states that a consuming task can start after the completion of the previous task. Constraint (6.23) takes care of proper sequencing time when a unit uses material which is previously stored, that is, when the producing task is active at time point $(p-2)$ and later produces and transfers the material to the storage at time point $(p-1)$. This available material in the storage at time point $(p-1)$ is then used by the consuming task in the following time points. This necessitates that the starting time of the consuming task must be later than the finishing time of the producing task at time point $(p-2)$. Consequently, constraint (6.23) together with constraint (6.21) results in a feasible sequencing time when the consuming task uses material, which is previously stored or/and material currently produced by the producing units.

$$t_u(s_{in,j'},p) \ge t_p(s_{in,j},p-2) - H(1 - y(s_{in,j},p-2)),$$

$$\forall\, j \in J,\ p \in P,\ s_{in,j} \in S_{in,J}^{sp},\ s_{in,j'} \in S_{in,j'}^{sc} \tag{6.23}$$

6.3.10   Sequence Constraints for FIS Policy

According to constraints (6.24) and (6.21), the starting time of a task that consumes state s at time point p must be equal to the finishing time of a task that produces state s at time point $(p-1)$, if both consuming and producing tasks are active at time point p and time point $(p-1)$, respectively.

$$t_u(s_{in,j'},p) \le t_p(s_{in,j},p-1) + H\Big(2 - y\,(s_{in,j'},p) - y(s_{in,j},p-1)\Big)$$

$$\forall\, j \in J,\ p \in P,\ s_{in,j} \in S_{in,J}^{sp},\ s_{in,j'} \in S_{in,J}^{sc} \tag{6.24}$$

6.3.11 TIME HORIZON CONSTRAINTS

The usage and the production of states should be within the time horizon of interest. These conditions are expressed in constraints (6.25) and (6.26).

$$t_u(s_{in,j}, p) \leq H, \quad \forall \, s_{in,j} \in S_{in,J}, \, p \in P, \, j \in J \tag{6.25}$$

$$t_p(s_{in,j}, p) \leq H, \quad \forall \, s_{in,j} \in S_{in,J}, \, p \in P, \, j \in J \tag{6.26}$$

6.3.12 OBJECTIVE FUNCTION

Constraint (6.27) is the objective function expressed as maximization of profit. This is obtained from revenue from the sale of products, less operating costs for tasks, raw material costs, and capital costs from piping and equipment.

$$\text{maximize} \left(\begin{array}{l} \displaystyle\sum_{s^p} price(s^p)\, q_s(s^p, p) - \sum_p \sum_{s_{inj} \in S_{rm}} \rho\left(s_{in,j}^{sc}\right) mu(s_{in,j}, p) \; - \\[2ex] \displaystyle\sum_p \sum_{s_{inj} \in S_{inJ}} \left(FOC * y(s_{in,j}, p) + VOC * mu(s_{in,j}, p) \right) \end{array} \right) *(AWH/H)$$

$$- \left(\begin{array}{l} \displaystyle\sum_{j \in J} \sum_{v \in V} \left(CNC * z(j,v) + VCN * pip(j,v) \right) \\[2ex] + \displaystyle\sum_{j' \in J} \sum_{j \in J} \left(FCNC * w(j,j') + VCNC * pipj(j,j') \right) \\[2ex] + \displaystyle\sum_{j \in J} \left(FEC * e(j) + VEQ * ss(j) \right) + \sum_{v \in V} \left(FECS * eu(v) + VEQS * s(v) \right) \end{array} \right) * CCF,$$

$$\forall \, p = P, \, s^p \in S^p$$

$$\tag{6.27}$$

Constraint (6.28) is the objective function expressed as minimization of capital and operating cost if the demand for the products is known beforehand within the specified time horizon.

$$\text{minimize} \left(\begin{array}{l} \displaystyle\sum_p \sum_{s_{inj} \in S_{rm}} mu(s_{in,j}, p) \\[2ex] + \displaystyle\sum_p \sum_{s_{inj} \in S_{inJ}} \left(FOC * y(s_{in,j}, p) + VOC * mu(s_{in,j}, p) \right) \end{array} \right) *(AWH/H)$$

$$\begin{pmatrix} + \displaystyle\sum_{j \in J} \sum_{v \in V} \left(CNC * z(j,v) + VCN * pip(j,v) \right) \\ + \displaystyle\sum_{j' \in J} \sum_{j \in J} \left(FCNC * w(j,j') + VCNC * pipj(j,j') \right) \\ + \displaystyle\sum_{j \in J} \left(FEC * e(j) + VEQ * ss(j) \right) + \displaystyle\sum_{v \in V} \left(FECS * eu(v) + VEQS * s(v) \right) \end{pmatrix} * CCF,$$

$$\forall \; p = P, \; s^p \in S^p$$

$$(6.28)$$

6.4 CASE STUDIES

In order to demonstrate the applicability of the proposed model, two published literature examples are presented and discussed. The results in all the case studies for the proposed model were obtained using CPLEX 9.1.2/GAMS 22.0 on a 2.4 GHz, 4 GB of RAM, Acer TravelMate 5740 G computer. The computational results for the literature models are taken directly from the literature for comparison.

6.4.1 CASE I

At first, this case was studied by Kondili et al. (1993) and it has become one of the most common examples appearing in literature. This is a typical multipurpose batch plant that produces two different products. Processing units are shared, a task can be conducted in multiple units, a unit can perform different tasks, and the products can take different production lines. The unit operations consist of preheating, three different reactions, and separation. The STN representation of the flowsheet is shown in Figure 6.1. Full connectivity between equipment is assumed. The plant superstructure

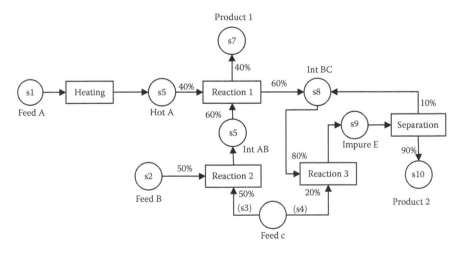

FIGURE 6.1 Recipe representation for case I.

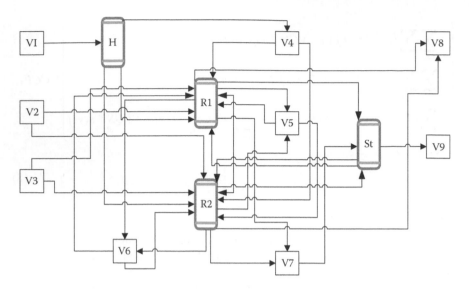

FIGURE 6.2   Superstructure for case I.

is shown in Figure 6.2. The data pertaining to equipment and materials can be found in Lin and Floudas (2001). The cost model for equipment is annualized capital cost.

6.4.2 RESULTS AND DISCUSSION

The results obtained for Case I are presented in Table 6.1. The proposed model requires 4 event points, 64 binary variables, 227 continuous variables, and 572 constraints and is solved in less than 1 s. Compared to other literature models the proposed model gave a smaller model size and required less computational time to solve. An objective value of 569.2 (k$) was obtained by this work, which is better than the objective values obtained by other formulations. The proposed model selected all the processing units; however, it avoided the necessity of all the intermediate storage units given

TABLE 6.1

Computational Statistics for Case I Using the Different Formulations

Model	Time Points	Integer Variables	Continuous Variables	Constraints	Obj MILP (k$)	CPU$_s$
Xia and Macchietto (1997)	8	288[o]	201[o]	425[o]	—	—
	8	62[t]	34[t]	122[t]	585.6	2407
Lin and Floudas (2001)	6	128	341	877	572.9	22.5
Castro et al. (2005)	7	74	294	395	572.7	6.6
This work	4	52	227	572	569.2	0.08

t = transformed formulation; o = original formulation.

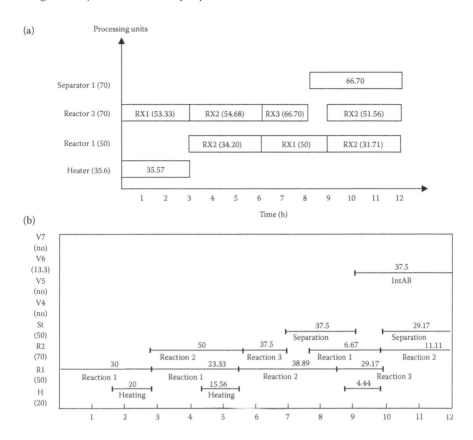

FIGURE 6.3 (a) Gantt chart using the proposed model, (b) Gantt chart using the formulation of Lin and Floudas. (Adapted from Lin, X., Floudas, C.A., 2001. *Comput. Chem. Eng.* 25, 665–674.)

in the superstructure and as a result led to less capital investment compared to other formulations, which required an intermediate storage for IntAB. Figure 6.3 shows the optimal Gantt chart in achieving the production requirement for the time horizon of 12 h, with the corresponding equipment sizes. The optimal plant structure obtained by the proposed model and by Lin and Floudas (2001) is depicted in Figure 6.4.

6.4.3 CASE II

This case study was taken from the petrochemical industry by Kallrath (2002) and later used as a benchmark problem in the scheduling environment for multipurpose batch plants. Pinto et al. (2008) adopted this case study for the design and synthesis problem. The model by Pinto et al. (2008) is the recent method for design and synthesis problems for multipurpose batch plants that also consider plant topology. The maximization of profit that results in the optimal design and synthesis of the plant is considered for the production time horizon of 120 h. The recipe representation for the plant is presented in Figure 6.5. It is a complex problem where 10 reactors, 15 storage units, and 18 states are considered in the superstructure presented in Figure 6.6.

(a)

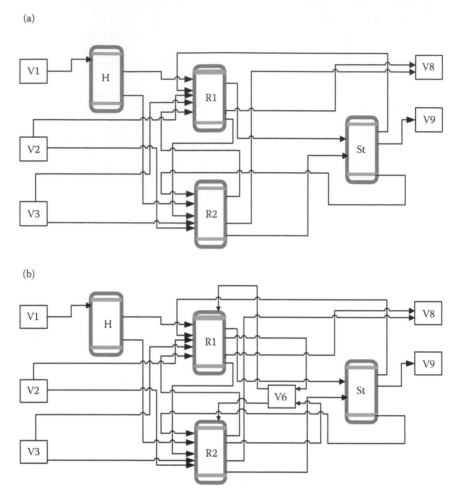

(b)

FIGURE 6.4 (a) Optimal structure using the proposed model, (b) optimal structure using the formulation of Lin and Floudas. (Adapted from Lin, X., Floudas, C.A., 2001. *Comput. Chem. Eng.* 25, 665–674.)

There are reactors dedicated to a specific task while there are also reactors conducting multiple tasks—some of them are capable of executing up to five different tasks. A mixed intermediate storage (MIS) policy is assumed. The piping cost from equipment connections is considered. The connections capacity range from 0 to 300 (mass unit [m.u.]/m²) with an associated fixed/variable cost of 0.1/0.01 (10^3 cost unit [c.u.]). Operating costs for the tasks are fixed/variable of 0.1/0.01 (10^3 c.u.). A raw material cost of 0.002 (10^3 c.u.) per m.u. is given. Data pertaining to equipment can be found in Pinto et al. (2008). The cost model for equipment is the annualized capital cost. The plant is required to satisfy a demand requirement of 30 tons for S14, S15, and S16; 20 tons for S17; and 40 tons for S18, in a time horizon of 120 h. The selling price for each product is 1000 c.u. per m.u. It is assumed that the plant operates for 366 days.

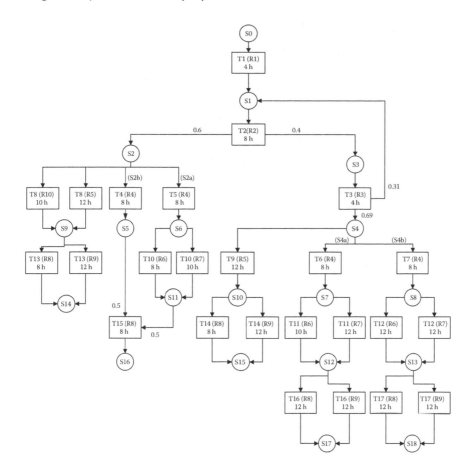

FIGURE 6.5 STN representation for Case II.

6.4.4 RESULTS AND DISCUSSION

Table 6.2 gives the computational statistics for Case II using the different formulations. The results for the literature models are taken directly from Pinto et al. (2008). The results obtained clearly differentiate the various formulation efficiencies. The adopted STN and RTN formulations of Pinto et al. (2008) led to a significant reduction in the model size from their classical STN and RTN representations (26,147 variables for STN-adopted vs. 31,547 variables for STN, 33,405 variables for RTN-adopted, 115,305 variables for RTN). The m-STN representation gives fewer variables and less computational time compared to the STN- and RTN-adopted models. The proposed formulation significantly reduced the number of binary and continuous variables required compared to other formulations, requiring 776 binary variables and 4485 total variables. The current formulation required 16 time points and was solved in a CPU time of 111 s to a 5% margin of optimality for fair comparison since the literature models were also solved with the same margin of optimality. It is worth mentioning that a new objective value of 6.92 (10⁷) c.u. was obtained, which is a

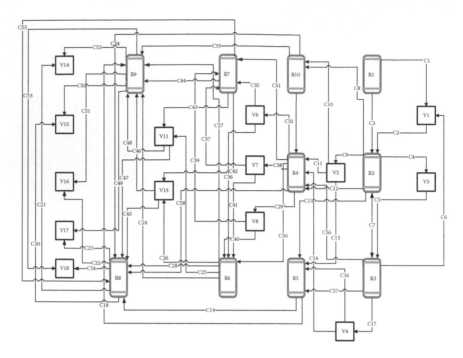

FIGURE 6.6   Superstructure of the plant for Case II.

great improvement (228.6%) compared to the objective value, 2.1 (10^7) c.u., obtained by other formulations. Within the time horizon of 120 h, the proposed formulation produced more products compared to other formulations (480 m.u. for S14, 243 m.u. for S15, 240 m.u. for S16, 20 m.u. for S17 and 40 m.u. for S18 obtained by this work vs. 60 m.u. for S14, 60 m.u. for S15, 50 m.u. for S16, 39 m.u. for S17 and 78 m.u. for S18 obtained by other formulation). The efficiency of the model is attributed to the use of the recent robust scheduling platform of Seid and Majozi (2012) for the design and synthesis problem. The optimal equipment selection and size obtained by this work is obtained in Table 6.3.

TABLE 6.2
Computational Statistics for Case II

Methodology	Objective Value	Total Variables	Binary Variables	Constraints	CPU Times (s)
		Pinto et al. (2008)			
STN	2.1 (10^7)	31,547	10,828	59,728	11,975.6
STN-adapted	2.1 (10^7)	26,147	9028	51,328	1445.4
m-STN	2.1 (10^7)	17,508	3147	29,822	1056.9
RTN	2.1 (10^7)	115,305	55,677	155,653	12,812.9
RTN-adapted	2.1 (10^7)	33,405	9074	54,136	1009.1
This work	6.92 (10^7)	4485	776	8895	111

TABLE 6.3

Optimal Equipment Design Capacity for Case II

Equipment	Capacity (m.u.)
V0	1088.4
R1, R2, V1, C2, C3	100
R3, V3, C5	280
R4, V11, C11, C47, C58	120
R5, R8, V2, V16, C15, C21, C22, C19	240
R6, V18, C4, C14, C25, C53, C24, C30	40
R9, C50, C18	110.4
V4	173.2
V6, C32, C36	26.7
V14	480
V15	243.6
V17, C49	20
C6	86.8
C9	60
C17	193.2
C16, C20	133.2

The formulation avoids having reactors 7 and 10 and storage units V7, V8, and V13, available in the superstructure. The optimal superstructure that depicts the selected equipment and the associated connectivity is given in Figure 6.7. The Gantt chart that details the amount of material processed, the type of task performed, the starting and finishing times for each piece of equipment using the proposed formulation is given in Figure 6.8. For comparison, the Gantt chart obtained for this case study using the formulation of Pinto et al. (2008) is also presented in Figure 6.9. This work also explicitly addressed the allocation of materials in the plant. Figure 6.10 depicts the amount of material and its location in the plant at any given time point p for state S6. The amount of material state S6 produced from unit R4 at time point $p4$, $p6$, and $p9$ was sent to storage at time 28 h, 44 h, and 60 h, respectively. The storage sent state S6 to the consuming unit R6 at time 28 h, 56 h, and 64 h. The storage holds the state S6 for the time interval from 44 h to 56 h and 60 h to 64 h only, during the entire time horizon. Direct transfer of material state S6 also occurred from the producing unit to consuming unit, where the amount of material processed by unit R4 at time point $p6$ produced state S6 at time 44 h and transferred to the consuming unit R6 at the next time point $p7$ where it started processing state S6 at the same time of 44 h. Consequently, the model allows one to know the exact location of materials and their amount in the plant at any given time.

6.5 CONCLUSIONS

A mathematical formulation for the synthesis and design of multipurpose batch plants has been presented. A recent robust scheduling formulation based on the continuous time representation is used as a platform for the design problem and achieves

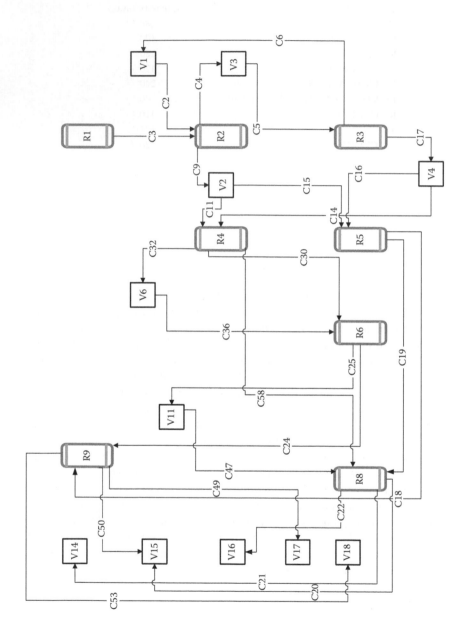

FIGURE 6.7   Optimal plant structure and associated pipe connectivity.

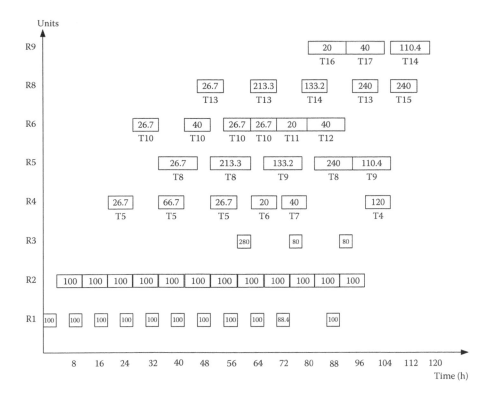

FIGURE 6.8   Optimal Gantt chart for Case II using the proposed formulation.

a better optimal design and requires less computational time. An improved objective value is obtained by this work compared to the recent published formulations for the design and synthesis problem. When this work is compared with other formulations, it gives a smaller size mathematical model that requires fewer binary variables, continuous variables, and constraints. The model explicitly considers the different locations of materials in the plant. The formulation also considers the costs arising from the pipe network and determines the optimal pipe network that should exist between equipment.

NOMENCLATURE

SETS

J $= \{j \mid j$ is a piece of equipment$\}$

J_s^{sc} $= \{j_s^{sc} \mid j_s^{sc}$ is a piece of equipment that consumes state $s\}$

J_s^{sp} $= \{j_s^{sp} \mid j_s^{sp}$ is a piece of equipment that produces state $s\}$

P $= \{p \mid p$ is a time point$\}$

S $= \{s \mid s$ is any state$\}$

S^p $= \{s \mid s$ is any state which is a product$\}$

S_{rm} $= \{s \mid s$ is any state which is a raw material$\}$

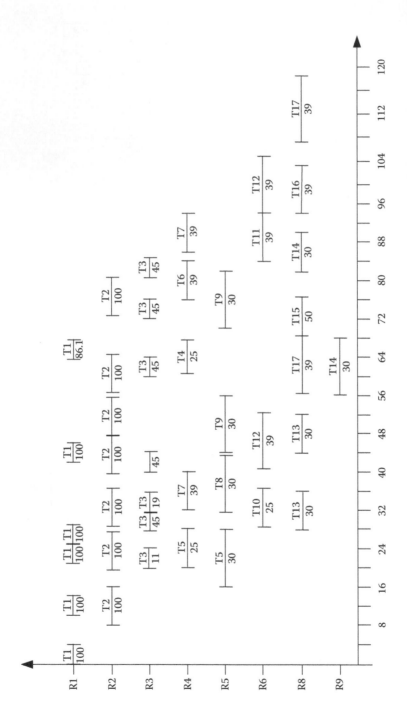

FIGURE 6.9 Optimal Gantt chart for Case II using the formulation of Pinto et al. (Adapted from Pinto, T., Barbosa-Povoa, A.P., Novais, A.Q., 2008. *Ind. Eng. Chem. Res.* 47, 6025–6044.)

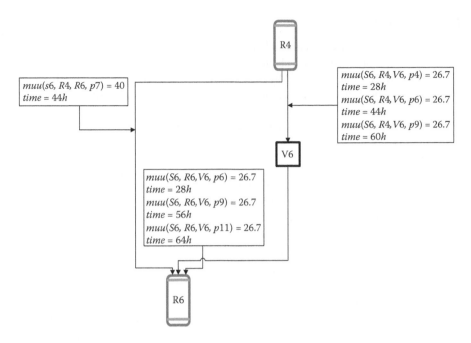

FIGURE 6.10 Material location in the plant for the state S6.

$S_{in,J}^{p} = \{s_{in,j}^{p} \mid s_{in,j}^{p}$ task which produce state s which is a product$\}$

$S_{in,J}^{sc} = \{s_{in,j}^{sc} \mid s_{in,j}^{sc}$ is a task which consumes state $s\}$

$S_{in,j}^{*} = \{s_{in,j}^{*} \mid s_{in,j}^{*}$ is a task performed in unit $j\}$

$S_{in,J} = \{s_{in,j} \mid s_{in,j}$ is an effective state representing a task$\}$

$S_{in,J}^{usc} = \{s_{in,j}^{usc} \mid s_{in,j}^{usc}$ is a task which consumes unstable state $s\}$

$S_{in,J}^{sp} = \{s_{in,j}^{sp} \mid s_{in,j}^{sp}$ is a task which produces state s other than a product$\}$

$S_{in,J}^{usp} = \{s_{in,j}^{usp} \mid s_{in,j}^{usp}$ is a task which produces unstable state $s\}$

$V = \{v \mid v$ is a storage$\}$

CONTINUOUS VARIABLES

$mu(s_{in,j},p)$	= amount of material processed by a task
$muu(s,j,p)$	= amount of state s transferred between unit j and storage unit at time point p
$mux(s,j,j',p)$	= amount of state s transferred between unit j and another unit j' at time point p
$pip(j,v)$	= design capacity of a pipe that connects unit j and storage unit v
$pipj(j,j')$	= design capacity of a pipe that connects unit j and another unit j'
$s(v)$	= design capacity of storage unit v
$ss(j)$	= design capacity of unit j
$t_u(s_{in,j},p)$	= time at which a task in unit j starts
$t_p(s_{in,j},p)$	= time at which a task in unit j finishes
$q_s(s,p)$	= amount of state s stored at time point p in storage unit

BINARY VARIABLES

$e(j)$ = 1 if unit j is selected; 0 otherwise

$eu(v)$ = 1 if storage unit v is selected; 0 otherwise

$t(j,p)$ = 1 if the state produced by unit j at time point p is consumed; 0 otherwise.

$w(j,j')$ = 1 if pipe that connects unit j and j' is selected; 0 otherwise

$y(s_{in,j},p)$ = 1 if state s is used in unit j at time point p; 0 otherwise

$z(j,v)$ = 1 if pipe that connects unit j and storage unit v is selected; 0 otherwise

PARAMETERS

$\beta(s_{in,j})$ = coefficient of variable term of processing time of a task

$\gamma^{L}_{s_{in,j}}$ = minimum percentage equipment utilization for a task

$\gamma^{U}_{s_{in,j}}$ = maximum percentage equipment utilization for a task

$\rho(s^{sc}_{in,j})$ = portion of state s consumed by a task

$\rho(s^{sp}_{in,j})$ = portion of state s produced by a task

$\tau(s_{in,j})$ = duration of a task conducted in unit j

AWH = annual working hour

CCF = capital charge factor

CNC = fixed capital cost for pipe connection between processing unit and storage

$FCNC$ = fixed capital cost for pipe connection between processing units

FEC = fixed capital cost of equipment

$FECS$ = fixed capital cost of storage

FOC = fixed operating cost for a task

H = time horizon of interest

M = any large number

$price(s^p)$ = selling price for a product

$QO(s)$ = initial amount of state s stored in unit

V^{L}_{j} = lower bound for unit j

V^{U}_{j} = upper bound for unit j

V^{L}_{v} = lower bound for storage unit v

V^{U}_{v} = upper bound for storage unit v

$V^{L}_{j,v}$ = lower bound for the capacity of the pipe connecting unit j and storage unit v

$V^{U}_{j,v}$ = upper bound for the capacity of the pipe connecting unit j and storage unit v

$V^{L}_{j,j'}$ = lower bound for the capacity of the pipe connecting unit j and another unit j'

$V^{U}_{j,j'}$ = upper bound for the capacity of the pipe connecting unit j and another unit j'

VCN = variable capital cost for pipe connection between processing unit and storage

$VCNC$ = variable capital cost for pipe connection between processing units

VEQ = variable capital cost of equipment

$VEQS$ = variable capital cost of storage

VOC = variable operating cost for a task

REFERENCES

Barbosa-Povoa, A.P.F.D., 2007. A critical review on the design and retrofit of batch plants. *Comput. Chem. Eng.* 31 (7), 833–855.

Castro, P.M., Barbosa-Povoa, A. P., Novais, A. Q., 2005. Simultaneous design and scheduling of multipurpose plants using resource task network based continuous-time formulations. *Ind. Eng. Chem. Res.* 44 (2), 343–357.

Glover, F., 1975. Improved linear integer programming formulations of nonlinear integer problems. *Manage. Sci.* 22, 455–460.

Kallrath, J., 2002. Planning and scheduling in the process industry. *OR Spectrum.* 24 (3), 219–250.

Kondili, E., Pantelides, C.C., Sargent, R.W.H., 1993. A general algorithm for short-term scheduling of batch operations. I. MILP formulation. *Comput. Chem. Eng.* 17, 211–227.

Lin, X., Floudas, C.A., 2001. Design, synthesis and scheduling of multipurpose batch plants via an effective continuous-time formulation. *Comput. Chem. Eng.* 25, 665–674.

Pinto, T., Barbosa-Povoa, A.P., Novais, A.Q., 2008. Design of multipurpose batch plants: A comparative analysis between the STN, m-STN, and RTN representations and formulations. *Ind. Eng. Chem. Res.* 47, 6025–6044.

Seid, E.R., Majozi, T., 2012. A robust mathematical formulation for multipurpose batch plants. *Chem. Eng. Sci.* 68, 36–53.

Xia, Q. S., Macchietto, S., 1997. Design and synthesis of batch plants MINLP solution based on a stochastic method. *Comp. Chem. Eng.* 21, S697–S702.

7 Process Synthesis Approaches for Enhancing Sustainability of Batch Process Plants

Iskandar Halim and Rajagopalan Srinivasan

CONTENTS

7.1 INTRODUCTION

Some of the most challenging issues facing the world today such as climate change, depletion of freshwater resources, and environmental pollution are a direct result of rapid industrialization that has taken place over the past hundred years. The evidence for climate change can be seen in the melting of polar ice, rising sea surface temperature, and extreme weather conditions such as heavy snow, storms, floods, and droughts in many parts of the world, which pose hazards to human lives and natural ecosystems [1]. Various studies also highlight that excessive water abstraction over the past decades has resulted in many water-stressed areas; this is expected to affect about half of the world's population by 2030 [2]. The spread of environmental pollution has further exacerbated the scarcity of usable water for domestic consumption.

With $4 trillion worth of products and 4.3% contribution to the world Gross Domestic Product (GDP), the chemical industry has become one of the most important sectors of the global economy today [3]. However, a recent survey conducted by the European Chemical Industry Council highlighted that the public perception of this industry was not so favorable. The survey showed that among the 10,000 respondents interviewed, only 49% gave a positive rating while 44% had a negative view [4]. This poor public perception can be attributed to the growing concerns over environmental damages caused by chemical pollutants (26%), health risks from chemical products (19%), and plant accidents (9%). Hence, there is a pressing need for the chemical industry to build a more positive image by addressing public concerns. One way of achieving this is by embracing sustainability.

7.1.1 Sustainability of Batch Chemical Processing

The chemical industry can be classified into commodity chemical and specialty chemical production. Commodity chemicals are characterized by large-scale production through continuous operation. Examples of commodity chemicals are petrochemicals and basic inorganics. On the other hand, specialty chemicals such as pharmaceuticals are produced in small amounts for niche markets. For example, consider an active chemotherapeutic drug called Paclitaxel. This drug was first commercially developed by Bristol-Myers Squibb and sold under the trademark Taxol. It has been proven effective for treatment of breast, lung, ovarian, and liver cancers [5]. Paclitaxel is produced by batch operation; this can be attributed to the following reasons:

- One of the routes for Paclitaxel production is a semisynthetic process involving an ingredient obtained from the bark of Pacific Yew tree [6]. Its complex chemistry involves multiple phases, synthetic steps, solvents, and reagents and is not fully understood. In this case, batch processing is more favored than the continuous one as the former is much more robust to the lack of accurate knowledge about the process [7].
- As the semisynthetic Paclitaxel technology is developed in the laboratory through batch experiments, batch operation is also preferred for the production scale process due to the ease of scale-up, which in turn reduces lead times.
- Furthermore, when the same pieces of equipment in the plant can be used to manufacture different products (i.e., the plant becomes multipurpose), it offers flexibility to the company to produce a large variety of products without incurring large investments for each production line [8].

While the scale of batch production plants is generally smaller than their continuous counterparts, batch operations typically exhibit a higher ratio of waste produced per unit of product (the E factor). For example, the typical E factor for a pharmaceutical plant has been reported to be between 50 and 100 [9]. To understand the reasons for the high E factor in a pharmaceutical process, consider the hierarchical chemical

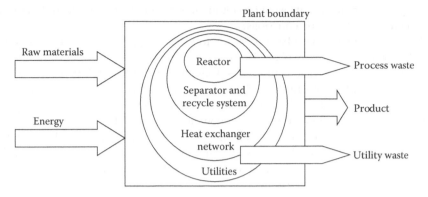

FIGURE 7.1 Onion model of waste origins in chemical process plants.

process plant design as represented by the "onion model" shown in Figure 7.1. It highlights two groups of waste arising in any chemical plant. The inner layers, which correspond to the activities of the reactor and separation and recycle systems, generate process wastes (waste by-products, purges, etc.) while the outer layers, which define the heat exchanger network and utility system, produce utility wastes (combustion emissions, wash water, etc.) [10]. The factors that result in high waste generation in a pharmaceutical process can be explained as follows. First, the high quality and purity requirements of the pharmaceutical product often call for huge quantities of solvents to be used during the reaction and/or separation process. Inadequate recovery and recycling of these solvents would lead to large volume of wastes. In addition, lack of a good understanding of the underlying complex phenomena of the pharmaceutical process prevents operating the plant at the most optimal condition to reduce the generation of waste by-products or off-quality products. Moreover, the individual pieces of equipment in the pharmaceutical plant can be used in a multiproduct environment or dedicated to individual products. Hence, product switch or batch change would require the production to be stopped, equipment washed, sterilized, reconfigured, and validated [11]. This would lead to additional wastes (waste wash water, waste solvent, etc.). Furthermore, the intermittent flows of streams during batch operation restrict opportunities for recovery, recycle, and reuse to reduce utility wastes.

From a design and operations perspective, the key issues around sustainable production are essentially about process efficiency and waste minimization [12]. These two are not mutually exclusive but two sides of the same coin. Consider the carbon emissions due to energy usage as an example. It is well-known that the pharmaceutical industry incurs substantial carbon emissions. In a study done in 2009, it was estimated that the U.S. healthcare sector, which accounts for activities such as hospital care and the manufacture and distribution of pharmaceutical products, contributes 8% of the country's total CO_2 emissions [13]. In the manufacturing context, a major portion of these emissions comes from the burning of fossil fuels to produce utilities (such as steam and hot water) for the plant. One way of minimizing the fuel consumption is by implementing energy recovery. This can be achieved by exchanging heat efficiently from streams that require cooling (hot streams) to those that require heating (cold streams). This would result in an overall reduction in the utilities consumption and thus carbon emissions.

Several techniques have been proposed to identify opportunities for waste minimization within a process plant. However, these techniques have been developed with the focus on continuous processes. Extending them to batch processes would necessitate modifications to the underlying methodologies to account for the time-dependent characteristics of batch operation. To illustrate, consider a stirred jacketed reactor during a batch process. This equipment can be used for multiple operation steps, such as heating up reactants, carrying out a reaction, boiling off solvent, or distilling a product. As a result, wastes differing in quantity and compositions may be generated from this equipment at different times. This temporal dimension needs to be accounted for when performing waste minimization analysis in a batch plant. This chapter will discuss the applications of various process synthesis methodologies—the knowledge-based technique, simulation-optimization, integrated scheduling, and the heat and water integration method—for sustainable batch production with emphasis on waste minimization. These will be illustrated using an industrial case study described next.

7.1.2   CASE STUDY DESCRIPTION: ORGANIC SALT PROCESS

This case study involves the production of an organic salt production that is commonly used for pharmaceutical processing. Figure 7.2 shows the flow diagram of the process. The operating procedure (process recipe) consists of the following operation steps [14]:

- Reaction: Initially, a stream of excess reactant A reacts with reactant B in a reaction tank. The reaction produces an acid product C and a waste by-product D. After the reaction is completed, the mixture is transferred to a neutralization tank.
- Neutralization: In the neutralization tank, the crude acid mixture is heated and reacted with an alkaline compound E. The reaction between acid C and alkaline E produces an organic salt S and a negligible amount of by-product. Next, the salt mixture is transferred to a crystallization unit.
- Crystallization and filtration: The salt mixture is cooled, causing salt crystals to form. From the crystallization unit, the crystal-bearing mixture is sent to a rotary vacuum filter unit. The salt crystals collected as "filter cake" are then sent to a dryer unit as product while the filtrate mixture is sent to a neutralization tank, where it is mixed further with alkaline E.
- Distillation: From the neutralization tank, the filtrate is sent to a distillation column. There, reactant A exits the top of the column and is recycled. The remaining liquid at the bottom of the column is finally sent to wastewater plant.

A team of experts has carried out a waste minimization review of this organic salt process and their results are available for comparison. In the next section, each of the process synthesis methodologies that can be used for this process will be discussed in detail.

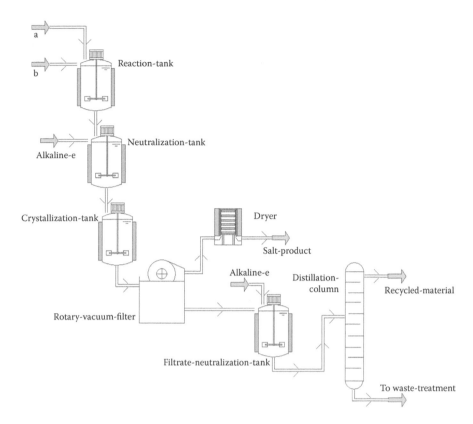

FIGURE 7.2   Organic salt process.

7.2   QUALITATIVE WASTE MINIMIZATION ANALYSIS

Waste minimization, defined as any technique, process, or activity which prevents, eliminates, or reduces waste at its source, or allows for reuse and recycling of waste, has become an important element of sustainability. In essence, the quest for waste minimization in any process plant involves the following two steps: identifying the origin of each material component that makes up the waste stream and finding ways to eliminate them through minimization and reuse/recycling options. Over the years, several methodologies have been proposed for waste minimization in batch processes including hierarchical design procedure, checklist approach, and the HAZOP (hazard and operability)-based technique. Houghton et al. [15] extended the hierarchical design procedure of Douglas [16] by listing a set of heuristic solutions to be implemented at each decision level—this covers input–output structure, reaction-recycle, separation, heat transfer, and cleaning process. Similarly, Mulholland and Dyer [17] developed a checklist to reduce waste generation during charging, reaction, discharging, and cleaning of a batch reactor. Another qualitative approach is an HAZOP-based technique, called Environmental Optimization (ENVOP) [18]. Like HAZOP, a combination of keywords (more, less, larger, smaller, other, etc.) and process variables (pressure, temperature, concentration, etc.) are used during an ENVOP analysis to evaluate each line and equipment in order to identify solutions for potential waste minimization. Due to its thorough analysis, ENVOP can be applied to both continuous and batch process plants. In this section, a waste minimization methodology that employs such a procedure for waste analysis in batch plants is described. The methodology, called BATCH-ENVOP *Expert* (BEE), applies process-graph (P-graph) analysis to diagnose the root cause of waste generation and a hierarchical design approach to derive broad waste minimization alternatives. BEE has been implemented as a knowledge-based system using Gensym's G2 expert system shell and has been successfully tested on a number of industrial-scale processes [19,20].

The first step in performing waste minimization analysis in any batch process is to identify each material component present in each process unit and stream of the process. This can be achieved using P-graph representation. P-graph originates from the work of Friedler et al. [21], who demonstrated a special directed bipartite graph for representing process structure to solve the synthesis problem in continuous processes. In their P-graph, an operating-unit is represented by a bar, a material stream by a circle, and connections between material streams and operating-units by directed arcs. For batch operations, the P-graph needs to be adapted to a recipe-centric viewpoint to account for the temporal dimension. Hence, an operation step is represented by a bar, a material stream by a circle, and connections between material streams and operations by directed arcs. Figure 7.3 shows the P-graph structure that describes the materials present at each unit and stream pertinent to the organic salt process.

Once the materials present in each stream and unit are fully established using P-graph, the next step is to decompose the P-graph into subgraphs comprised of

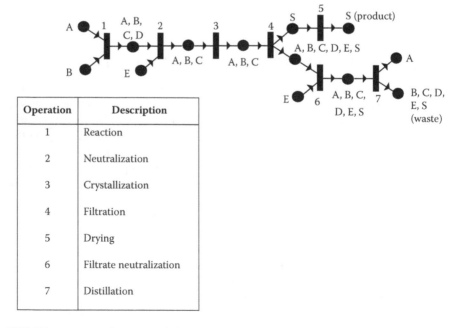

Operation	Description
1	Reaction
2	Neutralization
3	Crystallization
4	Filtration
5	Drying
6	Filtrate neutralization
7	Distillation

FIGURE 7.3   P-graph representation for organic salt production.

operation steps that contribute to the presence of each material component in the waste stream (or unit). This is done by tracing each component in the waste stream upstream through the P-graph model to identify the operations that lead to the escape of valuable materials into waste streams. This procedure is explained in Figure 7.4. By applying ENVOP, the following waste origins can be identified in the organic salt process:

- Unreacted reactants B (operation-1) and E (operation-2 and operation-6).
- Unrecovered intermediate C.
- Formation of by-product D (operation-1).
- Inefficiency in separating product S (operation-4).

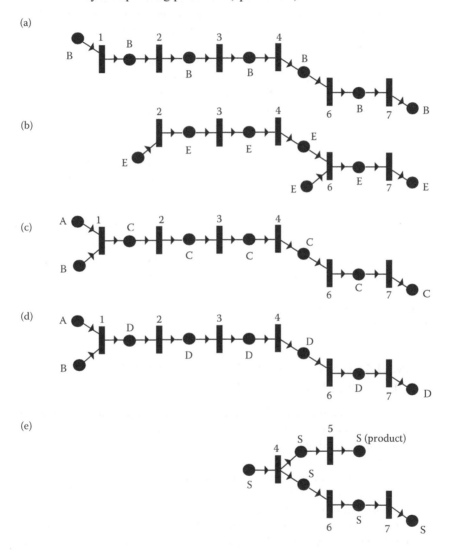

FIGURE 7.4 P-graph diagnosis of waste origins (a) reactant A, (b) alkaline E, (c) intermediate C, (d) waste by-product D, and (e) product S.

Once the waste origins are identified, alternatives to eliminate them can be proposed. This is done using a set of waste minimization heuristics that have been formulated following the hierarchical decision levels of Houghton et al. [15], which can be expressed using the following axioms:

- Input–output structure: (a) if an impurity exists in a feed stream, then purify that stream to prevent the impurity from entering the process; (b) if a useful material in the feed stream exits through a waste stream, then recover that material to prevent it from exiting as waste stream; (c) if a useful material exists in a waste stream, then recover and recycle that material.
- Reaction system: (a) if a reaction operation generates waste by-product, then eliminate or reduce that by-product formation; (b) if an excess reactant exits through a waste stream, then minimize the feed of that reactant.
- Separation process: if a separation operation is inefficient thus causing an escape of valuable materials into a waste stream, then improve the separation process or include another separation operation.
- Cleaning system: if a cleaning agent is used during a washing operation, then reuse it for another washing operation.

Table 7.1 lists the set of waste minimization alternatives identified by BEE for each operation step in the process recipe along. It also shows the results from the team of experts. A comparison shows that all the sources of waste identified and diagnosed by BEE are correct and in-line with those from the team of experts. However, the comparison between the two results shows that the team derives more detailed alternatives. For example, the team's suggestions "use evaporative cooling technique to crystallize salt" and "install larger cooling coils within crystallizer" are more detailed than the alternatives "use alternative crystallization technology or improve the current technology" identified by BEE. For BEE to propose such detailed options, it requires detailed technology-specific knowledge to be embedded inside its knowledge database.

7.3 SIMULATION-OPTIMIZATION METHODOLOGY FOR WASTE MINIMIZATION

The waste minimization methodology in BEE provides only broad-level qualitative alternatives that can be made to improve the sustainability of the process. The analysis is capable of suggesting "how" a particular waste source is to be solved, but not of specifically determining the process variables that need to be manipulated and the extent of manipulation that is required to reduce the waste streams. For example, consider the alternative "optimize the column control point" identified by the team. There are many ways to optimize the column variables such as by manipulating the reboiler temperature, column pressure, or distillate concentration. All these cannot be evaluated qualitatively as it requires a detailed model of the process.

The use of process simulators is prevalent in the chemical process industry. Using process simulators, various plant modifications can be evaluated easily and

TABLE 7.1
Comparison between Team's Results and That Generated by BATCH-ENVOP *Expert*

Operation	Team's Suggestions	BATCH-ENVOP *Expert* Alternatives
Reaction	• Use new technology to make acid C • Optimize the reactant ratio during acid production • Add water to dilute the excess reactant A	• Consider alternative process chemistry • Optimize the reaction condition to eliminate by-product D and improve reactant B conversion • Use reaction agent to suppress forming by-product D
Neutralization	• Decrease the amount of alkaline E • Optimize addition and mixing of alkaline	• Prevent excessive feed of E • Optimize the neutralization condition
Crystallization	• Add agent to enhance the crystallization of organic salt S • Use evaporative cooling technique to crystallize salt • Install chiller for crystallizer • Install larger cooling coils within crystallizer • Install automatic temperature controller • Use brine as cooling	• Optimize the crystallization operating condition • Use alternative crystallization technology or improve the current technology
Filtration	• Improve the filter cloths in vacuum filter • Cool filter wash stream to reduce dissolving and washing away of crystals • Isolate filter wash stream and recirculate it to the neutralizer • Cool the filter filtrate and recirculate it to the neutralizer	• Optimize the filtration operating condition • Use alternative filtration technology or improve the current technology • Use further separation after the filtration process to avoid salt S from becoming waste
Filtrate neutralization	• Decrease the amount of alkaline E • Optimize addition and mixing of alkaline	• Prevent excessive feed of E • Optimize the neutralization condition
Distillation	• Redesign the distillation column • Optimize the column control point • Recirculate the column bottoms during start-ups and shut-downs	• Use alternative separation technology • Optimize the distillation operating condition

in a short time without the need for extensive experimentation or pilot plant testing. In recent years, they have been extensively used for environmental studies; for instance, to evaluate the environmental impacts from various plant design and modifications [22–25]. An integrated knowledge-based simulation-optimization framework has been developed by combining the waste minimization analysis of BEE with a gPROMS simulator and multiobjective optimization [20]. The procedure starts with BEE extracting the material and energy balance information from the plant simulation model. Then, using P-graph analysis, BEE derives broad waste alternatives to the plant that cover basic suggestions such as: materials substitution, change of process chemistry, reuse and recycling, and optimizing certain unit operation's variables. This is followed by detailed evaluation using a simulation-optimization technique to evaluate the economic and environmental implications due to the modifications.

To illustrate the simulation-optimization framework, let us revisit the organic salt case study, particularly the acid C production process. Let us consider the team's suggestion "use a new technology to make acid C." Figure 7.5 shows the flow diagram of the new process for making acid C. It involves the following steps [26]:

- Fed-batch reaction: Initially, 2500 kmol of excess reactant A is charged from tank T101 into reaction-tank R101. This is followed by the slow addition of B at rate 1.25 kmol/s from tank T102 into R101. The reaction, which fully consumes B, also produces C and waste by-product D according to the following kinetics:

$$\text{Reaction 1 : A + B} \rightarrow \text{C; } r_c(\text{mol/m}^3\text{s}) = 0.0333 \, \exp\left(\frac{-2400}{T(K)}\right) C_a C_b$$

$$\text{Reaction 2: B + C} \rightarrow \text{D; } r_d(\text{mol/m}^3\text{s}) = 0.0833 \, \exp\left(\frac{-3000}{T(K)}\right) C_b C_c$$

As the reaction is highly exothermic, cooling is provided through a cooling jacket to maintain the temperature at 325 K. When the maximum yield of C is reached, the vessel is cooled rapidly to bring the temperature down to 298 K. The mixture is then drained into intermediate storage T104. At the end of draining operation, about 20 kmol of mixture is left in the vessel.

- Reactor washing: The reactor is washed with approximately 1080 kmol of cleaning agent L from tank T103 to bring the concentration of trace material C in the reactor to less than 0.01 (mole fraction). The used liquid L containing traces of reaction mixture is sent to tank T105 as waste. This combined reaction and washing step is repeated for three cycles to produce approximately 3400 kmol of C in T104.
- Distillation: From the intermediate storage T104, the reaction mixture is sent to tank T106 for boiling. This is done using steam that is passed through a heating coil. The vapor is sent for distillation in column T107 with the pressure drop along the column controlled by manipulating the steam flow rate. The distillation is performed through three set-point changes to the column pressure drop to separate the mixture into three cuts. The first cut, which contains mainly A, is transferred to tank T108 to be recycled. This is done until the end point of the first cut reaches about 0.9 (mole fraction) as measured using a concentration measurement. Once the concentration

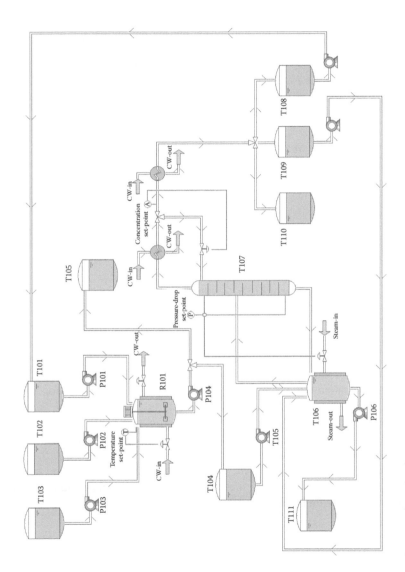

FIGURE 7.5   Modified acid C production.

of A falls below 0.9, the condensate is diverted to tank T109. Since this second cut contains a mixture of A and C, it is recycled back to T106 for further distillation. When the concentration of C in the condensate reaches a purity level of around 0.92, the third cut is obtained. This condensate is sent to tank T110 as product. As by-product D is the heaviest component in the mixture, it remains at the bottom of T106. This residue is later sent to T111 as waste.

Qualitative diagnosis of the waste streams of this process has been performed using BEE. It reveals the following waste sources: use of cleaning liquid L, waste by-product D generation, and inefficient separation (distillation). Based on the diagnosis, the following waste minimization alternatives are investigated further:

- Optimize the reaction-operating condition to minimize the formation of by-product D. This is done by manipulating the cooler setpoint. The specified range is 310–330 K.
- Improve the efficiency of the distillation process to minimize the amount of useful materials (A and C) from becoming waste. This can be done by manipulating the reboiler pressure drop setpoints and the purity setpoints of A and C. The variable ranges are as follows: first stage pressure drop: 0.060–0.070 kPa, second stage: 0.070–0.075 kPa, third stage: 0.075–0.080 kPa, purity of A: 0.85–0.92 and purity of C: 0.90–0.96.
- Recycle the cleaning liquid L from the first wash operation to the second and from the second washing to the third.

Once the variables to be manipulated are identified, the next step is to perform quantitative assessment to these variables. The objective here is to evaluate the economic and environmental implications from the modification using the process simulator. The economic impacts can be measured in terms of production rate and batch cycle time. For the environmental assessment, the measures are the amount of cleaning liquid, total energy consumption, and environmental impact, which can be calculated using WAste Reduction (WAR) algorithm. The WAR algorithm was proposed by Cabezas et al. [22], who introduced the concept of potential environmental impact balance of pollutants. In WAR, each process material is assigned an index value to indicate its potential environmental impact. Using this index value, the total impact of a waste stream in the plant ($\dot{I}^{waste}$) can be calculated as

$$\dot{I}^{waste} = \dot{M}_m \times x_{km}^{waste} \times \psi_k \qquad (7.1)$$

where $\dot{M}_m$ is the mass flowrate of the waste stream m, x_{km}^{waste} is the mass fraction of chemical k in the stream m, and ψ_k is the potential environmental impact of chemical k.

The economic and environmental aspects of the proposed alternatives may be in conflict with one another. Consider the alternative "optimize the column control point" as an example. While the waste material could be reduced by manipulating the control variables, implementing this alternative could lead to increased operating costs. Hence, the next step of the analysis is to perform multiobjective optimization to identify and resolve such trade-offs. Among different optimization techniques that can be used, simulated annealing algorithm has been shown to be one of the promising tools. Simulated annealing is a stochastic optimization algorithm, which mimics the thermodynamic phenomena in metal annealing [27]. In contrast to a greedy search, which is susceptible to local optima, the advantage of simulated annealing lies in its ability to escape from local optima by temporarily accepting poorer solutions. The algorithm has been extended to multiobjective optimization problems as follows [28]:

1. Start by identifying the elements of the solution vector x, whose elements are the various decision variables that impact the sustainability profile of the batch process. Specify also the minimum (x_{min}) and maximum (x_{max}) values for each of the decision variables. Meanwhile, the variable values from the base-case design are used as the initial solution x'.

2. Populate the Pareto set with $\langle x', f_{env}(x'), f_{eco}(x') \rangle$ as its initial element. Here, f_{env} is the environmental indicator of the process while f_{eco} is the plant economics measure.

3. Specify an initial annealing temperature T. In the beginning, T is set to a large number ($T_{initial}$) and then gradually reduced.

4. Perform random perturbation to generate a new solution vector x'' in the neighborhood of x' using the following expression:

$$x'' = x' + \frac{T}{T_{initial}}[r_1(x_{max} - x') - r_2(x' - x_{min})] \tag{7.2}$$

where r_1 and r_2 are random numbers between 0 and 0.5. Using the candidate solution vector x'', simulate the process to obtain $f_{env}(x'')$ and $f_{eco}(x'')$.

5. Compare objective values of x'' with all solutions in the Pareto set. If x'' dominates any element of the Pareto set, replace that element with x''. A solution is deemed as nondominated, if no other solution is superior when all objectives are considered. If x'' is Pareto-optimal with all the elements in the set, then include it in the set.

6. If x'' is dominated, temporarily accept it as the current solution vector with a probability P, where

$$P = \min \left\{ 1, \exp\left(\frac{-\Delta S_{env}}{T}\right) \cdot \exp\left(\frac{-\Delta S_{eco}}{T}\right) \right\} \tag{7.3}$$

$$\begin{aligned} \Delta S_{env} &= f_{env}(x'') - f_{env}(x') \\ \Delta S_{eco} &= f_{eco}(x'') - f_{eco}(x') \end{aligned} \tag{7.4}$$

At the same time, a random number P_{rand} is generated between 0 and 1.

7. x'' is accepted as the new solution, that is, $x' \leftarrow x''$ when $P > P_{rand}$ (a random number in [0, 1]); otherwise the earlier solution vector (x') is retained.
8. Periodically, reduce the annealing temperature T with a reduction factor R_t:

$$T \leftarrow T \cdot R_t \tag{7.5}$$

9. To escape from local optima, periodically, x' is replaced with a randomly selected solution from the Pareto set.
10. Steps (4) to (9) are repeated for a predefined total number of iterations N_{Total}.

The multiobjective simulated annealing algorithm described above has been successfully coded in BEE and linked to the gPROMS simulator via a connecting bridge. This two-way connection is used to send values of the decision variables that have been plotted by the simulated annealing algorithm in BEE to gPROMS to simulate their effects and then return the results back to BEE. In this case, the simulated annealing parameters were set as follows: number of iterations = 1500, $T_{initial}$ = 150, and $R_T = 0.99$. Table 7.2 shows the optimization results with 13 Pareto solution sets. The optimization results highlight a potential improvement of 3.3% of the environmental impact, 63.2% of the cleaning agent, 2.8% in C production, 1.8% of the production time, and 3.8% of the energy consumption from the base case scenario. The high savings in the cleaning liquid comes from recycling of cleaning effluent from the first wash to second and the second wash to third. The results also highlight the trade-offs between the objectives. No solution is found to dominate all the objectives. Further, they show that the optimization is mostly sensitive to changes in cooler setpoints and purity specifications. This conclusion is not surprising as changes in these variables greatly impact the amount of reactant consumed and product (and by-product) formed during the reaction and also the component purity during the distillation process. In turn, this would affect the objective values (i.e., the amount of cleaning agent, environmental impacts, product flow, energy consumption, and production time).

7.4   INTEGRATED SCHEDULING AND WATER AND ENERGY MINIMIZATION

One key issue that arises in batch process operations is optimal scheduling, which is effectively arranging tasks to produce a set of products with limited resources (raw materials, equipment, utilities, and manpower). Alongside is the need to minimize the utility (energy and water) consumption. However, the inherent nature of batch operations complicates the optimization of the schedule, along with minimization of energy and water usage. For example, much of the heating and cooling in batch processes is performed through external jackets or internal coils in vessels. Hence, the availability of hot and cold streams in the batch process is tightly integrated with the processing steps and consequently temporal and with time-varying temperatures [29]. This complicates heat recovery and heat integration between streams. Similarly, wash water minimization through reuse and recycling is constrained by the timings at which the water is utilized in the operation.

TABLE 7.2
Pareto Solution for Acid Production Case Study

		Decision Variables						Objective Functions				
Status	Solution Index	Cooler (K)	Reboiler-1st Stage Δp (kPa)	Reboiler-2nd Stage Δp (kPa)	Reboiler-3rd Stage Δp (kPa)	Purity A	Purity C	Environmental Impact	Cleaning Liquid (kmol)	Product (kmol)	Batch Time (s)	Total Energy (10^6 kJ)
Base case	0	313.9	0.063	0.073	0.078	0.908	0.933	680.7	3236.4	3007.3	13,008	3.926
Pareto solution	1	314.7	0.063	0.073	0.077	0.878	0.913	658.4	1191.0	3092.7	12,851	3.826
	2	319.3	0.063	0.073	0.076	0.877	0.912	660.7	1211.9	3063.9	12,903	3.825
	3	318.7	0.063	0.074	0.078	0.863	0.931	682.3	1209.0	2937.0	12,835	3.794
	4	322.3	0.065	0.072	0.077	0.855	0.925	677.5	1554.0	2945.9	12,857	3.775
	5	317.2	0.064	0.072	0.078	0.858	0.940	693.1	1202.4	2869.4	12,801	3.786
	6	319.6	0.064	0.072	0.077	0.860	0.927	677.7	1213.1	2956.8	12,842	3.784
	7	319.0	0.065	0.071	0.077	0.857	0.932	683.9	1210.1	2920.3	12,812	3.781
	8	319.7	0.063	0.072	0.077	0.857	0.930	681.7	1213.2	2929.6	12,869	3.777
	9	320.3	0.065	0.072	0.078	0.862	0.927	678.1	1216.2	2955.6	12,814	3.791
	10	318.6	0.064	0.073	0.078	0.858	0.931	681.8	1208.5	2934.6	12,800	3.783
	11	320.5	0.066	0.073	0.077	0.857	0.93	682.7	1216.7	2923.8	12,811	3.781
	12	315.9	0.065	0.072	0.078	0.867	0.923	671.2	1196.5	3011.7	12,772	3.801
	13	320.1	0.064	0.072	0.077	0.858	0.922	672.1	1215.3	2983.4	12,838	3.779

Two of the most important techniques that have been successfully developed for minimizing energy and water consumption in batch process plants are heat integration and water reuse synthesis. The former involves matching the hot streams with the cold streams so as to minimize the external utilities (cooling water and steam) consumption. In the latter, a network of water flows is to be synthesized over a set of processing units subject to contaminant mass load, maximum allowable contaminant concentration of process units, and timing of the operations. The goal is to maximize the reuse of water between operations so as to minimize the fresh water consumption and hence wastewater generation. In recent years, these methodologies have been extended to the batch domain. A graphical pinch analysis approach was developed by Kemp and MacDonald [30,31], Wang and Smith [32], and Uhlenbruck et al. [33] for design of heat recovery network in batch process. In mathematical-based approaches, Vaselenak et al. [34], Corominas et al. [35], Vaklieva-Bancheva et al. [36], and Bozan et al. [37] proposed a sequential method by decomposing the problem into two parts—the scheduling part that needs to be specified *a priori* followed by heat integration synthesis. Other mathematical techniques include the work of Papageorgiou et al. [38], Lee and Reklaitis [39], Zhao et al. [40], Pinto et al. [41], and Majozi [42], who developed an integrated mathematical model to simultaneously optimize the schedule and heat integration for maximizing profit. Methodologies for synthesizing the water reuse network in batch processes have also been proposed. Some noteworthy contributions in graphical techniques include the work of Wang and Smith [32], Foo et al. [43], and Majozi et al. [44]. While these techniques are useful, they share a common drawback—the schedule needs to be determined *a priori* or solved first. Further, their application is limited to single contaminant cases. On the other hand, mathematical programming-based techniques are capable of handling the drawbacks of the graphical approaches. A continuous-time scheduling framework was proposed by Majozi [45] and Gouws et al. [46] to simultaneously optimize the schedule and water reuse within the constraints of a single contaminant. Majozi and Gouws [47] extended the methodology of Gouws et al. (2008) [46] and applied linearization techniques for simultaneous scheduling the optimization and water reuse network involving multiple contaminants.

Though commonly encountered, problems entailing simultaneous scheduling, heat integration, and water reuse network synthesis have been avoided in literature [47]. This may be due to complexities both in terms of modeling and optimization. As a result, there is lack of a unified methodology that solves them within a single framework. A novel framework for integrated scheduling, heat integration, and water reuse optimization in batch processes has been proposed [48]. The framework is based on a sequential framework, where the overall problem is decomposed into a sequential solution of scheduling, heat integration, and water reuse optimization. Figure 7.6 shows the sequence of steps employed in the framework. First, optimization of the schedule is performed to meet the economic objective such as minimizing makespan or maximizing profit. The output from the optimization is an optimal schedule, which can be represented in the form of a Gantt chart. As the schedule solution is not necessarily unique, the next step is to explore alternative solutions. The goal is to find alternate schedules that optimize other objectives (e.g., energy and water consumption) without deteriorating the primary objective function

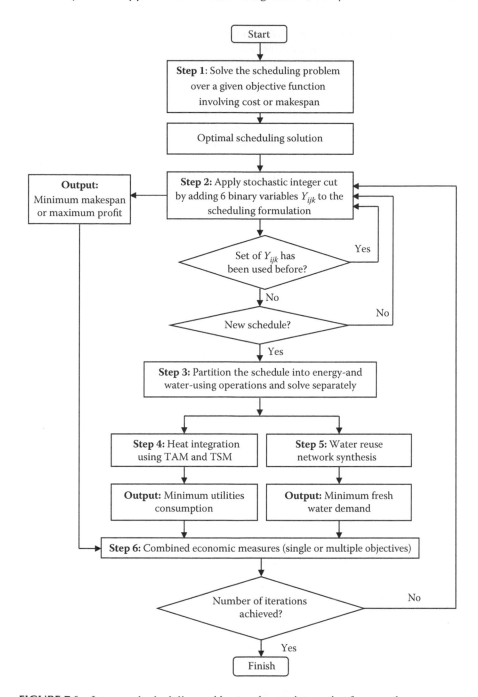

FIGURE 7.6 Integrated scheduling and heat and water integration framework.

(e.g., scheduling makespan). This is done using a stochastic search-based integer cut procedure that adds further constraints to the formulation. Subsequently, heat integration analysis to each of the resulting schedules is performed with the objective of establishing the minimum utility targets. Simultaneously, water reuse optimization is performed to establish the minimum fresh water targets and the water reuse network that meets this target. Compared to simultaneous methodologies published in the literature, the advantages of this sequential framework include its ability to solve multiobjective problems involving scheduling and heat and water reuse optimization with multiple contaminant cases.

To illustrate the sequential framework, consider again the organic salt case study. Let us assume that a modified technology to make salt S has been implemented. Figure 7.7 shows the new recipe that comprises the following steps [49]:

- Heating: Initially, an alkaline solution E is heated from 50°C to 70°C inside unit HR. Steam is used as the heating medium.
- Reaction 1: A mixture of 50% (mass basis) reactant A and 50% reactant F is reacted in two plausible units, RR1 and RR2. The reaction produces an intermediate INT. Throughout the operation, cooling water is supplied to bring down the mixture temperature from 100°C to 70°C. Once the reaction is completed, the unit is washed to prepare it for the next batch.
- Reaction 2: A mixture of 40% hot alkaline E and 60% INT is reacted in two plausible units, RR1 and RR2, to form an intermediate SD (60%) and useful by-product PB (40%). The reaction requires steam for heating up the mixture from 70°C to 100°C. This is then followed by a washing operation.
- Reaction 3: A mixture of 20% F and 80% SD is reacted further in two plausible units, RR1 and RR2. The reaction, which takes place from 100°C to 130°C, produces a solution mixture I. Again, washing of the unit is performed after the reaction is completed.

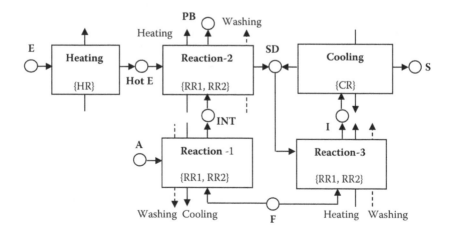

FIGURE 7.7   A new recipe for organic salt S production.

- Cooling: Solution mixture I is cooled in CR to separate the intermediate SD (10%) from the salt solution S (90%). This is followed by recycling the intermediate SD for the Reaction 3 process. Cooling water is used to bring down the temperature from 130°C to 100°C.

Figure 7.8 shows the flow diagram of the process. Here, both steam and cooling water can be used as external utilities for RR1 and RR2. In addition, water that is used for washing RR1 and RR2 would be contaminated with four contaminants (i.e., ar, br, cp, and dw).

7.4.1 BATCH SCHEDULING FORMULATION

Short-term scheduling of multipurpose batch plants is a challenging problem that has received considerable interest during the past decades. It is a decision-making problem that involves the following activities [50]:

- Assignment of process units and resources to tasks.
- Sequencing of tasks inside units.
- Determining the starting and ending times for the execution of each task.

In this framework, the continuous-time formulation of Sundaramoorthy and Karimi [51] is used as the scheduling model.

In the model, the batch horizon H is segmented into K ($k = 1, 2, ..., K$) number of slots which are synchronized on all units ($j = 1, 2, ..., J$). A time point T_k is defined as the time at which slot k ends; in this case, T_0 signifies the beginning of the horizon ($T_0 = 0$). A task starting at T_{k-1} can finish before, at, or after time T_k according to the following relation:

$$T_k = T_{k-1} + SL_k \tag{7.6}$$

where, SL_k is the variable slot length whose sum cannot exceed the time horizon H,

$$\sum_{k=1}^{K} SL_k \leq H \tag{7.7}$$

The objective function can be measured as total profit over either a given scheduling horizon or the required makespan to fulfill the production demand. This is expressed as follows:

$$\text{Min makespan} = \sum_{k=1}^{K} SL_k \quad \text{or} \quad \text{Max profit} = \sum_{m} g_m I_{mK} \tag{7.8}$$

where K is the total number of slots, I_{mK} is the inventory of material m at the final time slot T_K, and g_m is the cost or value of unit mass of material m. This objective

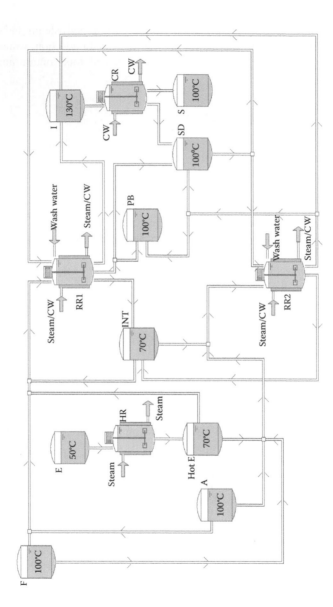

FIGURE 7.8 A new flowsheet of organic salt process.

function is subject to a set of constraints, involving the status of unit, processing time, mass balance, and material inventories. This is explained next.

A balance on the status of a unit j can be written as follows:

$$y_{ijk} = y_{ij(k-1)} + Y_{ij(k-1)} - YE_{ijk} \qquad (7.9)$$

where Y_{ijk}, y_{ijk}, and YE_{ijk} are binary variables described as

$$Y_{ijk} = \begin{cases} 1 & \text{if unit } j \text{ begins task } i \text{ at time } T_k \\ 0 & \text{otherwise} \end{cases} \quad i \in Ij, \ 0 \le k < K$$

$$y_{ijk} = \begin{cases} 1 & \text{if unit } j \text{ is continuing to perform task } i \text{ at time } T_k \\ 0 & \text{otherwise} \end{cases} \quad i \in Ij, \ 0 \le k \le K$$

$$YE_{ijk} = \begin{cases} 1 & \text{if unit } j \text{ ends task } i \text{ and releases its batch at time } T_k \\ 0 & \text{otherwise} \end{cases} \quad i \in Ij, \ 0 \le k \le K$$

Here, I_j represents the set of tasks that unit j can perform. Equation 7.9 states that whenever a task i is not under progress on unit j at time T_k, the value of y_{ijk} is zero. It becomes one only after a task has begun on the unit and zero when it ends at T_k. Based on the binary Y_{ijk}, binary variable Z_{jk} can be introduced as follows:

$$Z_{jk} = \begin{cases} 1 & \text{if unit } j \text{ begins a task (including } i = 0 \text{) at time } T_k \\ 0 & \text{otherwise} \end{cases} \quad 0 \le k \le K$$

In this case, if a task i merely continues on unit j at T_k, then Z_{jk} is 0. Since only one task can start at any time T_k on a unit j, the following expression can be written as

$$Z_{jk} = \sum_{i \in I_j} Y_{ijk}, \quad 0 \le k < K \qquad (7.10)$$

A time balance on task i as it progresses from T_k to T_{k+1} on unit j can be expressed as

$$t_{j(k+1)} \ge t_{jk} + \sum_{i \in I_j} (\alpha_{ij} Y_{ijk} + \beta_{ij} B_{ijk}) - SL_{(k+1)}, \quad k < K \qquad (7.11)$$

where t_{jk} is the time remaining at T_k to complete the task that was in progress during slot k on unit j, B_{ijk} is the batch size of task i that unit j begins at T_k, α_{ij} is the fixed processing time of task i on j, and β_{ij} is the variable processing time of task i on j.

For constant processing time, the parameter α_{ij} can be replaced with τ_{ij}, which is the constant processing time of task i on unit j, as follows:

$$t_{j(k+1)} \geq t_{jk} + \sum_{i \in I_j} \tau_{ij} Y_{ijk} - SL_{(k+1)}, \quad k < K \tag{7.12}$$

A mass balance on unit j can be expressed using the following equation:

$$b_{ijk} = b_{ij(k-1)} + B_{ij(k-1)} - BE_{ijk}, \quad i > 0, \ k > 0 \tag{7.13}$$

where b_{ijk} is the amount of material m that resides in unit j just before T_k and BE_{ijk} is the amount that task i discharges at its completion at T_k on unit j. Equation 7.13 states that whenever a unit j is not performing a task i at T_k then b_{ijk} is set to zero, and vice versa. An inventory balance of material m can be described as follows:

$$I_{mk} = I_{m(k-1)} + \sum_{i \in OI_m, i \neq 0} \sum_{j \in J_i} \sigma_{mi} BE_{ijk} + \sum_{i \in II_m, i \neq 0} \sum_{j \in J_i} \sigma_{mi} B_{ijk} \tag{7.14}$$

where I_{mk} is the inventory of material m at T_k, OI_m is the set of tasks that produce material m, II_m is the set of tasks that consume material m, J_i is a set of units that can perform task i, and σ_{mi} is the stoichiometric yield coefficient of material m in the mass balance of task I (σ_{mi} is set to be negative for the raw materials of task i and positive for its products). The scheduling model above is a mixed integer linear program (MILP) and can be solved for global optima using GAMS software by imposing good upper and lower bounds on the variables.

Let us now consider the organic salt scheduling case study, which has been solved using the formulation. Tables 7.3 through 7.5 show information on the processing times of the tasks and other information pertinent to heating/cooling and washing operations for the organic salt case study. The production demand is set to be 200 kg for both by-product PB and salt solution S. The objective function here is

TABLE 7.3

Task Information and Heating/Cooling Requirement of Organic Salt Production

Task (i)	T_{in} (°C)	T_{out} (°C)	Unit (j)	C_p (kJ/ kg°C)	Max Batch Size (kg)	Fixed Processing Time α_{ij} (h)	Variable Processing Time β_{ij} (h)	Washing Time (h)
Heating (H)	50	70	HR	2.5	100	0.667	0.007	0
Reaction-1 (R1)	100	70	RR1	3.5	50	1.334	0.027	0.25
			RR2	3.5	80	1.334	0.017	0.30
Reaction-2 (R2)	70	100	RR1	3.2	50	1.334	0.027	0.25
			RR2	3.2	80	1.334	0.017	0.30
Reaction-3 (R3)	100	130	RR1	2.6	50	0.667	0.013	0.25
			RR2	2.6	80	0.667	0.008	0.30
Cooling (C)	130	100	CR	2.8	200	1.334	0.007	0

TABLE 7.4

Material Inventory Information

Material State (m)	Initial Inventory (kg)	Max Storage (kg)
E	1000	1000
A	1000	1000
F	1000	1000
Hot E	0	100
SD	0	200
INT	0	150
I	0	200
PB	0	1000
S	0	1000

optimization with respect to makespan—this problem is known to be more complex than the profit maximization problem. To solve the problem, first the batch horizon was divided into $K = 8$ time slots ($k = 1, 2, 3, ..., 8$). Using the formulation, a minimum makespan of 19.96 h was obtained—this was presumed to be a global optima. The optimal schedule (Gantt chart) corresponding to this optimal solution is shown in Figure 7.9. It highlights the start and end times of different tasks in each unit and also the amount being processed.

7.4.2 INTEGER CUT GENERATION PROCEDURE TO GENERATE ALTERNATE SCHEDULES

The outcome of scheduling optimization is an optimal schedule which specifies what tasks to start/end at what times, on which units, and their batch size. This solution is not unique however. In general, multiple (alternate) scheduling solutions can exist when solving the MILP formulation of the scheduling problems. While this can

TABLE 7.5

Washing Information

Task (i)	Unit (j)	Max Inlet Concentration (ppm)				Max Outlet Concentration (ppm)				Contaminants Loading (g Contaminant/ kg Batch)
		ar	br	cp	dw	ar	br	cp	dw	
Reaction-1 (R1)	RR1	300	500	800	400	700	800	1200	900	0.2
	RR2	300	500	800	400	700	800	1200	900	0.2
Reaction-2 (R2)	RR1	700	600	300	400	1200	1000	600	800	0.2
	RR2	700	600	300	400	1200	1000	600	800	0.2
Reaction-3 (R3)	RR1	500	200	400	300	800	500	700	900	0.2
	RR2	500	200	400	300	800	500	700	900	0.2

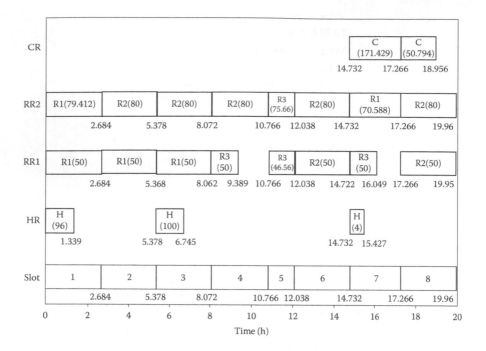

FIGURE 7.9 Optimal schedule for organic salt production process.

be considered a common drawback of the linear programming model, such traits can be exploited to our advantage. Hence, in the next step of the procedure, finding the alternate solutions of the scheduling problem is explored. This is done using an integer cut method that is invoked through the stochastic search method that adds further constraints to the original scheduling formulation. To illustrate the procedure, consider the optimal schedule shown in Figure 7.9. It comprises a binary variable $Y_{ijk} = Y_{H,HR,3} = 1$ (i.e., task H takes place in unit HR at slot (3)) in its solution. Using a stochastic search-based procedure, additional constraints in the form of a set of binary variable Y_{ijk} can be added to the formulation and a new schedule can be obtained from solving this "extended" formulation. Such a schedule is termed alternate optima if it has the same optimal value of the objective function. When this procedure is iterated over N different sets of Y_{ijk} cuts, many feasible alternate schedules can be generated.

Figure 7.10 shows one alternate optima of the scheduling formulation. This schedule has been obtained by adding a set of Y_{ijk} constraints of size $L = 6$ in the formulation. The schedule has tasks in units differing in size and timing compared to the original schedule in Figure 7.9, but the same makespan value of 19.96 h.

7.4.3 PARTITIONING THE SCHEDULE INTO ENERGY AND WATER-USING OPERATIONS

The subsequent step is to partition the schedule into processing (i.e., material transfer, reaction, separation, etc.) and nonprocessing (i.e., washing) parts. The reason for

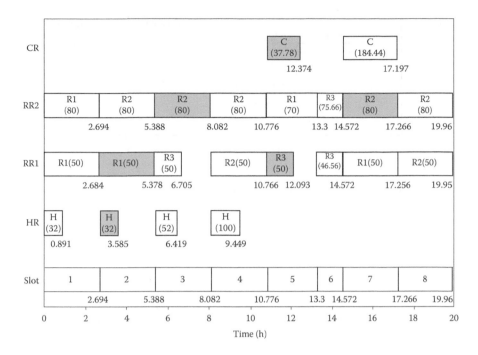

FIGURE 7.10 Alternate schedule of organic salt process. The filled boxes are the specified integer cuts.

this is that in the scheduling formulation, both the processing and washing times have been lumped into the task-processing time. Hence, in order to analyze opportunities for heat integration and water reuse between operations, the partitioning steps are necessary. The following conditions are also presumed:

- The time for "material transfer" in and out of the process unit is negligible compared to the "material processing" time in that unit.
- Heat exchange between two operations can only take place during the "material processing" operation.
- The amount of leftover materials to be washed is proportional to the size of materials that had been processed inside the unit.

Alternate schedules differ in the timing and duration of tasks and hence their capability for temporal clustering, heat recovery, and water reuse. This is illustrated in Figure 7.11 which shows the operations that take place inside two unit operations, A and B. In Figure 7.11a, which is a base case involving two tasks, there are no possibilities for heat integration and water reuse. Figure 7.11b shows an alternate schedule where the start time of the operation in unit B has been moved forward. In this case, heat integration is possible for a duration of 1 h. Altering both the timing for processing and washing tasks is exemplified in Figure 7.11c. Under this scenario, the heat integration is possible for as long as the processing

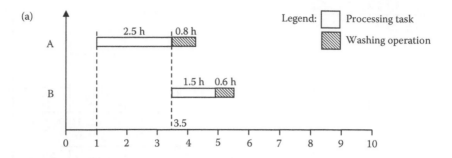

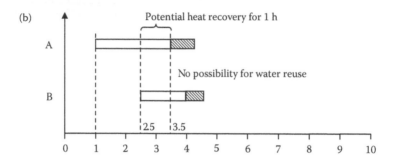

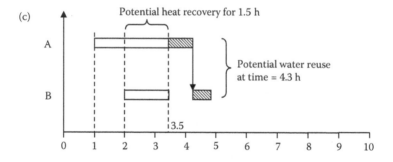

FIGURE 7.11   Rescheduling of tasks for heat integration and water reuse. (a) base case, (b) with heat integration, and (c) with heat integration and water reuse.

task of unit B, which is 1.5 h. Water reuse is also possible since the starting time of the washing operation in unit B has been delayed to match the finish washing time of unit A at 4.3 h.

7.4.4   HEAT INTEGRATION ANALYSIS

In the simplest form, the goal of heat integration is to find matches between hot streams and cold streams so as to minimize the use of external utilities (cooling water and steam). The opportunities for heat integration are thus higher in the optimal (or near optimal) schedules where tasks are temporally clustered together in a more "compact" form. In this framework, the heat integration analysis procedure for continuous processes has been extended to batch processes. This is done by incorporating time average model (TAM) and time slice model (TSM) analysis [52]. The former assumes a pseudo-continuous behavior of the batch process by averaging the total heat load of each stream over its availability period. The latter involves dividing the batch period into different time intervals and allowing heat exchange between the hot and cold streams within each interval in the same way as in a continuous process.

Consider the task scheduling shown in Figure 7.11b. Let us assume the following scenario: Vessel A contains 200 kg material ($C_p = 3$ kJ/kg°C) that needs to be cooled down from 100°C to 40°C while vessel B has 250 kg material ($C_p = 4$ kJ/kg°C) that needs to be heated up from 20°C to 60°C. One possible heat exchange set-up between the two streams is based on the cocurrent configuration shown in Figure 7.12. In this set-up, both the hot and cold streams are drawn from their vessels and then returned to after passing through a heat exchanger. In principle, such a heat exchange mode is analogous to heating/cooling of the vessel content through the vessel jacket. However, unlike in continuous processes, calculating the potential heat recovery under this scenario is very difficult. The reason is that the contents in both vessels would gradually change in temperature.

To allow for heat recovery calculation, TAM and TSM are applied. Using TAM, the average heat load of each unit can be calculated as follows:

Vessel A: total heat flow = batch size $\times C_p \times \Delta T = 200 \times 3 \times (100-40) = 36$ MJ;

average heat flow = 36/(3.5–1.0) = 14.4 MJ/h;

Vessel B: total heat flow = $250 \times 4 \times (60-20) = 40$ MJ;

average heat flow = 40/(4.0–2.5) = 26.7 MJ/h;

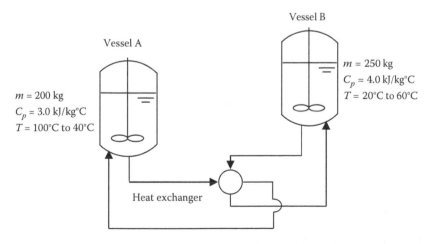

FIGURE 7.12 Cocurrent heat exchange configuration.

To explore the opportunity for heat integration, TSM is applied—this is done by splitting the batch period into different time intervals by associating the boundaries of these intervals with the starting and finishing times of tasks. In this example, the batch period can be split into three intervals: 1–2.5 h, 2.5–3.5 h, and 3.5–4 h and analyzed for the possibility of heat exchange within each interval. Analysis based on this figure reveals that, in total, a supply of $14.4 \times 1.5 = 21.6$ MJ of cold utility should be made available during the first interval. In the same way, the amount of hot utility to be supplied in the third interval is $26.7 \times 0.5 = 13.4$ MJ. In the second interval, however, where the two streams overlap, heat recovery is possible to the extent of 14.4 MJ—this is the "theoretical maximum" amount of heat that can be recovered through integration.

In this framework, a heat transshipment model of a continuous process has been extended to batch processing. It involves creating a set of temperature interval compartments and synthesizing an optimum network of heat transfer from sources (hot streams and utilities) to sinks (cold streams and utilities) through the compartments by taking into account the minimum temperature difference for heat flow (i.e., $\Delta T_{\min}$ approach). This procedure is illustrated in Figure 7.13, which shows five possible heat flows in each compartment [53]. Both the hot streams and hot utilities can transfer heat to the cold streams and cold utilities. The hot utilities can also transfer heat to the cold streams. Furthermore, the residual heat from the hot streams of higher temperature can be cascaded to the cold streams and cold utilities of lower temperature intervals.

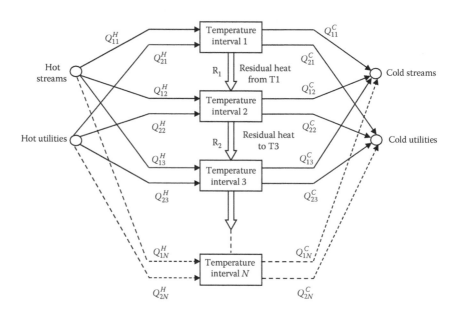

FIGURE 7.13   Heat-transshipment model.

Let Q_{hv}^H be the heat flow of hot stream h entering a temperature interval v. The heat equation describing this stream can be expressed as

$$Q_{hv}^H = F_h \times Cp_{hv} \times \Delta T_v^h \tag{7.15}$$

where F_h is the flowrate of hot stream h, Cp_{hv} is the heat capacity of hot stream h at temperature interval v, and ΔT_v^h is the temperature change of stream h at interval v. In the same way, the heat equation for cold stream c exiting from temperature interval v is defined as

$$Q_{cv}^C = F_c \times Cp_{cv} \times \Delta T_v^c \tag{7.16}$$

The following equations can be similarly written for hot utility p and cold utility n at temperature interval v:

$$Q_{pv}^S = F_p^S \times \Delta H_{pv} \tag{7.17}$$

$$Q_{nv}^W = F_n^W \times \Delta H_{nv} \tag{7.18}$$

where F_p^S, F_n^W, F_n^W, ΔH_{pv}, and ΔH_{nv} are the flowrate and the enthalpy change of the hot utility p and cold utility n, respectively. The energy balance over the interval v is defined as

$$R_v - R_{v-1} - Q_{pv}^S + Q_{nv}^W = Q_{hv}^H - Q_{cv}^C \tag{7.19}$$

where R_{v-1} and R_v are the residual heat load entering and exiting interval v, respectively. Here, the objective function is minimization of the utilities consumption and this can be expressed as

$$\text{Min Utilities} = \sum_{p \in S, v \in V} Q_{pv}^S + \sum_{n \in W, v \in V} Q_{nv}^W, \quad v = 1, 2, \ldots, V \tag{7.20}$$

The resulting mathematical formulation is an MILP problem and can be solved for global optima using GAMS [54].

Heat integration analysis has been performed for the schedule of Figure 7.10 to establish the minimum utility targets. The minimum temperature difference (ΔT_{min}) for heat transfer was specified as 10°C. The output from the heat integration was the amount of utilities posed by this schedule. In this case, the total amount of utilities that need to be supplied are 96.75 MJ (i.e., 61.36 MJ steam and 35.39 MJ cooling water).

7.4.5 WATER REUSE SYNTHESIS

In tandem with heat integration analysis, water reuse synthesis is performed on the schedule to establish the minimum fresh water targets and the water network design that meets this target. However, unlike in continuous processes, water reuse in a batch process is constrained by time as well as concentration. As previously mentioned, retiming of the washing operations may be necessary so that water can be reused optimally between the operations. The following procedure describes the retiming procedure to address the time constraint in batch water reuse:

1. Start by identifying the time interval of interest. This can be referred from the specified slot of the scheduling solution. Specify also the washing operations that take place within the interval.
2. Set the maximum number of reuse (N_R) that can be accommodated by the set of washing operations within that time slot. For instance, we can fix the maximum number of reuse to be three stages (i.e., $N_R = 3$).
3. For each specified number of reuse ($N_R \leq 3$), perform a permutation analysis to the set of washing operations to achieve the reuse target. Retiming of the starting time of the washing operation within the given time slot is needed to achieve the number of reuse targets.
4. If the reuse scheme is feasible, then collect it as a member of a solution set R_S.
5. For each candidate in R_S, perform the water reuse optimization problem for minimum fresh water demand.
6. Steps (2) through (5) are repeated for another time interval of interest of the schedule.

Figure 7.14 depicts the schematic flow of the washing operation in a batch process [45]. A contaminant j with a mass load of $\Delta m_{i,j,tot}$ is to be washed away from operation i. The maximum concentrations of contaminant j at the inlet and outlet water

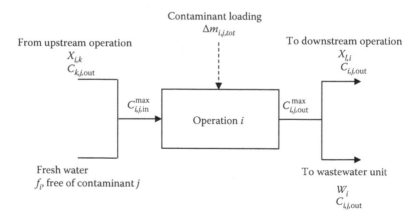

FIGURE 7.14 Schematic of batch washing operation.

stream of this operation are limited to $C_{i,j,\text{in}}^{\max}$ and $C_{i,j,\text{out}}^{\max}$, respectively. The inlet stream to the operation can be sourced from a fresh water stream f_i ($i = 1, 2, 3, ..., n_{\text{operations}}$) with zero contaminants, reuse water stream $X_{i,k}$ from the upstream operation k (provided its contaminant loading $C_{k,j,\text{out}}$ is less than the maximum allowable contaminant concentration) or mixing of the two streams. Here, the term upstream is used to describe any operation k ($k = 1, 2, 3, ..., n_{\text{operations}}$ with $k \neq i$) with the finishing time matching the starting time of the operation i. The outlet stream from operation i consists of flowstream W_i, which can be sent to wastewater treatment and/or flowstream $X_{l,i}$ to downstream operation l ($l = 1, 2, 3, ..., n_{\text{operations}}$ with $l \neq i$). To solve the concentration constraints of batch water reuse, the superstructure mathematical formulation of Mann and Liu [55] of continuous process has been extended to batch operations. This is done by introducing a binary variable $Yr_{i,k}$, which describes water reuse between washing operation i and k. In this case, $Yr_{i,k} = 1$ if the ending time of operation k matches with the starting time of i and $Yr_{i,k} = 0$ otherwise.

The objective function for minimizing the total fresh water flows to all water-using operations i between starting time t_s and finishing time t_f can be expressed as:

$$\text{Min} \sum_{t_s}^{t_f} \sum_i f_i \tag{7.21}$$

An overall water balance around each process unit can be defined as

$$f_i + \sum_{k \neq i} X_{i,k} Yr_{i,k} - W_i - \sum_{k \neq i} X_{k,i} Yr_{k,i} = 0 \tag{7.22}$$

where $Yr_{i,k}$ is a binary variable which describes the connectivity of upstream operation k to downstream i. The contaminant mass balance around each operation can be formulated as

$$\sum_{k \neq i} C_{k,j,\text{out}} X_{i,k} Yr_{i,k} + \Delta m_{i,j,\text{tot}} = \left(W_i + \sum_{l \neq i} X_{l,i} Yr_{l,i} \right) C_{i,j,\text{out}} \tag{7.23}$$

The constraint on the maximum inlet concentration of contaminant j to operation i can be expressed as

$$\sum_{k \neq i} X_{i,k} Yr_{i,k} C_{k,j,\text{out}} \leq \left(f_i + \sum_{k \neq i} X_{i,k} Yr_{i,k} \right) C_{i,j,\text{in}}^{\max} \tag{7.24}$$

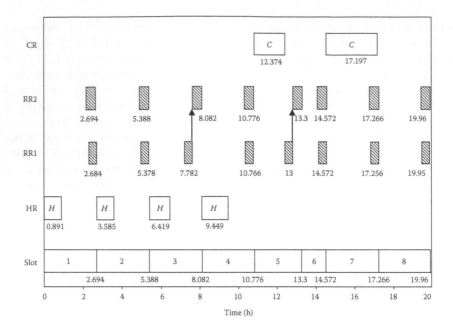

FIGURE 7.15   Optimal direct water reuse scheme for schedule of Figure 7.10.

The resulting formulation is an LP problem for a single contaminant case and can be easily solved for global optima. For a multiple contaminant case, the resulting formulation is an NLP and can be solved for global optima over suitable initial values of variables.

The above formulation has been applied to solve the water reuse problem for each time interval of the schedule in Figure 7.10. The total fresh water demand and thus the wastewater generation for this schedule is 275.09 kg. Figure 7.15 shows the possible water reuse scheme. As shown in the figure, the reuse of water between the washing operation of RR1 and RR2 is possible at time 7.782 h and 13 h.

7.4.6   COMBINED ECONOMIC AND ENVIRONMENTAL ASSESSMENT

The outcomes from the scheduling optimization, heat integration, and water reuse optimization together yield the complete economic and environmental measures of the process that can be represented in several forms depending on the nature of the application. In the context of scheduling for maximum profit, the measure is a total profit. For scheduling problems involving makespan minimization, the measure may include up to three distinct objectives of minimizing makespan and the energy and water demand. Figure 7.16 shows the three-dimensional plot of Pareto optimal results for this case study. This was obtained after 3500 runs. As shown in the figure, about 90% of the optimal results are based on the alternate optima of the scheduling solution. The result also highlights the trade-offs between the three objectives—no solution is found to dominate all the objectives. Among the optimal solutions is an alternate

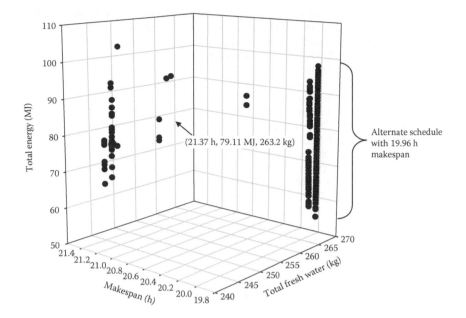

FIGURE 7.16   Three-dimensional plot of multiple objective scheduling solution.

schedule with a makespan of 21.37 h, total energy demand of 79.11 MJ, and fresh water consumption of 263.2 kg. Comparison between this schedule with the optimal schedule in Figure 7.10 (19.96 h, 96.75 MJ, and 275.09 kg) reveals that increasing the makespan by 7% could lead to 18.2% energy reduction and 4.3% water saving.

7.5   CONCLUSIONS

Recent emphasis on sustainability has challenged the batch chemical process industries to improve its process efficiency while at the same time reduce their waste generation. This chapter presented different process synthesis techniques—the knowledge-based method, the process simulation-optimization approach and integrated scheduling, and the heat and water integration method—for sustainable batch process design and operation. The rationale behind this is, since sustainable production is a multifaceted problem, different techniques are needed, each providing a different perspective to solve the synthesis problem. The techniques have been successfully tested using several interconnected case studies with convincing outcomes.

REFERENCES

1. IPCC 2007. Climate change 2007: Synthesis report. http://www.ipcc.ch/pdf/assessment-report/ar4/syr/ar4_syr.pdf (accessed April 23, 2014).
2. 2030 Water Resource Group. 2009. Charting our water future economic framework to inform decision-making. http://www.2030wrg.org/wp-content/uploads/2012/06/Charting_Our_Water_Future_Final.pdf (accessed April 23, 2014).

3. OECD 2011. 40 years of chemical safety at the OECD: Quality and efficiency. http://www.oecd.org/env/ehs/48153344.pdf (accessed April 23, 2014).

4. Uctas, R. 2007. Chemical industry struggles with public image. http://www.icis.com/Articles/2007/09/17/9062166/chemical-industry-struggles-with-public-image.html (accessed April 23, 2014).

5. Priyadarshini, K. and Keerthi Aparajitha, U. 2012. Paclitaxel against cancer: A short review. *Medicinal Chemistry* 2: 139–141.

6. Ritter, S. 2004. Green innovations. *Chemical & Engineering News* 82(28): 25–30.

7. Sharratt, P.N. 1997. Chemicals manufacture by batch processes. In *Handbook of Batch Process Design*, ed. P.N. Sharratt, 1–23. London: Blackie Academic and Professional.

8. Kossik, J. 2002. Think small: Pharmaceutical facility could boost capacity and slash costs by trading in certain batch operations for continuous versions. http://www.pharmamanufacturing.com/articles/2002/6/(accessed April 23, 2014).

9. Sheldon, R.A. 1997. Catalysis: The key to waste minimization. *Journal of Chemical Technology and Biotechnology* 68: 381–388.

10. Smith, R. 1995. *Chemical Process Design*. New York: McGraw-Hill.

11. Jiménez-González, C. and Constable, D.J.C. 2011. *Green Chemistry and Engineering a Practical Design Approach*. Hoboken: John Wiley & Sons.

12. Dickson, M.J. 2011. Plant design. In *Process Understanding for Scale-Up and Manufacture of Active Ingredients*, ed. I. Houson, 283–305. Weinheim: Wiley-VCH.

13. Chung, J.W. and Meltzer, D.O. 2009. Estimate of the carbon footprint of the US health care sector. *The Journal of the American Medical Association* 302: 1970–1972.

14. Du Pont de Nemours, E.I. 1993. *DuPont Chambers Works Waste Minimization Project*. New York: U.S. Environmental Protection Agency.

15. Houghton, C., Sowerby, B., and Crittenden, B. 1996. Clean design of batch processes. In *Case Studies in Environmental Technology*, ed. P. Sharratt, and M. Sparshott, 59–71. Rugby: Institution of Chemical Engineers.

16. Douglas, J.M. 1992. Process synthesis for waste minimization. *Industrial Engineering and Chemistry Research* 31: 238–243.

17. Mulholland, K.L. and Dyer, J.A. 1999. *Pollution Prevention: Methodology, Technologies and Practices*. New York: American Institute of Chemical Engineers.

18. Isalski, W.H. 1995. ENVOP for waste minimization. *Environmental Protection Bulletin* 34: 16–21.

19. Halim, I. and Srinivasan, R. 2006. Systematic waste minimization in chemical processes: Part III. Batch operations. *Industrial and Engineering Chemistry Research* 45: 4693–4705.

20. Halim, I. and Srinivasan, R. 2008. Designing sustainable alternatives for batch operations using an intelligent simulation-optimization framework. *Chemical Engineering Research and Design* 86: 809–822.

21. Friedler, F., Varga, J.B., and Fan, L.T. 1994. Algorithmic approach to the integration of total flowsheet synthesis and waste minimization. In *Pollution Prevention via Process and Product Modifications*, ed. M.M. El-Halwagi, and D.P. Petrides, 86–97. New York: American Institute of Chemical Engineers.

22. Cabezas, H., Bare, J.C., and Mallick, S.K. 1999. Pollution prevention with chemical process simulators: The generalized waste reduction (WAR) algorithm—Full version. *Computers and Chemical Engineering* 23: 623–634.

23. Fu, Y., Diwekar, U. M., Young, D., and Cabezas, H. 2000. Process design for the environment: A multi-objective framework under uncertainty. *Clean Products and Processes* 2: 92–107.

24. Chen, H. and Shonnard, D. 2004. Systematic framework for environmentally conscious chemical process design: Early and detailed design stages. *Industrial Engineering Chemistry and Research* 43: 535–552.

25. Othman, M.R., Repke, J.U., Wozny, G., and Huang, Y. 2010. A modular approach to sustainability assessment and decision support in chemical process design. *Industrial and Engineering Chemistry Research* 49: 7870–7881.
26. von Watzdorf, R., Naef, U.G., Barton, P.I., and Pantelides, C.C. 1994. Deterministic and stochastic simulation of batch/semicontinuous processes. *Computers and Chemical Engineering* 18(supplement): 343–347.
27. Kirkpatrick, S., Gelatt, C.D., and Vecchi, M.P. 1983. Optimization by simulated annealing. *Science* 220: 671–680.
28. Suppapitnarm, A., Seffen, K.A., Parks, G.T., and Clarkson, P.J. 2000. A simulated annealing algorithm for multiobjective optimization. *Engineering Optimization* 33: 59–85.
29. Kemp, I.C. 2007. *Pinch Analysis and Process Integration a User Guide on Process Integration for the Efficient Use of Energy*. Oxford: Elsevier.
30. Kemp, I.C. and MacDonald, E.K. 1988. Energy and process integration in continuous and batch processes. *IChemE Symposium Series* 105: 185–200.
31. Kemp, I.C. and MacDonald, E.K. 1988. Application of pinch technology to separation, reaction and batch processes. *IChemE Symposium Series* 109: 239–257.
32. Wang, Y.P. and Smith, R. 1995. Time pinch analysis. *Chemical Engineering Research and Design* 73: 905–914.
33. Uhlenbruck, S., Vogel, R., and Lucas, K. 2000. Heat integration of batch processes. *Chemical Engineering and Technology* 23: 226–229.
34. Vaselenak, J.A., Grossmann, I.E., and Westerberg, A.W. 1986. Heat integration in batch processing. *Industrial and Engineering Chemistry Research* 25: 357–366.
35. Corominas, J., Espuña, A., and Puigjaner, L. 1994. Method to incorporate energy integration considerations in multiproduct batch processes. *Computers and Chemical Engineering* 18: 1043–1055.
36. Vaklieva-Bancheva, N., Ivanov, B.B., Shah, N., and Pantelides, C.C. 1996. Heat exchanger network design for multipurpose batch plants. *Computers and Chemical Engineering* 20: 989–1001.
37. Bozan, M., Borak, F., and Or, I. 2001. A computerized approach for heat exchanger network design in multipurpose batch plants. *Chemical Engineering and Processing* 40: 511–524.
38. Papageorgiou, L.G., Shah, N., and Pantelides, C.C. 1994. Optimal scheduling of heat-integrated multipurpose plants. *Industrial and Engineering Chemistry Research* 33: 3168–3186.
39. Lee, B. and Reklaitis, G.V. 1995. Optimal scheduling of a cyclic batch processes for heat integration—I. Basic formulation. *Computers and Chemical Engineering* 19: 883–905.
40. Zhao, X.G., O'Neill, B.K., Roach, J.R., and Wood, R.M. 1998. Heat integration for batch processes Part 1: Process scheduling based on cascade analysis. *Chemical Engineering Research and Design* 76: 685–699.
41. Pinto, T., Novais, A.Q., and Barbosa-Póvoa, A.P.F.D. 2003. Optimal design of heat-integrated multipurpose batch facilities with economic savings in utilities: A mixed integer mathematical formulation. *Annals of Operations Research* 120: 201–230.
42. Majozi, T. 2006. Heat integration of multipurpose batch plants using a continuous-time framework. *Applied Thermal Engineering* 26: 1369–1377.
43. Foo, D.C.Y., Manan, Z.A., and Tan, Y.L. 2005. Synthesis of maximum water recovery network for batch process systems. *Journal of Cleaner Production* 13: 1381–1394.
44. Majozi, T., Brouckaert, C.J., and Buckley, C.A. 2006. A graphical technique for wastewater minimisation in batch processes. *Journal of Environmental Management* 78: 317–329.
45. Majozi, T. 2005. An effective technique for wastewater minimisation in batch processes. *Journal of Cleaner Production* 13: 1374–1380.

46. Gouws, J.F., Majozi, T., and Gadalla, M. 2008. Flexible mass transfer model for water minimization in batch plants. *Chemical Engineering and Processing* 47: 2323–2335.

47. Majozi, T. and Gouws, J.F. 2009. A mathematical optimisation approach for wastewater minimisation in multipurpose batch plants: Multiple contaminants. *Computers and Chemical Engineering* 33: 1826–1840.

48. Halim, I. and Srinivasan, R. 2011. Sequential methodology for integrated optimization of energy and water use during batch process scheduling. *Computers and Chemical Engineering* 35: 1575–1597.

49. Kondili, E., Pantelides, C.C., and Sargent, R.W.H. 1993. A general algorithm for short-term scheduling of batch operations—I. MILP formulation. *Computers and Chemical Engineering* 17: 211–227.

50. Reklaitis, G.V., Rekny, J., and Joglekar, G.S. 1997. Scheduling and simulation of batch processes. In *Handbook of Batch Process Design*, ed. P.N. Sharratt, 24–60. London: Blackie Academic and Professional.

51. Sundaramoorthy, A. and Karimi, I.A. 2005. A simpler better slot-based continuous-time formulation for short-term scheduling in multipurpose batch plants. *Chemical Engineering Science* 60: 2679–2702.

52. Linnhoff, B., Ashton, G.J., and Obeng, E.D.A. 1988. Process integration of batch processes. *IChemE Symposium Series* 109: 221–237.

53. Papoulias, S.A. and Grossmann, I.E. 1983. A structural optimization approach in process synthesis– II. Heat recovery networks. *Computers and Chemical Engineering* 7: 707–721.

54. Shenoy, U.V. 1995. *Heat Exchanger Network Synthesis: Process Optimization by Energy and Resource Analysis*. Houston: Gulf Publishing Company.

55. Mann, J.G. and Liu, Y.A. 1999. *Industrial Water Reuse and Wastewater Minimization*. New York: McGraw-Hill.

8 A Mixed-Integer Linear Programming Model for Optimal Synthesis of Polygeneration Systems with Material and Energy Storage for Cyclic Loads

Kathleen B. Aviso, Aristotle T. Ubando,
Alvin B. Culaba, Joel L. Cuello,
Mahmoud M. El-Halwagi, and Raymond R. Tan

CONTENTS

8.1 INTRODUCTION

Polygeneration plants offer the prospect of efficient utilization of fuel inputs, as well as minimal generation of carbon emissions, relative to separate generation of individual streams of electricity, heat, cooling, and other products (Chicco and Mancarella, 2009; Serra et al., 2009). They are also particularly well-suited to applications requiring stand-alone facilities to provide energy self-sufficiency, for instance, in the case of residential buildings, hospitals and hotels, and so on (Lozano et al., 2009b, 2011). To some extent, the topology of polygeneration plants can be designed to allow some degree of operational flexibility; however, it has also been noted that the interdependence among process units may also lead to vulnerability to failure, as process unit inoperability may cascade through a process network via stream linkages (Tan et al., 2012; Kasivisvanathan et al., 2013).

Various mathematical programming methods have been proposed for the synthesis and design of polygeneration plants. Hemmes et al. (2007) proposed a generic optimization approach for multiinput, multioutput energy hubs. The framework allows for flexible coproduction of multiple streams. Process units within this framework were assumed to be "black boxes" with fixed efficiencies or coefficients of performance. Alternatively, "gray box" representations based on empirical equations for the material and energy balances of the process units were also considered based on empirical efficiency curves. Lozano et al. (2009a) proposed a linear programming (LP) model to optimize operations of a trigeneration plant producing electricity, heat, and cooling. Liu et al. (2007) proposed a multiperiod mixed-integer nonlinear programming (MINLP) model for the design of polygeneration plants with projected changes in product mix over a long-term planning horizon. This model was then linearized to yield a mixed-integer linear program (MILP) for which global optimum can be readily determined. An MINLP model specifically for the synthesis of coal-based polygeneration plants producing electricity and methanol was then proposed by Liu et al. (2009). Since the optimization of polygeneration plants needs to account for both economic and environmental performance, a multiobjective modeling approach has also been proposed (Kavvadias and Maroulis, 2010; Carvalho et al., 2012a). Varbanov and Friedler (2008) also proposed an alternative synthesis approach based on graph theoretic methodology, which nevertheless applies the same linearity assumptions for component process units.

Due to the joint production of multiple streams, analysis of polygeneration plants is inherently complicated from an accounting standpoint, especially if the streams are intended to be purchased by different customers. Carvalho et al. (2012b) developed a model based on life-cycle assessment (LCA) and thermoeconomic principles to account for system emissions. A cost allocation model for systems with a variable load was proposed by Lozano et al. (2011) based on the principle of avoided expenses. A fuzzy fractional programming model was later proposed by Ubando et al. (2013) to design polygeneration plants, while simultaneously allocating the system carbon footprint to the different product streams. Variable loads have been considered in different polygeneration design models at different time scales (Liu et al., 2007; Lozano et al., 2009b, 2011). In most cases, flexibility is introduced via the option to import or export power as demands shift (Lozano et al., 2009a,b, 2011). Tan et al. (2012) proposed an algebraic approach for identifying bottlenecks in polygeneration systems as a result of demand changes. A subsequent paper proposed an MILP model for the optimal allocation of streams under abnormal operating conditions resulting from partial or complete loss of capacity in some process units (Kasivisvanathan et al., 2013). This framework was then extended to the design of robust energy systems given multiple anticipated scenarios (Kasivisvanathan et al., 2014). In this series of works, all process units were assumed to have a known feasible part-load operating range. In addition, all methodologies were also shown to be applicable to biorefineries under similar assumptions. Then, Chen et al. (2014) proposed a transshipment MILP model for grid-connected hybrid renewable energy systems. Lee et al. (2014)

then developed a model with energy storage options for periodic loads; their formulation also considered energy losses incurred during storage.

In this chapter, we propose an MILP model for the optimal design of polygeneration systems with storage for cyclic loads. The overall approach is based on the early MILP work of Grossmann and Santibanez (1980). The succeeding sections are organized as follows. First, a formal problem statement is given. This is followed by a description of the model formulation. A case study is then presented to illustrate the methodology. Finally, conclusions and prospects for further research are given.

8.2   PROBLEM STATEMENT

The formal problem statement is as follows:

- We consider a polygeneration system to be comprised of n candidate process units and with a total of m streams.
- Each process unit is modeled as a black box characterized by a fixed set of ratios of direct inputs and output streams. The signs of the coefficients follow the input–output convention used in previously published work (e.g., Carvalho et al., 2012a; Ubando et al., 2013) while the ratios represent process unit efficiency or yield levels that are assumed to be scale-invariant. Furthermore, each unit is assumed to have a linear capital cost function (i.e., investment cost for a process unit is proportional to its size or capacity). Finally, each process unit has a predefined minimum fractional part-load operating limit.
- The streams within the plant are grouped into m_1 streams whose flow-rates are exogenously specified (e.g., as product demand) and m_2 streams whose values are endogenously determined within the model (e.g., fuel or feedstock inputs), such that $m = m_1 + m_2$. The unit price of each stream is assumed to be fixed.
- For each of the m_1 streams with specified net flowrates, there is an option to provide a storage or buffer unit, for which there is a linear capital cost function.
- The plant operates cyclically over p periods, each with a unique duration and demand for product streams.
- The problem is to determine the polygeneration plant that meets the desired product demand while giving the maximum annual profit.

8.3   MODEL NOMENCLATURE

SETS

M	Set of product streams with m elements
M_1	Product streams with exogenously defined flowrates
M_2	Product streams with endogenously determined flowrates
N	Set of processes with n elements
P	Set of time periods in a day with p elements

PARAMETERS

δ_k	Duration of period k in h/day
a_{ij}	Amount of stream i produced/required by process unit j
AF	Annualizing factor
β_{jk}	Fractional part-load operating limit
$CAPEX_j$	Variable capital cost for process unit j
$COST_i$	Unit cost for product stream i
D	Any big number
$SCOST_i$	Variable capital cost for storage unit i
T	Annual operating hours of the system
y_{ik}	Exogenously defined demand for product stream i in period k (or endogenously determined requirements for stream i in period k)

VARIABLES

B_{jk}	Binary variable which indicates operation of process j in time period k
B'_{jk}	Linearization parameter for the product of B_{jk} and x_{jk}
ACC	Annualized capital cost for process units and storage unit facilities
NSV	Net stream value or economic potential
OC	Annual operating cost
Profit	Total profit achieved by the system
s_i^0	Amount of product stream i stored in storage facility during the start of the operating cycle
s_{ik}	Amount of product stream i stored in storage facility during period k
x_j	Capacity of process j
x_{jk}	Required capacity of process j in time period k
Z_i	Capacity of storage facility i
Z_{ik}	Cumulative amount of product stream i stored in the storage facility in period k

8.4   MODEL FORMULATION

The overall objective is to maximize the annual profit of the system as shown in Equation 8.1. The profit is the difference between the net stream value (NSV) less the annualized capital cost (ACC) and the annual operating cost (OC) as shown in Equation 8.2. The NSV is obtained from the annual sale of the products and by-products less the cost of raw materials (Douglas, 1985).

$$\text{Maximize Profit} \tag{8.1}$$

$$\text{Profit} = \text{NSV} - \text{ACC} - \text{OC} \tag{8.2}$$

The system operates a total of T hours in one year such that each day is divided into P periods, where each period k has duration of δ_k (h/day). Equation 8.3 is used to calculate the NSV on an annual basis. Stream i has a net flowrate of y_{ik} (product units) in period k with a unit cost of $COST_i$. Note that a positive value for y_{ik} indicates a net output of stream i, meaning that it can be sold as a product, while a negative value indicates a net input, which is then a raw material or fuel cost to the system; thus, the summation corresponds to the NSV as defined by Douglas (1985). This term also includes most of the major operating costs in the form of purchased inputs such as fuel and feedstocks. ACC, on the other hand, is calculated using Equation 8.4, where AF is the annualizing factor (calculated based on an assumed plant life and discount rate), $CAPEX_j$ is the variable capital cost of process j, which is proportional to its capacity x_j and $SCOST_i$ is the variable capital cost for the storage unit for stream I, which is proportional to its capacity Z_i. The annual operating cost (OC), on the other hand, will include annual recurring expenses (such as labor and maintenance) other than the raw material costs that have already been considered in the NSV. The equation for OC cannot be generalized because it varies with each case study. A more detailed discussion on cost estimation in chemical process plants may be found in El-Halwagi (2012).

$$NSV = T \sum_{i=1}^{m} COST_i \sum_{k=1}^{p} \frac{y_{ik}\delta_k}{24} \qquad (8.3)$$

$$ACC = AF\left(\sum_{j=1}^{n} CAPEX_j x_j + \sum_{i=1}^{m_1} SCOST_i Z_i \right) \qquad (8.4)$$

Each process unit (j) can be linearly scaled such that the ratios of the input and output streams (i) are fixed. The parameter a_{ij} represents the amount of stream i needed or produced by process unit j while the scaling factor x_j indicates the capacity of each process. The net total flowrate of stream i in period k is given by Equation 8.5, where s_{ik} corresponds to the flowrate of stream I, which is stored in period k. Note that if s_{ik} is negative this suggests that the stream is released from the storage device or buffer unit. In the case of streams for which no storage options are considered, the storage flowrate is equal to zero in all periods, as given by Equation 8.6.

$$y_{ik} = \sum_{j=1}^{n} a_{ij}x_{jk} - s_{ik} \quad \forall i \in M, \ k \in P \qquad (8.5)$$

$$s_{ik} = 0 \quad \forall i \in M_2, \ k \in P \qquad (8.6)$$

The process units function only within exogenously defined operating capacities, where the upper limit is set at x_j and the lower limit is a fraction β_{jk} (Equation 8.8) of the upper limit as shown in Equation 8.7. B_{jk} is a binary variable (Equation 8.9), which indicates whether process j is operational in time period k.

$$\beta_{jk}B_{jk}x_j \leq x_{jk} \leq B_{jk}x_j \quad \forall j \in N, \ k \in P \qquad (8.7)$$

$$0 \leq \beta_{jk} \leq 1 \quad \forall j \in N,\ k \in P \tag{8.8}$$

$$B_{jk} \in \{0,1\} \quad \forall j \in N,\ k \in P \tag{8.9}$$

Note that Equation 8.7 is a nonlinear constraint. However, it can be linearized by transforming Equation 8.7 into Equations 8.10a through 8.10e using a standard linearization procedure (Chen et al., 2011), where B'_{jk} is introduced to replace the nonlinear product of B_{jk} and x_j and D is an arbitrary large number.

$$B'_{jk} \leq x_j \quad \forall j \in N,\ k \in P \tag{8.10a}$$

$$B'_{jk} \geq x_j - D(1 - B_{jk}) \quad \forall j \in N,\ k \in P \tag{8.10b}$$

$$B'_{jk} \leq DB_{jk} \quad \forall j \in N,\ k \in P \tag{8.10c}$$

$$x_{jk} \leq B'_{jk} \quad \forall j \in N,\ k \in P \tag{8.10d}$$

$$x_{jk} \geq \beta_{jk} B'_{jk} \quad \forall j \in N,\ k \in P \tag{8.10e}$$

The storage or buffer unit has the capability of storing excess amounts of the product streams which can be used up at a later time period in order to meet the demand requirement y_{ik}. The total amount stored in the storage or buffer unit in any time period k is Z_{ik}, which is equal to the cumulative amount stored or withdrawn into or from the unit since the start of the time cycle until current period of interest, k (Equation 8.11) where s_i^0 indicates the amount in the storage unit at the start of the time cycle. If the storage facility is empty, $Z_{ik} = 0$ and thus Z_{ik} can never be negative (Equation 8.12). For streams that cannot be stored, the initial amount in the storage unit at the beginning of the time cycle will always be zero (Equation 8.13). The size of the storage facility should be bigger than the total amount stored at any given time period (Equation 8.14) and since the operation is cyclical, the amount that remains inside the storage unit at the end of the time cycle should be equal to the amount at the start of the next time cycle (Equation 8.15).

$$Z_{ik} = \sum_{v=1}^{k} s_{iv} \delta_v + s_i^0 \quad \forall i \in M,\ \forall k,\ v \in P \tag{8.11}$$

$$Z_{ik} \geq 0 \quad \forall i \in M,\ k \in P \tag{8.12}$$

$$s_i^0 = 0 \quad \forall i \in M_2 \tag{8.13}$$

$$Z_i \geq Z_{ik} \quad \forall i \in M, \ k \in P \qquad (8.14)$$

$$Z_{ik} = s_i^0 \quad \forall i \in M, \ k = p \qquad (8.15)$$

The resulting MILP model may be readily solved without significant computational difficulties. In this work, the model is implemented using the commercial optimization software LINGO 13.0 (Lindo Systems, 2010). The LINGO code for the linearized model is shown in the chapter appendix; the next section shows the application of the model to a realistic case study.

8.5 CASE STUDY

The formulated model is demonstrated using a polygeneration case study with two periods ($p = 2$). The first period lasts 16 h, while the second has duration of 8 h. The schematic diagram of the polygeneration plant is shown in Figure 8.1. It consists of four main process units, which are the boiler, combined heat and power (CHP) unit, electric chiller, and reverse osmosis (RO) module. Moreover, a water tank is added into the system to enable the storage of purified water. Each process in the polygeneration plant is treated as a black box where the material and energy follow scale-invariant ratios (based, for instance, on unit efficiencies or coefficients of performance). Each process unit is shown separately in Figure 8.2, along with the respective input and output streams. Treating each process together with its material and energy flows as a column vector, the process matrix is then determined as

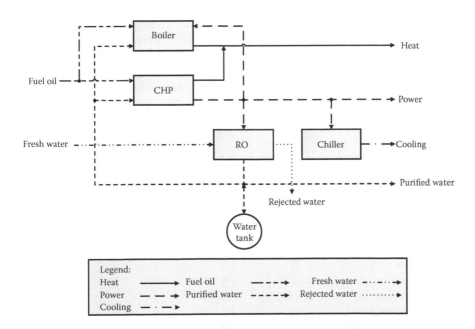

FIGURE 8.1 Polygeneration plant process flow diagram.

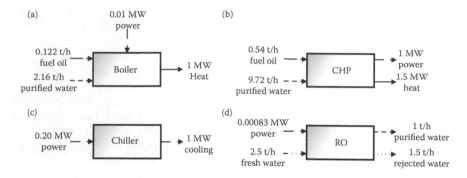

FIGURE 8.2 Polygeneration plant process units: (a) boiler, (b) combined heat and power module, (c) electric chiller, and (d) reverse osmosis unit.

shown in Table 8.1. All negative values in the process matrix are material or energy inputs into a process column vector, while all positive values are material or energy outputs. Each element in the technology matrix, a_{ij}, corresponds to the entry in the ith row (ith product stream) and in the jth column (jth process unit). For example, the chiller unit, illustrated in Figure 8.2c (process unit 3), operates using a vapor compression cycle. The black-box representation shows an input power (product stream 2) of 0.20 MW and a cooling output (product stream 3) of 1 MW. Under the column vector of the chiller unit in Table 8.1, the required power is expressed as a negative value ($a_{23} = -0.20$ MW) while the cooling product is expressed as a positive value ($a_{33} = 1$ MW). Note that the ratio of the absolute values of the two flowrates is 5, which is the coefficient of performance (COP) of the chiller. This COP is assumed to be fixed even if the chiller unit is scaled up or down.

The minimum partial operational load at any time period (β_{jk}), for the process units is 40% for the boiler, 50% for the CHP module, 60% for the chiller, and 0% for the RO module. The unit prices of the product streams are shown in Table 8.2 and are based on figures used in Carvalho et al. (2012a). The variable capital costs of the process and storage units are shown in Tables 8.3 and 8.4, and are based on literature values (Seider et al., 2009; Carvalho et al., 2012a). An annualizing factor of 0.06 y^{-1}

TABLE 8.1
Process Matrix of the Polygeneration System

Process Matrix	Boiler	CHP	Chiller	RO
Heat (MW)	1	1.5	0	0
Power (MW)	−0.01	1	−0.20	−0.00083
Cooling (MW)	0	0	1	0
Purified water (t/h)	−2.16	−9.72	0	1
Fuel oil (t/h)	−0.122	−0.54	0	0
Fresh water (t/h)	0	0	0	−2.5
Rejected water (t/h)	0	0	0	1.5

TABLE 8.2
Unit Prices of Product Streams

Product Stream	Unit Price	Unit
Heat	50	US$/MWh
Power	120	US$/MWh
Cooling	60	US$/MWh
Purified water	25	US$/t
Fuel oil	900	US$/t
Fresh water	1	US$/t
Rejected water	0	US$/t

TABLE 8.3
Variable Capital Cost Coefficients of Process Units

Process Unit	Cost Coefficient	Units
Boiler	1896	Thousand US$/MW
CHP	350	Thousand US$/MW
Chiller	536	Thousand US$/MW
Reverse osmosis	13.33	Thousand US$/(t/h)

TABLE 8.4
Variable Capital Cost Coefficient for Storage Unit

Storage Units	Cost Coefficient	Units
Water tank	1	Thousand US$/t

is used to convert the capital costs for the process equipment and storage units to an annual amount. For this case study, annual operating costs associated with labor and utilities are assumed to be negligible. It is also assumed that the polygeneration plant operates 8000 h/y. The demands for the product streams for each period are shown in Table 8.5.

TABLE 8.5
Product Demands per Period

Product Stream	Period 1	Period 2
Heat (MW)	20	24
Power (MW)	10	8
Cooling (MW)	4	6
Purified water (t/h)	900	430

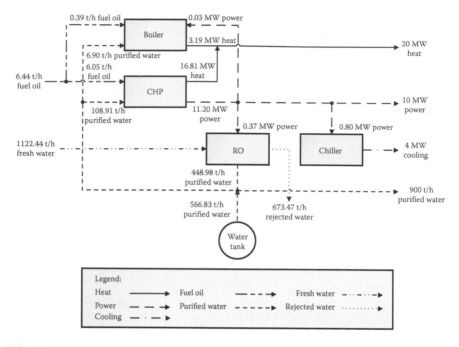

FIGURE 8.3   Resulting optimal polygeneration configuration for period 1.

Solving the model results in a maximum profit of US$100,866,000/y. The resulting annual NSV (i.e., sale of products minus purchase of inputs) is US$ 104,094,000/y, while the ACC amounts to US$3,228,000/y. The polygeneration plant operates in Period 1 as shown in Figure 8.3. The cycle begins with the storage tank containing 9069 t of purified water (accumulated from the previous cycle). During this 16-h period, the water tank dispenses an additional 566.83 t/h of this previously stored purified water to meet the high demand during this period. The polygeneration plant thus supplies the required demand of 20 MW of heat, 10 MW of power, 4 MW of cooling, and 900 t/h of purified water, while consuming 6.44 t/h of fuel oil and 1122.44 t/h of fresh water. The configuration for Period 2 is shown in Figure 8.4. The water tank stores 1133.66 t/h of purified water for 8 h, which results in an accumulation of 9069 t for the next cycle. The polygeneration plant produces the required demand of 24 MW of heat, 8 MW of power, 6 MW of cooling effect, and 430 t/h of purified water while consuming 6.74 t/h of fuel oil and 4211.73 t/h of fresh water for this period. The water profile in the storage tank is shown in Figure 8.5.

In summary, the developed model has been demonstrated using a polygeneration plant that operates in two periods during one operating cycle. A storage unit (water tank) is integrated into the design to store purified water and to accommodate the changes in the product demand during the operating cycle. It should be noted that this case study has no feasible solution if storage options are excluded.

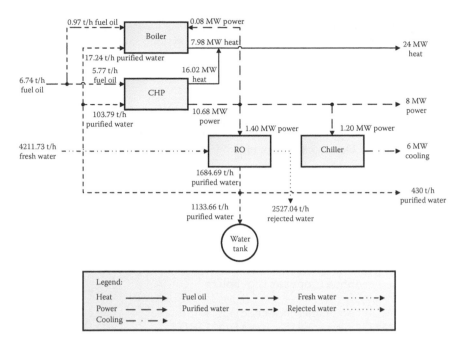

FIGURE 8.4   Resulting optimal polygeneration configuration for period 2.

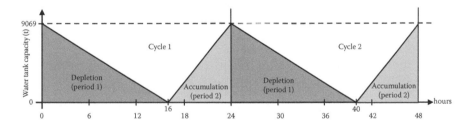

FIGURE 8.5   Water tank capacity profile for two 24-h cycles.

8.6   CONCLUSION

An optimization model for the design of polygeneration systems with cyclic loads has been developed. The model considers material or energy storage options for various product streams, and is formulated as an MILP, which ensures that the global optimum can be readily determined. A polygeneration case study has been solved to illustrate the model. Results show that the optimal operating condition of the polygeneration system differs between the two periods. Using a storage unit, which serves as a buffer, allows load variations throughout the process cycle to be balanced. In some cases, no solution may exist without the inclusion of storage units; in other cases, storage units may improve the overall profitability of the design. Future work on polygeneration will focus on the inclusion of cyclic loading with fuzzy loads or demands, and other energy storage devices such as batteries, ice storage systems, and so on.

ACKNOWLEDGMENTS

The PhD work of the second author was supported in part by The Fulbright Foundation, the Philippine Commission on Higher Education, and the De La Salle University Faculty Development Program.

APPENDIX—LINGO CODE

```
!     I        index for product streams
      J        index for process units
      K        index for time period
      N        time period counter
      T        annual operating hours

      A        technical coefficients
      BETA     fractional operating limit
      CAPEX    variable capital cost of process units
      COST     unit cost of product streams
      DELTA    duration of time period
      Y        product output demand
      SCOST    Variable capital cost of storage unit
      Z        size of storage unit

      ACCUM    amount of accumulated product in storage unit
      AINIT    initial load in storage unit
      B        binary variable
      S        flowrate of stored/released product
      XF       scaling factor for process unit;

SETS:
PROCESS: X, BETA, CAPEX;
PRODUCT: AINIT, COST;
TIME: DELTA;
STORAGE(PRODUCT): Z, SCOST;
NONSTORAGE(PRODUCT);
AMATRIX (PRODUCT, PROCESS): A;
XMATRIX (PROCESS, TIME):XF, B, BX;
SMATRIX (PRODUCT, TIME): Y, S, AX, ACCUM;

ENDSETS

DATA:
PROCESS = BOILER      CHP     CHILLER         RO;
PRODUCT = STEAM ELECTRICITY REFRIGERATION PUREWATER OIL FWATER
REJWATER;
```

```
STORAGE = PUREWATER;
NONSTORAGE = STEAM ELECTRICITY REFRIGERATION OIL FWATER REJWATER;

A =     1        1.5      0       0              !Steam (MW);
        -0.01    1        -0.2    -0.00083       !Power (MW);
        0        0        1       0              !Cooling (MW);
        -2.16    -9.72    0       1              !Purified H2O (t/h);
        -0.122   -0.54    0       0              !Fuel Oil (t/h);
        0        0        0       -2.5           !Fresh H2O (t/h);
        0        0        0       1.5;           !Reject H2O (t/h);
BETA =      0.4  0.5      0.6     0;  !Minimum fractional partial
load operation for process units;
T = 8000;
TIME =      P1       P2;
DELTA =     16       8;                   !Duration of time intervals (h);
Y =     20,      24,      ! Steam (MW);
        10,      8,       ! Power (MW);
        4,       6,       ! Cooling (MW);
        900,     430,     ! Purified Water (t/h);
        ,        ,
        ,        ,
        ,        ;
COST =      50   120      60      25      900     1     0;     !Unit cost
of streams in $ per MWh or per t;
CAPEX =     1896 350      536     13.33 ;                      !Variable
capital cost of process units in Thousand $/t;
SCOST =                   1    ;        !Variable capital cost of
storage tank in Thousand $/t;
ENDDATA

MAX = PROFIT;              !In Thousand $ per year;
@FREE(PROFIT);

PROFIT = @SUM(PRODUCT(I): COST(I)*@SUM(TIME(K): DELTA(K)*
Y(I,K)/(24))*T)/1000
- AF * (@SUM(PROCESS(J): CAPEX(J)*X(J)) + @SUM(STORAGE(I):
SCOST(I)*Z(I)));

AF = 0.06;
@FOR(STORAGE(I): AINIT(I) = ACCUM(I,2));
@FOR(NONSTORAGE(I): AINIT(I) = 0);
@FOR(AMATRIX(I,J):@FREE(A(I,J)));
@FOR(SMATRIX(I,K): Y (I,K) = (@SUM(PROCESS(J): A(I,J)*XF(J,K))
- S(I,K));
@FOR(SMATRIX(I,K): AX(I,K) = (@SUM(PROCESS(J): A(I,J)*XF(J,K)));
@FOR(SMATRIX(I,K):@FREE(AX(I,K)));
!- - - - - - - - - - - - - - - - - - - - - - - - - - - - - - - -
- - - - - - - - - - - -;
!LS***;
D = 1000000;
! Use linear constraints to force BX(J,K) = B(J,K)*X(J);
!BX(J,K) corresponds to B'(J,K);
```

```
@FOR (XMATRIX (J,K);
    BX(J,K) <= X(J);
    BX(J,K) >= X(J) - D*(1 - B(J,K));
    BX(J,K) <= D*B(J,K);
    );
!LS*** @FOR(XMATRIX(J,K): XF(J,K) <= B(J,K)*X(J));
    @FOR(XMATRIX(J,K): XF(J,K) <= BX(J,K));
!LS***@FOR(XMATRIX(J,K): XF(J,K) >= BETA(J) * X(J)*B(J,K));
    @FOR(XMATRIX(J,K): XF(J,K) >= BETA(J)*BX(J,K));

@FOR(XMATRIX(J,K): X(J) >= XF(J,K));
@FOR(XMATRIX(J,K): @BIN(B(J,K)));
@FOR(SMATRIX(I,K): @FREE(S(I,K)));
@FOR(SMATRIX(I,K): @FREE(Y(I,K)));

!ACCUM(I,K) corresponds to Z_ik;
@FOR(SMATRIX(I,L): ACCUM(I,L) = @SUM(TIME(K)|K#LE#L:S(I,K)*DEL
TA(K)) + AINIT(I));

@FOR(NONSTORAGE(I):@FOR(TIME(K): S(I,K) = 0));
@FOR(STORAGE(I): @FOR(TIME(K): Z(I) >= ACCUM(I,K)));
```

REFERENCES

Carvalho, M., Lozano, M. A., Serra, L. M. 2012a. Multicriteria synthesis of trigeneration systems considering economic and environmental aspects. *Applied Energy* 91: 245–254.

Carvalho, M., Lozano, M. A., Serra, L. M., Wohlgemuth, V. 2012b. Modeling simple trigeneration systems for the distribution of environmental loads. *Environmental Modelling & Software* 30: 71–80.

Chen, C.-L., Lai, C.-T., Lee, J.-Y. 2014. Transshipment model-based MILP (mixed-integer linear programming) formulation for targeting and design of hybrid power systems. *Energy* 65: 550–559.

Chen, D. S., Batson, R. G., Dang, Y. 2011. *Applied Integer Programming: Modeling and Solution*. John Wiley & Sons, New Jersey.

Chicco, G., Mancarella, P. 2009. Distributed multi-generation: A comprehensive view. *Renewable & Sustainable Energy Reviews* 13: 535–551.

Douglas, J. M. 1985. A hierarchical decision procedure for process synthesis. *AIChE Journal* 31: 353–362.

El-Halwagi, M. M. 2012. *Sustainable Design Through Process Integration—Fundamentals and Applications to Industrial Pollution Prevention, Resource Conservation, and profitability Enhancement*. Elsevier, Amsterdam.

Grossmann, I. E., Santibanez, J. 1980. Applications of mixed-integer linear programming in process synthesis. *Computers & Chemical Engineering* 4: 205–214.

Hemmes, K., Zachariah-Wolff, J. L., Geidl, M., Andersson, G. 2007. Towards multi-source multi-product energy systems. *International Journal of Hydrogen Energy* 32: 1332–1338.

Kasivisvanathan, H., Ng, D. K. S., Tan, R. R. 2013. Optimal operational adjustment in multi-functional energy systems in response to process inoperability. *Applied Energy* 102: 492–500.

Kasivisvanathan, H., Ubando, A. T., Ng, D. K. S., Tan, R. R. 2014. Robust optimisation for process synthesis and design of multi-eunctional energy systems with uncertainties. *Industrial & Engineering Chemistry Research* 53: 3196–3209.

Kavvadias, K. C., Maroulis, Z. B. 2010. Multi-objective optimization of a trigeneration plant. *Energy Policy* 38: 945–954.

Lee, J.-Y., Chen, C.-L., Chen, H.-C. 2014. A mathematical technique for hybrid power system design with energy loss considerations. *Energy Conversion and Management* 82: 301–307.

Lindo Systems. 2010. *LINGO: The Modelling Language and Optimizer*. Lindo Systems, Inc., Chicago.

Liu, P., Gerogiorgis, D. I., Pistikopoulos, E. N. 2007. Modeling and optimization of polygeneration energy systems. *Catalysis Today* 127: 347–359.

Liu, P., Pistikopoulos, E. N., Li, Z. 2009. A mixed-integer optimization approach for polygeneration energy systems design. *Computers & Chemical Engineering* 33: 759–768.

Lozano, M. A., Carvalho, M., Serra, L. M. 2009a. Operational strategy and marginal costs in simple trigeneration systems. *Energy* 34: 2001–2008.

Lozano, M. A., Carvalho, M., Serra, L. M. 2011. Allocation of economic costs in trigeneration systems at variable load conditions. *Energy and Buildings* 43: 2869–2861.

Lozano, M. A., Ramos, J. C., Carvalho, M., Serra, L. M. 2009b. Structure optimization of energy supply systems in tertiary sector buildings. *Energy and Buildings* 41: 1063–1075.

Seider, W. D., Seader, J. D., Lewin, D. R., Widagdo, S. 2009. *Product and Process Design Principles: Synthesis, Analysis and Design*. 3rd ed. Wiley, NJ, USA.

Serra, L. M., Lozanao, M. A., Ramos, J., Ensinas, A. V., Nebra, S. A. 2009. Polygeneration and efficient use of natural resources. *Energy* 34: 575–586.

Tan, R. R., Lam, H. L., Kasivisvanathan, H., Ng, D. K. S., Foo, D. C. Y., Kamal, M., Hallale, N., Klemes, J. J. 2012. An algebraic approach to identifying bottlenecks in linear process models of multi-functional energy systems. *Theoretical Foundations of Chemical Engineering* 46: 642–650.

Ubando, A. T., Culaba, A. B. Aviso, K. B., Tan, R R. 2013. Simultaneous carbon footprint allocation and design of trigeneration plants using fuzzy fractional programming. *Clean Technologies and Environmental Policy* 15: 823–832.

Varbanov, P., Friedler, F. 2008. P-graph methodology for cost-effective reduction of carbon emissions involving fuel cell combined cycles. *Applied Thermal Engineering* 28: 2020–2028.

9 Scheduling and Design of Multipurpose Batch Facilities
Periodic versus Nonperiodic Operation Mode through a Multi-Objective Approach

Tânia Pinto-Varela

CONTENTS

9.1　INTRODUCTION

In multipurpose batch facilities, a wide variety of products can be produced through different processing recipes by sharing all available resources, such as equipment, raw material, intermediates, and utilities. These facilities may operate through different operating modes, based on market demand characteristics.

If the facility faces stable demand periods, it becomes more profitable to operate in a periodic mode. However, if facing is a constant change of production, operating in a nonperiodic mode is more adequate. Like most real-world problems, the multipurpose design batch facilities involve multiple objectives and most of the existing literature on the design problem has been centered on a mono-criterion objective (Barbosa-Povoa 2007).

However, some works have been appearing in the scientific community addressing such a problem. On the basis of a nonperiodic operating mode, Dedieu et al. (2003) developed a two-stage methodology for the multi-objective batch plant design and retrofit, according to multiple criteria. A master problem characterized as a multi-objective genetic algorithm defines the problem design and proposes several plant structures. A subproblem characterized as a discrete event simulator evaluates the technical feasibility of those configurations. Later on, Dietz et al. (2006) presented a multi-criteria cost-environment design of multiproduct batch plants. The approach used consists of coupling a stochastic algorithm, defined as a genetic algorithm, with a discrete event simulator. A multi-objective genetic algorithm was developed with a Pareto optimal ranking method. The same author proposed the problem of the optimal design of batch plants with imprecise demands using fuzzy concepts (Dietz et al. 2008). The author extended the previous work applying a multi-objective approach using a genetic algorithm to take into account simultaneously maximization of the net value and two performance criteria, that is, the production delay/advance and flexibility. Mosat et al. (2007) presented a novel approach for solving different design problems related to single products in multipurpose batch plants. A new concept of super equipment is used and requires an implicit definition of a superstructure. Essentially the optimization is made on the transfers between different equipment units in a design. The Pareto optimal solutions are generated by a Tabu search algorithm. Chen et al. (2009) integrated the scheduling and heat recovery problems into a unified framework for multipurpose batch processes. The batch scheduling formulation is based on a continuous resource-task network (RTN) formulation. In the same work, the author extended the formulation to consider the heat-integrated periodic scheduling for batch processes.

As the problem complexity increases, multi-objective optimization may become lengthy, requiring an alternative optimization approach. Pinto-Varela et al. (2010) proposed an adapted form of the symmetric fuzzy linear programming (SFLP) approach, which is applied to the design and scheduling of multipurpose facilities. An example is used to show the methodology application and results are compared to those located on the efficient frontier, which were obtained with the mono-objective model through the ε-constraint approach.

Nowadays, the nature and dimension of these problems usually leads to large mixed-integer linear program (MILP) formulations that are associated with a high computational burden. To overcome this difficulty, Chibeles-Martins et al. (2011) proposed a meta-heuristic approach, based on Simulated Annealing, which is compared with an exact approach. More recently, Seid and Majozi (2013) presented a robust scheduling formulation for the synthesis and design of multipurpose batch plants. The formulation performance was compared with some benchmark examples and a better computational performance was presented.

Beyond the nonperiodic operating mode, the multipurpose batch facilities operate under conditions of relatively stable production demands, over an extended period of time. In such situations, it is profitable to establish a regular periodic operating schedule, in which the same sequence of operations is carried out repeatedly, thus simplifying the operation and control of such facilities; this is obtained by operating under a periodic mode. The periodic mode applied to the design/schedule of multipurpose batch plants has been, so far, centered on mono-criterion objectives (Barbosa-Povoa 2007). Despite the fact that some work has been already presented, very few works focus on the periodic operation mode application.

Fuchino et al. (1994, 1995) studied the design problem of multipurpose batch plants under a periodic operation through an evolutionary design method. Voudouris and Grossmann (1996) proposed a method for integrating scheduling and design for a special class of multipurpose batch processes. A new representation for periodic schedules is proposed that aggregates the number of batches for each product. Heo et al. (2003) addressed the periodic scheduling, planning, and design of batch plants. A nonlinear model was proposed that was linearized using a separable programming method. Finally, Castro et al. (2005) presented a continuous general mathematical formulation based on the RTN process representation. A periodic operation is assumed where the cycle time value is optimized simultaneously with the rest of the design variables. More recently, Nonyane and Majozi (2012) extended the periodic scheduling operation mode to the wastewater minimization concept.

Frequently, real-world problems involve multiple objectives. The multi-objective approach is a form of modeling such as that realistic situations conferring the resulting models characteristics will allow them to act as potentially powerful decision-making tools. On the basis of the Pareto optimal surface, decision makers will be able to select any given solution depending on the relative worth of each objective.

We realize that the existing literature does not completely explore the application of the multi-objective optimization to the design of multipurpose facilities, where detailed design aspects are considered. Therefore, the multi-objective optimization is still a modeling approach that requires further study when applied to the design of batch plants. By this reason, the work of Pinto et al. (2005, 2008) is generalized here to account for a multi-objective decision context where more than one economic objective is considered. These early works present generic and detailed formulations for the optimal design and retrofit of multipurpose batch facilities operating in a nonperiodic and periodic mode for a single mono-criterion objective. A multi-objective approach allows different values of different objectives at the Pareto optimal surface and the decision makers will be able to select any solution depending on how much one objective is worth in relation to the other. The method ε-constraint is explored, which presents the advantage of suiting any arbitrary problem, that is, with either convex or nonconvex objective spaces.

The final results allow the identification of a range of plant topologies, facilities design, and storage policies associated with a scheduling operating mode that minimizes the total cost of the system while maximizing the production, subject to total product demands and operational restrictions.

In this way, the resulting models would be able to act as potentially powerful decision-making tools where different decisions are accounted for.

Section 9.2 provides a brief discussion about concepts and classical multi-objective methodologies. The modeling framework to be addressed is introduced in Section 9.3. The nonperiodic and periodic formulation is presented in Section 9.4. A brief discussion and results analysis is given in Section 9.5, followed by conclusion in Section 9.6.

9.2 MULTI-OBJECTIVE CONCEPTS AND CLASSICAL METHODS OVERVIEW

In general, these models requiring a single objective have a clear, quantitative way to compare alternative feasible solutions. In many applications, a single objective is sufficient to realistically model the actual decision process. However, decisions become much more problematic when dealing with complex engineering designs, where more than one objective is at stake. For such cases, as previously referred, a multi-objective optimization model is required to capture all the possible perspectives. This is the case of the scheduling and design of batch facilities, addressed in this work, where two objectives are under consideration—maximization of revenue and minimization of cost. Contrary to common practice, where these are usually assigned equal weights and merged into a single index such as profit, revenue, and cost, are now handled separately since the desired trade-off between them is meant to be open to the decision maker. The multi-objective optimization can be generically represented as

$$\text{Maximize } f_m(x) \quad m = 1, 2, \ldots, M;$$

$$s.t.$$

$$g_j(x) \leq 0 \qquad\qquad j = 1, 2, \ldots, J; \qquad\qquad (9.1)$$

$$h_k(x) = 0 \qquad\qquad k = 1, 2, \ldots, K;$$

$$x_i^{(L)} \leq x_i \leq x_i^{(U)} \qquad i = 1, 2, \ldots, n$$

where M identifies the objective function $f(x) = (f_1(x), f_2(x), \ldots, f_m(x))^T$ and J and K are, respectively, the number of inequality and equality constraints. A solution will be given by a vector X of n decision variables: $X = (x_1, x_2, \ldots, x_{n-1}, x_n)^T$.

9.2.1 CONCEPTS

When addressing the multi-objective problem, different concepts need to be presented. To recall, a description of the most important ones is given below.

9.2.1.1 Variable Space versus Objective Space

Within the multi-objective optimization problems (MOOP), two cases may exist: the nonlinear and the linear problems. The latter case is the theme of the present work and is denoted as multi-objective linear problem (MOLP).

One of the important differences between single-objective and multi-objective optimization is that the multi-objective optimization constitutes a multi-dimensional space, in addition to the usual decision variable space. This additional space is called

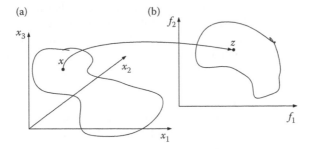

FIGURE 9.1   (a) Variables decision space and (b) corresponding objective space. (Adapted from Deb, K. 2004. *Multi-Objective Optimization Using Evolutionary Algorithms*. West Sussex: John Wiley & Sons.)

the objective space Z. For each solution X (vector X) in the decision variable space, there exists a point in the objective space, denoted by $f(x) = z = (z_1, z_2, z_3, \ldots, z_M)^T$. The mapping takes place between the solution space and the objective space (Figure 9.1).

Another important difference is that in a multi-objective optimization no solution vector X exists that maximizes all objective functions simultaneously. A feasible vector X is called an optimal solution if there is no other feasible vector that increases one objective function without causing a reduction in at least one of the other objective functions. It is up to the decision maker to select the best compromising solution among a number of optimal solutions in the efficient frontier. Multi-objective optimization is sometimes referred to as a vector optimization, because a vector of objectives, instead of a single objective, is optimized.

9.2.1.2 Nondominated Solutions

Within the MOLP problem, one important concept is the nondominated solution. To illustrate this concept, Figure 9.2 adapted from Deb (2004) will be used. From the feasible objective space, many solutions can be obtained and compared between them. This comparison allows the identification of the best solution in both objectives.

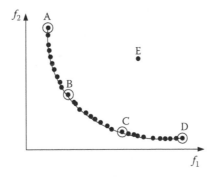

FIGURE 9.2   Nondominated solutions vs. dominated solutions. (Adapted from Deb, K. 2004. *Multi-Objective Optimization Using Evolutionary Algorithms*. West Sussex: John Wiley & Sons.)

However in certain cases, one solution is better than the other in one objective, but it is worse in the other objective. Figure 9.2 illustrates the nondominated solution.

The minimum from F1 is the solution A, but the minimum of F2 is the solution D. It is visible that solution A has a smaller value of F1, but a larger F2 value. Hence, none of these solutions can be said to be better than the other with respect to both objectives. These solutions are called nondominate solutions. The same happens when solutions B and C are compared. All the solutions lying on the curve are known as nondominated solutions. The curve formed by joining these solutions is known as the efficient frontier. Comparing E with C, it is visible that the latter is better than solution E in both objectives. When this happens, solution E is dominated by solution C. Thus, solution E is not of interest.

9.2.1.3 Supported versus Unsupported Nondominated Solutions

The efficient set of solutions resulting in MOLP can be classified as supported or unsupported solutions. If the solution is on the boundary of the feasible objective space, the solution is a supported nondominated solution. Otherwise, it is an unsupported (convex dominated) nondominated vector (i.e., those solutions whose objective value vector does not lie in the border of the convex hull). The vector unsupported nondominated is dominated by some convex combination of other nondominated solutions. In Figure 9.3, these concepts are illustrated using eight solutions defined from A to J.

A dashed line in Figure 9.3a shows that solution B results from a convex combination of solutions A and C. Based on that, it is possible to define three classes of optimal solution, which are

- Supported-extreme solution: A, C, and I
- Supported nonextreme: B
- Unsupported: D, E, and F

Through a bold line visible in Figure 9.3b, the definition of the objective space is made, using the supported extreme solutions.

The inverse images of the supported nondominated solution are said to be supported efficient points (in the decision space) and inverse images of the unsupported nondominated solution are said to be unsupported efficient points (in the decision space).

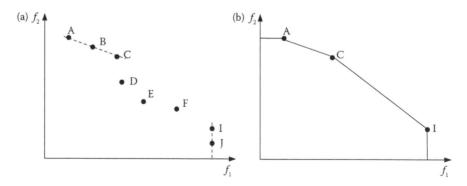

FIGURE 9.3 Supported vs. unsupported nondominated solutions.

9.2.2   Classical Methods Overview

Classical multi-objective optimization methods have been used for at least the past four decades. During this period, many algorithms have also been suggested (Steuer 1986). We will outline here some of the classical methods and highlight its advantages and disadvantages.

9.2.2.1   Weighted Sum Method

As the name suggests, this method converts a set of objectives into a single normalized objective by multiplying each objective by a weight. The weight of an objective is usually chosen based on its importance in the problem. Different objectives have results with different magnitude orders requiring its normalization, to characterize the composite objective function. Only after that, the MOOP is converted into a single-objective optimization problem (Deb 2004):

$$\text{Minimize } F(x) = \sum_{m=1}^{M} w_m f_m(x)$$

s.t.

$$
\begin{aligned}
g_j(x) &\geq 0 & j &= 1,2,\ldots,J \\
h_k(x) &= 0 & k &= 1,2,\ldots,K \\
x_i^L &\leq x_i \leq x_i^U & i &= 1,2,\ldots,n \\
w_m &\in [0,1]
\end{aligned}
$$

(9.2)

9.2.2.2   ε-Constraint

To overcome the difficulties/limitation faced by the weighted sum approach, the ε-constraint method is used. Haimes et al. (1971) suggested reformulating the MOOP by just keeping one of objectives and restricting the rest of the objectives within user-specified values. The problem is as follows:

$$\text{Maximize } f_u(x)$$

s.t.

$$
\begin{aligned}
f_m(x) &\leq \varepsilon_m & m &= 1,2,\ldots,M \text{ and } m \neq u \\
g_j(x) &\leq 0 & j &= 1,2,\ldots; \\
h_k(x) &= 0 & k &= 1,2,\ldots,K; \\
x_i^{(L)} &\leq x_i \leq x_i^{(U)}
\end{aligned}
$$

(9.3)

where ε_m represents an upper bound of the value of f_m. This technique suggests handling one of the objectives and restricting the others within user-specified values.

Once obtained, the efficient frontier allows the decision maker to select any solution depending on the relative worthiness of each objective.

9.2.2.3 Weighted Metric Methods

This method defines another way to combine several objectives into a single objective. This approach is based on weighted distance functions, as metrics. The most used metrics are L_1, L_2, and L_∞, its generalization is L_p. The L_p metric measures the distance between solution X and the ideal solution Z^* or another point. If different importance is given to each objective, a weighted metric should be used with the family weighted metrics characterization, L_p^λ, where the λ_m is a set of weight, which can be minimized as follows:

$$\text{Minimize } L_p(x) = \left(\sum_{m=1}^{M} \lambda_m \left| f_m(x) - z_m^* \right|^p \right)^{1/p}$$

s.t.   (9.4)

$$g_j(x) \ge 0 \qquad j = 1,2,\ldots,J$$
$$h_k(x) = 0 \qquad k = 1,2,\ldots,K$$
$$x_i^L \le x_i \le x_i^U \quad i = 1,2,\ldots,n$$

The parameter p can take any value between 1 and ∞. Some of the value of p defines a *well-known metric*, with $p=1$ the problem obtained is a weighted sum approach, with $p=2$ a weighted Euclidean distance of any point in the objective space from the ideal point is defined. When a large p is used, the above problem reduces to a one of minimizing the largest deviation $|f_m(x) - z_m^*|$ denoted by the weighted *Tchebycheff* problem:

$$\text{Minimize } l_\infty(x) = \max_{m=1}^{M} w_m \left| f_m(x) - z_m^* \right|$$

s.t.

$$g_j(x) \ge 0 \qquad j = 1,2,\ldots,J$$ (9.5)
$$h_k(x) = 0 \qquad k = 1,2,\ldots,K$$
$$x_i^L \le x_i \le x_i^U \quad i = 1,2,\ldots,n$$

9.2.2.4 Benson's Method

This procedure uses as a reference value a random solution, z^0, from the feasible non-Pareto-optimal solution. Therefore, the nonnegative difference $(z_m^0 - f_m(x))$ of each objective is calculated and their sum is maximized (Deb 2004):

$$\text{Minimize } \sum_{m=1}^{M} \max(0,(z_m^0 - f_m(x))$$

s.t.   (9.6)

$$f_m(x) \le z_m^0 \qquad m = 1,2,\ldots,M$$
$$g_j(x) \ge 0 \qquad j = 1,2,\ldots,J$$
$$h_k(x) \ge 0 \qquad k = 1,2,\ldots,K$$
$$x_i^L \le x_i \le x_i^U \quad i = 1,2,\ldots,n$$

9.2.2.5 Value Function Method

In this method, a utility function is used, $U : R^M \rightarrow R$ relating M objectives. The value function must be valid over the entire feasible search space. The method goal is to maximize the value function as follows (Deb 2004):

$$\text{Maximize } U(f(x))$$

$$s.t.$$

$$g_j(x) \geq 0 \qquad j = 1,2,\ldots,J \qquad (9.7)$$

$$h_k(x) = 0 \qquad k = 1,2,\ldots,K$$

$$x_i^L \leq x_i \leq x_i^U \quad i = 1,2,\ldots,n$$

where $f(x) = (f_1(x), f_2(x), \ldots, f_M(x))^T$.

In Table 9.1, all the above methods are summarized and the main advantages and disadvantages are enumerated.

Having these characteristics in mind and taking into account the fact that we will be working with only two objectives, the ε-constraint method is chosen to be used with the present work. This method has several characteristics that will satisfy the final work purpose (Table 9.1):

- It is able to produce supported and unsupported solutions;
- Every run produces a different efficient solution;
- The number of efficient solutions can be adjusted by $\Delta\varepsilon$ in the objective function range.

9.3 MODELING FRAMEWORK

The optimal scheduling and design of multipurpose plants are addressed using the RTN representation methodology (Pantelides, 1994). Such framework appears as one of the most general and conceptually simpler representations to deal with process facility problems where two types of entities, tasks, and resources are defined.

A task is an operation that consumes and/or produces a specific set of resources, whereas a resource describes all different entities involved in the facility. Each resource is classified within a type, based on its functionality (i.e., two items of processing equipment belong to the same resource type if they have the same capacity and connectivity, and are able to perform the same set of tasks). Resources are produced or consumed at discrete times during the execution of the task, and can be classified depending on whether or not their availability is restored to the original state after being used by a task. A nonrenewable resource represents raw materials, utilities, manpower, and so on, whereas a renewable resource represents all types of equipment such as processing, storage, piping, and so on.

Each task has a fixed duration τ_k and the execution of task k starting at time t is characterized by the pair of variables N_{kt}, ξ_{kt}. N_{kt} is an integer variable whereas ξ_{kt} is continuous. These variables can be interpreted, respectively, as the number

TABLE 9.1

Classical Multi-Objective Methods Characteristics Summary

Method	Advantages	Disadvantages
Weighted sum method	– Is simplest way to solve a MOOP – Guarantees solutions on the entire Pareto-optimal set, only in problems having convex Pareto-optimal front – Only identify extreme solutions	– The objective functions should be normalized – All objectives have to be converted into same type (max or min objective) – Different weight vectors combinations may lead to the same Pareto-optimal solutions – It is not able to find unsupported solutions
ε-Constraint	– Different Pareto-optimal solutions can be found by using different ε values – Can be used in problems with convex or nonconvex objective spaces – It overcomes duality gaps in convex sets	– The solution to the problem depends on the chosen ε vector – As the number of objectives increases, there exist more elements in the ε vector, requiring more information from the user
Weighted metric methods	– The weighted Tchebysheff metric guarantee to find each and ever Pareto-optimal solution when z^* is a utopian objective vector (Miettinen 1999). – Some improvement can be made using: rotated weighted metric or dynamically changing the ideal solution – These improvements may be useful in solving problems with a nonconvex objective space	– The objective functions should be normalized – This method requires the ideal solution z^* sometimes difficult to reach
Benson's method	– To avoid scaling problems, individual differences can be normalized before the summation – To obtain different Pareto-optimal solutions, the differences can be weighted before summation – Changing the weight vector, different Pareto-optimal solutions can be obtained	– Requires a random solution Z^0 – Only find solution in nonconvex region if z^0 is chosen appropriately
Value function method	– Simple to use, if adequate value function information is available. – Mainly used in practice to multi-attribute decision analysis problems with discrete set of feasible solution (Keeney and Raiffa 1976).	– The solution depends on the chosen value function – Requires the uses of a value function which is globally applicable over the entire search space – There is a danger of using an oversimplified value function

Source: Adapted from Deb, K. 2004. *Multi-Oobjective Optimization Using Evolutionary Algorithms.* West Sussex: John Wiley & Sons.

of instances (e.g., batches) of task k starting simultaneously at time t, and the total amount of material being processed by all these instances.

The amount of resource type r produced at time θ relative to the start of task k at time t is given by

$$\mu_{kr\theta} N_{k,t-\theta} + \nu_{kr\theta} \xi_{k,t-\theta}$$

where $\mu_{kr}\theta$ and $\nu_{kr}\theta$, $\theta = 0, \ldots, \tau_k$ are known constants. Negative values for the latter indicate consumption of the resource, while positive values denote production.

These coefficients allow the modeling of different type of tasks such as those with time-dependent manpower or utility requirements, as well as those that involve a simple transformation from a feed to a product, using an equipment item.

Changes on the resource utilization can only occur at interval boundaries. The variable R_{rt} denotes the amount of excess resource r over time interval t. The change in the excess resource level for each resource type from one time interval to the next is given by excess resource balance constraints.

In this work, a discretization of time over a nonperiodic and periodic operation mode is used.

The nonperiodic operation mode is defined over a given time horizon, H, where the time horizon is divided into intervals of equal duration. The market is satisfied at the end or along the horizon with multideliveries. Every time the market demand requirements change over the horizon, a different scheduling can be defined. However, if the facility faces a stable market, demand is more suitable to operate in periodic mode using the cycle time period (T). The latter is taken as the shortest interval of time at which a cycle is repeated, where the cycle represents a sequence of operations involving the production of all desired products and the utilization of all available resources. Since all cycles are equal for a given campaign, the problem is formulated over a single cycle where it is guaranteed that the operation of the facility is the same at the beginning and at the end of the cycle. The execution of a task is allowed to overlap successive cycles and, since a cycle is repeated over the time of planning, its execution is modeled by wrapping around to the beginning of the same cycle. To do so, the wrap-around operator defined by Shah et al. (1993) is used:

$$\Omega(t) = \begin{cases} t & \text{if } t \geq 1 \\ \Omega(t + T) & \text{if } t \leq 0 \end{cases} \tag{9.8}$$

When this is applied, for instance, the variable $N_{k,\Omega(t-\theta)}$ for $t - \theta \leq 0$ leads to an identical equipment resource allocation that will start at time $t - \theta + T$.

The start of the cycle is defined as time $t = 1$, and the end as $t = T + 1$. The latter coincides with the starting point of the next cycle. A planning horizon (H) is assumed which is divided into N equal cycles of duration (T). A cycle is divided into a number of elementary time steps of fixed duration (δ), as shown in Figure 9.4.

The scheduling and design problem of multipurpose batch facilities can be characterized as follows.

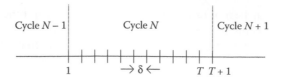

FIGURE 9.4 Time discretization for a single cycle.

Given:

- The process/plant description, including the plant topology
- Resources availability, characteristics and costs
- Time horizon of scheduling and mode of operation
- Demand over the time horizon (production range) and cost data

Determine:

- The optimal plant configuration (i.e., number and type of equipment units and their connections, as well as their sizes)
- The optimal process schedule (i.e., timing of all tasks, storage policies, batch sizes, amounts transferred, allocation of tasks and consumption of resources)

Mixed storage policies, shared intermediate states, material recycles, and multi-purpose batch plant equipment units with continuous sizes, are allowed.

9.4 MATHEMATICAL FORMULATION

In this section both formulations, nonperiodic and periodic operation mode, presented by Pinto et al. (2005, 2008) are compared.

To enable a better understanding between formulations, the indices, parameters, and decision variables nomenclature are presented. However, detailed explanation will be omitted.

Sets:

$D = \{r:$ *set of all equipment resources*$\}$

$D_{fp}/D_{rm} = \{r \in D:$ *set of dedicated storage vessels for final product/raw material*$\}$

$C = \{r:$ *set of all material resources*$\}$

$C_{is} = \{r:$ *set of intermediate material with dedicated storage*$\}$

$C_r = \{r:$ *set of material resources with dedicated storage*$\}$

$T_p = \{k:$ *set of processing tasks in an equipment resource, r*$\}$

$T_s = \{k:$ *set of storage task for raw and product material in dedicated vessel*$\}$

$D_c/D_v = \{r \in D:$ *set of all connections/storage vessels*$\}$

$D_p = \{r \in D:$ *set of process equipment resources*$\}$

$C_p/C_f = \{r:$ *set of final products/raw materials*$\}$

$T_{kr} = \{k:$ *set of all tasks requiring an equipment resource, r*$\}$

$T_{tv} = \{k, r \in D_v:$ *transfer task, k, for an intermediate storage vessel, r, which is a sink*$\}$

$T_t = \{k:$ *set of all transfer tasks*$\}$

Parameters:

H planning horizon	T cycle time
CCF capital charge factor	v_r/p_r price of resource
α_k^0/α_k^1 fixed and variable cost coefficients	φ_r^{max} size factor
R_r0 resource r available initially	R_r^{min}/R_r^{max} min/max at H
$\mu_{kr\theta}/v_{kr\theta}$ consumption or production of a	Δ_r^{max} the max available of r
renewable $(-1,1)$/nonrenewable $(-1,0)$ resource	
r, at the start/end of θ	

Variables:

R_{rt} excess of resource at t	ξ_{kt} batch size of task k at time t
Δ_r amount of resource required	V_r capacity of resource r
N_{kt} number of processing tasks k at instant t	$\Omega(t)$ wrap-around time operator
$Et_r = 1$ if r is installed; 0 otherwise	$Ec_r = 1$ if r is installed; 0 otherwise

Subject to:
Excess resource balance for the renewable resource
Pinto et al. (2008)

$$R_{rt} = R_{r_0|t=1} + R_{r,t-1|t\geq 2} + \sum_k \sum_{\theta=0}^{\tau_k} \mu_{kr\theta} N_{k,t-\theta} \quad \forall r \in D_p, \ t = 1...H \qquad (9.9)$$

Pinto et al. (2005)

$$R_{rt} = R_{r,\Omega(t-1)} + \sum_k \sum_{\theta=0}^{\tau_k} \mu_{kr\theta} N_{k,\Omega(t-\theta)} \quad \forall r \in D_p, \ t = 1...T \qquad (9.10)$$

Excess resource balance for the nonrenewable resource
Pinto et al. (2008)

$$R_{rt} = R_{r_0|t=1} + R_{r,t-1|t\geq 2} + \sum_k \sum_{\theta=0}^{\tau_k} \upsilon_{kr\theta} \xi_{k,t-\theta} \quad \forall r \in C, \ t = 1...T \qquad (9.11)$$

Pinto et al. (2005)

$$R_{rt} = R_{r_0|t=1, r\in R_f} + R_{r,\Omega(t-1)|t\geq 2} + \sum_k \sum_{\theta=0}^{\tau_k} \upsilon_{kr\theta} \xi_{k,\Omega(t-\theta)} \quad \forall r \in C, \ t = 1...T \qquad (9.12)$$

The constraints (9.13) through (9.15) are only used by periodic formulation.

Excess resource balance for the intermediate storage

$$R_{rt|t=1} = R_{rt|t=T+1} \quad \forall r \in C_{is} \tag{9.13}$$

Excess resource balance for the product resource

$$R_{rt} = R_{r,\Omega(t-1)|1 \prec t \leq T+1} + \sum_k \sum_{\theta=0}^{\tau_k} v_{kr\theta} \xi_{k,\Omega(t-\theta)|t \prec T+1} \quad \forall r \in C_p, t = 1...T+1 \tag{9.14}$$

Wrap-around operator (Shah et al. 1993)

$$\Omega(t) = \begin{cases} t & \text{if } t \geq 1 \\ \Omega(t+T) & \text{if } t \leq 0 \end{cases} \tag{9.15}$$

The resource existence, design, boundaries characterization, connectivity, and production requirement constraints are similar for both formulations, characterized by constraints (9.16) through (9.37).

Processing equipment resource existence constraints: These guarantee that each equipment resource is idle or processing a task, which cannot be pre-empted once started

$$\sum_{t'=t-pi+1}^{t} \sum_{k \in T_{kr}} N_{k,t'} \leq \Delta_r \quad \forall r \in D_p, t = 1...T \tag{9.16}$$

Equipment resource constraints: The equipment resource required must be available with Δ_r^{max} as an upper bound

$$0 \leq R_{rt} \leq \Delta_r \quad \forall r \in D_p, t = 1...T+1 \tag{9.17}$$

$$0 \leq \Delta_r \leq \Delta_r^{max} \quad \forall r \in D_p \tag{9.18}$$

Equipment in continuous size ranges: The equipment resource is available in a continuous size range.

$$V_r^{min} \leq V_r \leq V_r^{max} \quad \forall r \in D_p \tag{9.19}$$

$$0 \leq V_r \leq \Delta_r (V_r^{max} - V_r^{min}) + V_r^{min} \quad \forall r \in D_p \tag{9.20}$$

$$\sum_t \sum_k \xi_{kt} \leq V_r \quad \forall k \in Ts, r \in D_v \tag{9.21}$$

Capacity and batch size constraints: The amount of material being processed must always be within the maximum and minimum equipment capacity.

$$\phi_{kr}^{min} V_r N_{kt} \leq \xi_{kt} \leq \phi_{kr}^{max} V_r N_{kt} \quad \forall k \in T_K, r \in C, t = 1...H \tag{9.22}$$

Considering a bilinear term in constraint (9.22), linearization must be made

$$\tilde{N}_{jkt} = \begin{cases} 1 & \text{if } N_{kt} = j \\ 0 & \text{otherwise} \end{cases}$$

where j varies from 0 to N_k^{max}, we have

$$N_{kt} = \sum_{j=1}^{N_k^{max}} j \tilde{N}_{j,k,t} \quad \forall k \in (T_p \wedge T_{kr}), t = 1...T \tag{9.23}$$

$$\sum_{j=0}^{N_k^{max}} \tilde{N}_{jkt} \leq 1 \quad \forall k \in (T_p \wedge T_{kr}), t = 1...T \tag{9.24}$$

By multiplying Equation 9.23 by V_r, we have

$$V_r N_{kt} = \sum_{j=1}^{N_k^{max}} j V_r \tilde{N}_{jkt} = \sum_{j=1}^{N_k^{max}} j \tilde{V}_{rjkt}$$

where $\tilde{V}_{rjkt} \equiv V_r \tilde{N}_{jkt}$ can be defined through the linear constraints.

$$0 \leq \tilde{V}_{rjkt} \leq V_r^{max} \tilde{N}_{jkt} \quad \forall k \in (T_p \wedge T_{kr}), r \in D_p, j = 1...N_k^{max}, t = 1...T \tag{9.25}$$

$$\sum_{j=0}^{N_k^{max}} \tilde{V}_{rjkt} = V_r \quad \forall k \in (T_p \wedge T_{kr}), r \in D_p, j = 1...N_k^{max}, t = 1...T \tag{9.26}$$

and finally the capacity constraint can be represented by

$$\varphi_{kr}^{min} \sum_{j=1}^{N_k^{max}} j \tilde{V}_{rjkt} \leq \xi_{kt} \leq \varphi_{kr}^{max} \sum_{j=1}^{N_k^{max}} j \tilde{V}_{rjkt} \quad \forall k \in (T_p \wedge T_{kr}),$$

$$r \in D_p, j = 1...N_k^{max}, t = 1...T \tag{9.27}$$

Storage constraints for a dedicated vessel: The storage capacities must account for the respective amounts of material stored

$$\sum_{r \in C_f} \left[Ro_f - R_{rt} \right] \le V_{r|r \in D_{rm}} \quad \forall t = 1 + T \tag{9.28}$$

$$\sum_{r \in C_p} R_{r,t} \le V_{r|r \in D_{fb}} \quad t = 1 + T \tag{9.29}$$

$$V_r^{\min} Et_r \le V_r \le V_r^{\max} Et_r \quad \forall r \in D_v \tag{9.30}$$

$$\sum_t \sum_k \xi_{kt} - M_r Et_r \le 0 \quad \forall r \in D_v, \, k \in T_{tv}, \, t = 1 \ldots T \tag{9.31}$$

$$\sum_k \xi_{kt} \le \varphi_r^{\max} V_r \quad \forall r \in D_v, \, k \in T_{tv}, \, t = 1 \ldots T \tag{9.32}$$

Connectivity constraints: The connection capacity must account for the batch transfer task:

$$\xi_{kt} \le \varphi_{kr}^{\max} V_r \quad \forall k \in T_t, \, r \in D_c \tag{9.33}$$

$$V_r^{\min} Ec_r \le V_r \le V_r^{\max} Ec_r \quad \forall r \in D_c \tag{9.34}$$

Production requirement constraints:

$$R_r^{\min} \le R_{r,t} \frac{H}{T} \le R_r^{\max} \quad \forall r \in C_p, \, t = 1 + T \tag{9.35}$$

Two objectives are considered in this work: the cost minimization by constraint 9.36 and revenue maximization by constraint (9.37). The economic costs include operational, raw material, and design.

$$
\begin{aligned}
F_1 = &\left(\sum_t \sum_{k \in Tp} \left(\alpha_k^0 N_{kt} + \alpha_k^1 \xi_{kt} \right) + \sum_{r \in C_f} \left[R_{r0} - R_{rt} \right] v_r \right) \frac{H}{T} \\
&+ \left(\begin{aligned} &\sum_{r \in D_p} \sum_s \left(\Delta_r \, CC_{rs}^0 + V_r \, CC_{rs}^1 \right) + \sum_{r \in D_C} \sum_s \left(Ec_r \, CC_{rs}^0 + V_r \, CC_{rs}^1 \right) \\ &+ \sum_{r \in D_v} \sum_s \left(Et_r \, CC_{rs}^0 + V_r \, CC_{rs}^1 \right) \end{aligned} \right) \times CCF
\end{aligned}
$$

$$\tag{9.36}$$

3. Boundaries' characterization
    3.1 Perform a single objective approach with revenue maximization
    3.2 Upper bound of the efficient solution set, (x_1, y_1) is defined
    3.3 Perform a single objective approach with cost minimization
    3.4 Lower bound of the efficient solution set, (x_2, y_2) is defined
4. Number of steps definition, *nstep*
5. ε interval dimension
    5.1 $\Delta\varepsilon = \dfrac{x_1 - x_2}{nstep}$
6. Efficient frontier definition through the efficient solution set characterization
    6.1 $k = 0$ to $k = nstep$
        6.1.1 perform a single objective approach, with revenue maximization, *s.t*
            $F_1 \le x_1 - k\,\Delta\varepsilon \quad k = 0,\ldots, nstep$
        6.1.2 $k = k + 1$
        6.1.3 go to step 4.1

FIGURE 9.5 ε-Constraint implementation.

$$F_2 = \frac{H}{T} \sum_{r \in C_P} R_{rt} p_r \qquad (9.37)$$

The multi-objective methodology used is the ε-constraint, which is applied to both models, periodic (PM) and nonperiodic model (NPM). An implementation template is presented in Figure 9.5.

9.5   COMPUTATION RESULTS

In this section, we present three examples and explore different operational/design conditions. The first one is based on Barbosa-Povoa and Macchietto (1994) and explores the nonperiodic operation mode, while the second example is based on the first one, however to be explored by the periodic operation mode some adaptations were made. The third example is explored through a nonperiodic and periodic mode as cases (a) and (b), respectively.

The General Algebraic Modeling System (GAMS) was used, coupled with the CPLEX 11.0 in a Pentium (R) Duo Core, T7300, 2.0 GHz, 2 GB RAM.

EXAMPLE 9.1

The formulation proposed above is applied to the Barbosa-Povoa and Macchietto (1994) example, where the maximization of the revenue with the production of three final products, S4, S5 and S6 is made. The market demand may vary between [0:80] tonnes for S4 and S5, and [0:60] tonnes for S6 using two raw materials, S1 and S2.

The facility operates in a nonperiodic mode over a time horizon of 12 h. In terms of equipment suitability, reactors R1, R2, and R3 are multipurpose equipment. Task T1 may process S1 during 2 h in R1 or R2 producing the unstable material S3; Task T2 may process S2 during 2 h in R2 or 1 h in R1 producing S4, which is both an intermediate material and a final product; Task T3 processes 0.5 of S3 and 0.5 of S4 in R3 during 3 h producing S5, which, like S4, is both an

TABLE 9.2

Capacities and Equipment Cost

	V1	V2	V4	V5, V6	R1	R2	R3	Ci
Capacity (u.m/m²) max:min	Unl	Unl	0:1000	0:1000	0:100	0:150	0:200	0:200
Cost (10³ c.u.) fix:var	0	0	1:0	0:0.1	20:0.5	55:0.5	30:1	0:0.1

Abbreviations: c.u. = currency units; u.m = mass units; unl. = unlimited; fix:var = fix and variable cost; max:min = maximum:minimum, the capacity range available for the design.

intermediate and a final product; Task T4 may process 0.5 of S3 and 0.5 of S5 in R3 during 2 h producing the final product S6.

Table 9.2 shows the maximum capacity allowed for each unit and its cost.

The efficient frontier obtained for the facility operating in a nonperiodic mode can be seen in Figure 9.6. Every efficient point lies along the depicted boundary because no further progress is possible in one objective function without degrading the other.

For the case under study, the objective value space is represented with axes for the cost and revenue objective functions. In the efficient frontier, some optimal plant topologies are visible. The points A, B, C, D, and E represent points where topology changes as a result of addition or withdrawal of equipment units to the previous topology. The optimal design for the main equipment is presented in Table 9.3, followed by Table 9.4 with the final product quantity produced, for each assigned point in Figure 9.6.

Between points A and B, the plant topology requires the reactor R1 (with capacity 31.33, Table 9.3) to produce S4 (Table 9.4), but in reaching point B, new topology is suggested until C. However, a new reactor R2 is used for producing the same final product, but this time with capacity 80 tonnes (Table 9.3). This results from the fact that reactors R1 and R2 present different capacities and costs. Between C and D points, the plant topology requires one more reactor R3 (with capacity 11.42, Table 9.3) and produces two products, 80 and 11.42 tonnes of S4

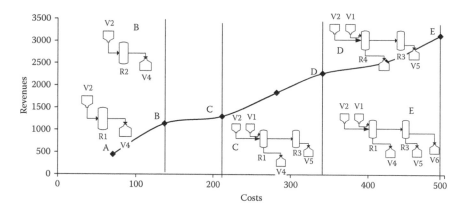

FIGURE 9.6   Efficient frontier for Example 9.1.

TABLE 9.3
The Optimal Design for the Main Equipment (Tonnes)

Equipment	A	B	C	D	E
R1	31.33	–	85.71	–	67.54
R2	–	80	–	120	–
R3	–	–	11.42	80	110

TABLE 9.4
Quantities Produced for Each Final Product (Tonnes)

Products	A	B	C	D	E
S4	31.33	80	80	80	80
S5	–	–	11.42	80	80
S6	–	–	–	–	60

and S5, respectively (Table 9.4). At point E maximum production is reached and a more complex topology is required.

The optimal scheduling is characterized for the revenue maximization, point E (Figure 9.7). As highlighted, R1 and R3 present multitask characteristics, performing Tasks T1 or T2 and T3 or T4, respectively. This model requires 238 binary variables, 919 variables, and 1466 equations, reaching 0% of the optimality gap.

EXAMPLE 9.2

Using the first example with slight adaptations, the periodic operation mode is explored. Revenue is maximized to produce three final products, S4, S5, and S6 with market demands between [0:11000] and [0:6000] tonnes for S4, S5, and S6, respectively. The facility operates in a periodic mode over a time horizon of 8 h/day in a campaign of 100 days. Table 9.5 shows the capacity allowed for each unit and its associated cost.

The efficient frontier characterization using the periodic mode is shown in Figure 9.8 and identifies four points where topology changed (A, D, F, and H). The optimal equipment design is shown in Table 9.6, followed by the final products characterization in Table 9.7.

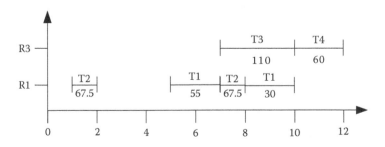

FIGURE 9.7 Optimal scheduling for the plant topology at point E.

TABLE 9.5

Capacities and Equipment Cost

	V1	V2	V4, V5, V6	R1	R2	R3	Ci
Capacity (u.m/m²)							
max:min	Unl	Unl	0:1000	0:100	0:250	0:200	0:200
Cost (10³ c.u.) fix:var	0	0	0:0.1	20:0.5	45:0.5	30:1	0:0.1

Abbreviations: c.u. = currency units; u.m = mass units; unl. = unlimited; fix:var = fix and variable cost; max:min = maximum:minimum, the capacity range available for the design.

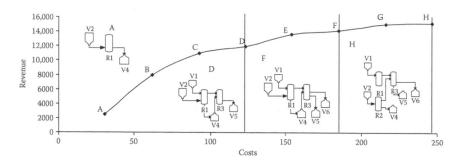

FIGURE 9.8   The efficient frontier for Example 9.2.

TABLE 9.6

The Optimal Design for the Main Equipment

Equipment	A	B	C	D	E	F	G	H
R1	25.01	79.43	82.21	58.82	65.67	67.57	71.2	81.81
R2	–	–	–	–	–	–	–	50.4
R3	–	–	–	19.08	53.35	62.85	80.99	84
V4	25.01	79.43	110	110	110	110	110	110
V5	–	–	–	19.08	53.35	60	60	60
V6	–	–	–	–	–	7.13	52.48	60

TABLE 9.7

Quantities Produced for Each Final Product

Final Products	A	B	C	D	E	F	G	H
S4	25.01	79	110	110	110	110	110	110
S5	–	–	–	58.82	53.35	60	60	60
S6	–	–	–	–	–	7.13	52.48	60

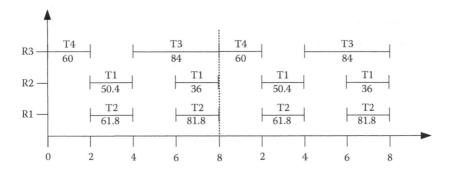

FIGURE 9.9 Optimal scheduling for the plant topology at point H.

The efficient frontier shows that the topology used between A and D only requires reactor R1, and is able to produce 25 tonnes of S4 (Table 9.7). At point D, the topology requires one more reactor, R3, and is able to produce 110 of S4 and 58.8 tonnes of S5 (Table 9.7), which is used between D and F. As the revenue increases, the design and operational costs also increase, mainly because of the equipment capacity expansion, allowing a higher production. Topology at F produces one more product S6, requiring the respective, V6, to storage of 7.1 tonnes. The topology that maximizes the revenue (H) requires all three reactors R1, R2, and R3, whose designs are presented in Table 9.6.

The revenue maximization (H) is used to characterize its scheduling in Figure 9.9. In the scheduling not only the cycle characteristics of the periodic operation mode is shown, but also the multipurpose characteristics of equipment R3, performing Tasks T3 and T4. The model requires 166 binary in a total of 649 variables and 1069 constraints and was solved until optimality is reached.

EXAMPLE 9.3

To highlight the characteristics of the formulation, a more complex example is explored. The example considers a design of a multipurpose batch plant where the maximization of revenue is undertaken for the production of five products (S5, S9, S10, S6, and S11) from three raw materials (S1, S2, and S7). Two of the products are also intermediated products (S5 and S6).

The facility has available six reactors (R1 to R6), and nine dedicated vessels. In terms of equipment suitability, only reactors R1 and R2 may carry out two processing tasks, T1 and T2, while each storage vessel and reactors R3, R4, R5 and R6 are dedicated to a single State/Task. Task T1 may process S1 during 2 h in R1 or R2 producing, S3, an unstable product; Task T2 may processes S2 during 2 h in R1 or R2 producing S4; Task T3 may process 0.5 of S3 and S4 during 4 h in R3 to produce S5; T4 processes 0.5 of S3 and S4 during 2 h in R4 to produce S6; Task T5 may process S6 during 1 h to produce the final product 0.3 of S11 and 0.7 of S8 in R5, and finally Task T6 processes during 1 h S8 in reactor R6 to produce the final products 0.5 of S9 and 0.5 of S10. The connections capacity ranges from 0 to 200 m.u./m^2 at a fix/variable cost of 0.1/0.01 [10^3 c.u.]. The capacity of R1, R2, R5, and R6 ranges from 0 to 150 m.u./m^2 while the others range from 0 to 200 m.u./m^2 (where m.u. and c.u. are, respectively, mass and currency units).

This example will be explored through a nonperiodic mode as case (a) and a periodic mode as case (b). The quantities produced for each case are shown in

TABLE 9.8
Final Products Quantities for Each Case

Tonnes (min:max)	S5	S6	S9	S10	S11
Case (a)	0:170	0:270	0:166	0:166	0:143
Case (b)	0:5100	0:8100	0:4980	0:4980	0:4290

Table 9.8, over 24 h and through a single campaign of 720 h, in a cycle of 24 h, for cases (a) and (b), respectively.

In this example, different stop criteria were used. In case (a) 5% of the optimality gap and 5000 s, and for case (b) 5% of the optimality gap and 8000 s.

Case a. Nonperiodic mode operation:
An approximated efficient frontier is characterized in Figure 9.10; points A, B, C, D, and E characterize the topology change, resulting from the production increase. The characteristics of each topology are shown, respectively, in Tables 9.9 and 9.10. The optimal scheduling is characterized for the last topology change (E), as shown in Figure 9.11.

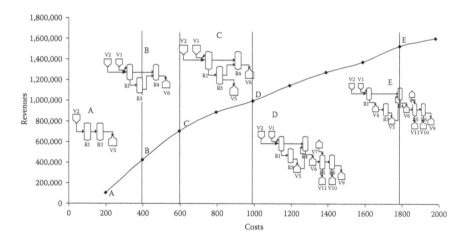

FIGURE 9.10   Approximation of the efficient frontier, case (a).

TABLE 9.9
Quantities Produced for Each Final Product, Case (a)

Final Products	A	B	C	D	E	F
S5	76.2	–	51.72	170	170	170
S6	–	155.5	258.6	270	270	270
S9	–	–	–	7.6	145.1	166
S10	–	–	–	7.6	145.1	166
S11	–	–	–	6.5	124.4	142

TABLE 9.10

The Design for the Main Equipment, Case (a)

Equipment	A	B	C	D	E	F
R1	76.2	93.3	103.4	141.2	120.3	124.4
R3	76.2	62.2	103.4	141.2	180.5	124.4
R4	–	155.5	129.3	140.5	159.1	169.4
R5	–	–	–	21.8	138.2	118.6
R6	–	–	–	15.3	96.8	83
V4	–	–	–	–	120.3	169.4
V5	76.2	–	51.72	170	170	170
V6	–	155.5	258.6	270	270	449
V9/V10	–	–	–	7.6	145.1	166
V11	–	–	–	6.5	124.4	143

As is shown in Figure 9.10, the topology between points A and B requires two reactors R1 and R3 with the same capacity (Table 9.10), which produce 76.5 tonnes of S5. The topology changes between B and C with an additional reactor, R4, whose capacity is 155.5 (Table 9.10) and produces 155.5 tonnes of S6 (Table 9.9). The topology between C and D is able to produce both products, S5 and S6, using one more tank. The following topologies (D, E) only differ in tank V4.

The scheduling shown in Figure 9.11 characterizes point E and shows the multipurpose characteristic of reactor R1, by performing T1 and T2. The remaining processing equipment is dedicated. The model requires 612 binary in 2582 variables and 4378 equations, and reached the optimality gap of 5%.

Case b. Periodic mode operation:
The periodic mode approximate efficient frontier is shown in Figure 9.12, where several topologies are characterized—A, B, F, and G—followed by the equipment design and final products quantities characterization in Tables 9.11 and 9.12, respectively.

The topology obtained at point A requires three reactors, R1, R3, and R4 and processes 170 tons of S5 and 270 tons of S6 (Table 9.11). The topology obtained in

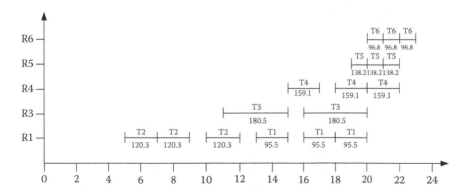

FIGURE 9.11 Scheduling for the plan topology assigned by E, case (a).

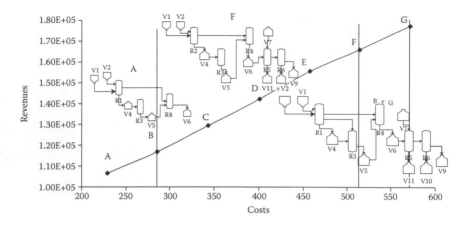

FIGURE 9.12   Approximation to the efficient frontier Example 9.3, case (b).

TABLE 9.11

Quantities Produced for Each Final Product, Case (b)

Final Products	A	B	C	D	E	F	G
S5	170	170	170	170	170	170	170
S6	270	270	270	270	270	270	270
S9/S11	–	24.15	52.97	83.4	115.12	139.56	166
S11	–	20.70	45.41	71.5	98.67	119.63	142.3

point B is similar to G topology, only differing in the design capacities (Table 9.12). All the topologies have the ability to produce all the products, except the topology defined in A (Table 9.11).

The scheduling characterized in Figure 9.13 (point G) highlights the periodic operation mode by overlapping two cycles with tasks T3 and T4.

This model requires 612 binary in 2609 variables and 4691 constraints and reached the optimality gap of 5%.

A comparison of the two operating modes shows an increase of complexity associated with the periodic operation mode, not only in the number of variables

TABLE 9.12

Design for the Main Equipment, Case (b)

Equipment	A	B	C	D	E	F	G
R1	71.75	72.95	77.07	65.13	68.76	–	124.3
R2	–	–	–	–	–	71.55	–
R3	69.50	58.36	61.65	54.28	68.76	59.63	103.52
R4	90.00	101.50	115.23	97.28	108.61	117.34	126.79
R5	–	34.50	50.45	79.42	82.23	99.69	118.57
R6	–	24.15	35.32	55.59	57.56	69.78	83

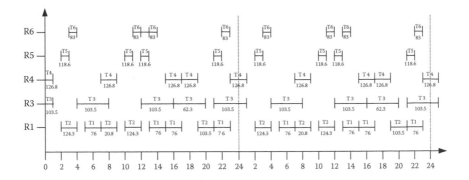

FIGURE 9.13   Scheduling for the plant topology at point G, case (b).

and constraints but also in computational time. The scheduling using the cycle concept takes advantage of allowing tasks overlapping over cycle boundaries for increasing their productivity. This is only possible if the facility is facing a stable demand period.

9.6   CONCLUSION

The work describes an analysis of the trade-off between capital and products costs, revenues, and operational flexibility for the scheduling and design of multipurpose batch facilities. The topology, equipment design, scheduling, and storage policies of multipurpose batch plants are addressed, considering a bi-objective approach, production maximization, and cost minimization; the multi-objective method used was the ε-constraint.

The efficient frontier characterizes the nondominated solutions, allowing the identification of a range of topologies, design, and storage policies that minimize the total cost of the system, while maximizing production, subject to total product demands and operational restrictions. The proposed methodology allows the decision makers to evaluate the relationship between revenue and cost of given batch facilities, thus enabling them to develop an adequate business strategy.

Since more realistic problems are associated with a higher complexity, higher computation requirements and more efficient solution tools are necessary to solve these cases. The implementation of decomposition approaches or meta-heuristics-based solutions should definitely be explored.

REFERENCES

Barbosa-Povoa, A. P. 1994. *Detailed design and retrofit of multipurpose batch plants*. PhD Thesis, University of London.

Barbosa-Povoa, A. P. 2007. A critical review on the design and retrofit of batch plants. *Computers & Chemical Engineering* 31(7): 833–855.

Barbosa-Povoa, A. P. and S. Macchietto. 1994. Detailed design of multipurpose batch plants. *Computers & Chemical Engineering* 18(11–12): 1013–1042.

Castro, P. M., A. P. Barbosa-Povoa, and A. Q. Novais. 2005. Simultaneous design and scheduling of multipurpose plants using resource task network based continuous-time formulations. *Industrial & Engineering Chemistry Research* 44(2): 343–357.

Chen, C.-L. and C.-Y. Chang. 2009. A resource-task network approach for optimal short-term/ periodic scheduling and heat integration in multipurpose batch plants. *Applied Thermal Engineering* 29(5–6): 1195–1208.

Chibeles-Martins, N., T. Pinto-Varela, A. P. Barbosa-Povoa, and A. Q. Novais. 2011. A simulated annealing approach for the biobjective design and scheduling of multipurpose batch plants. *21st European Symposium on Computer Aided Process Engineering*, Vol. 29, pp. 865–869.

Deb, K. 2004. *Multi-Objective Optimization Using Evolutionary Algorithms*. West Sussex: John Wiley & Sons.

Dedieu, S., L. Pibouleau, C. Azzaro-Pantel, and S. Domenech. 2003. Design and retrofit of multiobjective batch plants via a multicriteria genetic algorithm. *Computers & Chemical Engineering* 27(12): 1723–1740.

Dietz, A., A. Aguilar-Lasserre, C. Azzaro-Pantel, L. Pibouleau, and S. Domenech. 2008. A fuzzy multiobjective algorithm for multiproduct batch plant: Application to protein production. *Computers & Chemical Engineering* 32: 292–306.

Dietz, A., C. Azzaro-Pantel, L. Pibouleau, and S. Domenech. 2006. Multiobjective optimization for multiproduct batch plant design under economic and environmental considerations. *Computers & Chemical Engineering* 30(4): 599–613.

Fuchino, T., M. Muraki, and T. Hayakawa. 1994. Scheduling method in design of multipurpose batch plants with constrained resources. *Journal of Chemical Engineering of Japan* 27(3): 363–368.

Fuchino, T., M. Muraki, and T. Hayakawa. 1995. Design of multipurpose batch plants with multiple production plans based on cyclic production. *Journal of Chemical Engineering of Japan* 28(5): 541–550.

Haimes, Y. Y., L. S. Lasdon, and D. A. Wismer. 1971. On a bicriterion formulation of the problems of integrated system identification and system optimization. *IEEE Transactions on Systems and Cybernetics* 1(13): 296–297.

Heo, S. K., K. H. Lee, H. K. Lee, I. B. Lee, and J. H. Park. 2003. A new algorithm for cyclic scheduling and design of multipurpose batch plants. *Industrial & Engineering Chemistry Research* 42(4): 836–846.

Keeney, R. L. and H. Raiffa. 1976. *Decisions with Multiple Objectives: Preferences and Value Tradeoffs*. New York: Wiley.

Miettinen, K. 1999. *Nonlinear Multiobjective Optimization*. Boston: Kluwer.

Mosat, A., U. Fischer, and K. Hungerbuhler. 2007. Multiobjective batch process design aiming at robust performances. *Chemical Engineering Science* 62(21): 6015–6031.

Nonyane, D. R. and T. Majozi. 2012. Long term scheduling technique for wastewater minimisation in multipurpose batch processes. *Applied Mathematical Modelling* 36(5): 2142–2168.

Pantelides, C. C. 1994. Unified frameworks for optimal process planning and scheduling. In: Rippin, D.W.T. and Hale, J., Editors, *Proc. Second Conf. on Foundations of Computer Aided Operations*, New York: Cache Publications, pp. 253–274.

Pinto, T., A. Barbosa-Povoa, and A. Q. Novais. 2005. Optimal design and retrofit of batch plants with a periodic mode of operation. *Computers & Chemical Engineering* 29(6): 1293–1303.

Pinto, T., A. Barbosa-Povoa, and A. Q. Novais. 2008. Design of multipurpose batch plants: A comparative analysis between the STN, m-STN, and RTN representations and formulations. *Industrial & Engineering Chemistry Research* 47(16): 6025–6044.

Pinto-Varela, T., A. P. F. D. Barbosa-Povoa, and A. Q. Novais. 2010. Fuzzy-like optimization approach for design and scheduling of multipurpose non-periodic facilities. *20th European Symposium on Computer Aided Process Engineering*, Vol. 28, pp. 937–942.

Seid, E. R. and T. Majozi. 2013. Design and synthesis of multipurpose batch plants using a robust scheduling platform. *Industrial and Engineering Chemistry Research* 52(46): 16301–16313.

Shah, N., C. C. Pantelides, and R. W. H. Sargent. 1993. Optimal periodic scheduling of multipurpose batch plant. *Annals of Operations Research* 42: 193–228.

Steuer, R. E. 1986. *Multiple Criteria Optimization: Theory, Computation and Application.* Malabar, FL: Robert E. Krieger Publishing Company, Inc.

Voudouris, V. T. and I. E. Grossmann. 1996. MILP model for scheduling and design of a special class of multipurpose batch plants. *Computers & Chemical Engineering* 20(11): 1335–1360.

Part III

Resource Conservation

10 Flexibility Analyses and Their Applications in Solar-Driven Membrane Distillation Desalination System Designs

Vincentius Surya Kurnia Adi and Chuei-Tin Chang

CONTENTS

Owing to the rapidly growing world population and the alarming effects of global warming, there appears to be an ever-increasing demand for freshwater almost everywhere. For this reason, considerable effort has been devoted in recent years to develop an efficient and sustainable desalination technology. Among various alternatives, the *air gap membrane distillation* (AGMD) is widely considered as a promising candidate since the energy consumed per unit of water generated by this method is the lowest (Cabassud and Wirth, 2003; Ben Bacha et al., 2007; Bui et al., 2010). Many researchers have already constructed rigorous mathematical models to simulate and analyze the underlying transport phenomena so as to identify the key variables affecting the water flux in an AGMD module (Koschikowski et al., 2003; Meindersma et al., 2006; Ben Bacha et al., 2007; Chang et al., 2010). Particularly, Ben Bacha et al. (2007) and Chang et al. (2010, 2012) have built models of all units embedded in a *solar-driven membrane distillation desalination system* (SMDDS), that is, (1) the solar absorber, (2) the thermal storage tank, (3) the counter-flow shell-and-tube heat exchanger, (4) the AGMD modules, and (5) the distillate tank, and then discussed various operational and control issues accordingly. The process flow diagram of a typical SMDDS design can be found in Figure 10.1. Gálvez et al. (2009) meanwhile designed a 50 m³/day desalination setup with an innovative solar-powered membrane, and Guillen-Burrieza et al. (2011) also assembled a solar-driven AGMD pilot plant. These two studies were performed with the common goal of minimizing energy consumption per unit of distillate produced. Note that SMDDS should be operated in *batch mode* since the solar energy can only be supplied intermittently and periodically. Furthermore, the freshwater demand of SMDDS is assumed to be time variant, thus the traditional continuous operation is almost out of the question in this study.

Obviously, a good SMDDS design should be not only cost optimal but also operable in a realistic environment. Although successful applications of the solar-driven

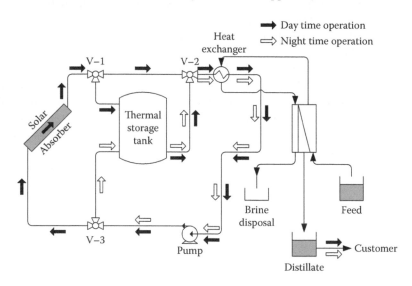

FIGURE 10.1   A typical SMDDS design—Configuration I.

AGMD modules were reported in the literature, these works focused primarily on thermal efficiency while the important issues concerning *operational flexibility* have not been addressed rigorously. Notice that the term "flexibility" is generally regarded as a system's capability of ensuring feasible operation over a specified region in the uncertain parameter space (Halemane and Grossmann, 1983). The sources of uncertainties may be either external or internal. Typical examples of the former case can be changes in throughput, feed quality, product demand, ambient conditions, and so forth, whereas the latter uncertainties are often associated with equipment deterioration, such as exchanger fouling and catalyst deactivation. Traditionally, the operational flexibility of a process is ensured in an ad hoc fashion by choosing conservative operating conditions, applying empirical overdesign factors, and introducing spare units. The major disadvantages of this approach can be summarized as follows:

i. Since the interactions among units are not considered, the actual flexibility level of the entire process cannot be accurately determined.
ii. Since the economic penalties of the heuristic design practices are not evaluated, their financial implications cannot be properly assessed.

A number of mathematical programming models have already been developed to facilitate quantitative flexibility analysis so as to provide the designers with the capabilities to (1) determine the performance index of any system design in relation to the expected operational requirements, (2) identify the bottleneck conditions which limit the flexibility in a design, and (3) compare alternative designs on an objective basis (Swaney and Grossmann, 1985a). Three performance measures for the steady-state, dynamic, and temporal flexibilities are discussed in this chapter. The first index is used primarily for gauging the continuous processes (Pistikopoulos and Grossmann, 1988a,b; 1989a,b; Petracci, et al., 1996), whereas the second is for the dynamic systems (Dimitriadis and Pistikopoulos, 1995). By considering the cumulative effects of uncertain disturbances over time, the programming model for computing the temporal flexibility index can be constructed by modifying its dynamic counterpart (Adi and Chang, 2013). Finally, note that the flexible SMDDS designs are identified in the present work on the basis of this last version of flexibility analysis.

10.1 STEADY-STATE FLEXIBILITY ANALYSIS

Design and control decisions are usually made in two consecutive steps over the life cycle of a *continuous* chemical process. In the design phase, the "optimal" operating conditions and the corresponding material- and energy-balance data are determined traditionally on the basis of economic considerations. Since it is often desirable to address the operability issues at the earliest possible stage, the systematic incorporation of flexibility analysis in process synthesis and design has received considerable attention in recent years (Grossmann and Halemane, 1982; Halemane and Grossmann, 1983; Swaney and Grossmann, 1985a,b; Dimitriadis and Pistikopoulos, 1995; Bansal et al., 2000, 2002; Floudas et al., 2001). As mentioned before, these so-called uncertainties may arise either from random exogenous disturbances (such as those in feed qualities, product demands, and environmental conditions, etc.) or from uncharacterizable variations in the internal parameters (such as heat transfer coefficients, reaction rate constants, and other physical properties) (Malcolm et al., 2007; Lima and Georgakis, 2008; Lima et al., 2010a,b). The ability of a chemical process to maintain feasible operation despite these uncertain deviations from the nominal states was referred to as its *operational flexibility*. The so-called *flexibility index* (FI_s) was first proposed by Swaney and Grossmann (1985a,b) to provide a quantitative measure of the feasible region in the parameter space. More specifically, FI_s can be associated with the maximum allowable deviations of the uncertain parameters from their nominal values, by which feasible operation can be assured with the proper manipulation of the control variables. The aforementioned authors also showed that, under certain convexity assumptions, critical points that limit feasibility and/or flexibility must lie on the vertices of the uncertain parameter space. Grossmann and Floudas (1987) later exploited the fact that sets of active constraints are responsible for limiting the flexibility of a design and developed a mixed integer nonlinear programming (MINLP) model accordingly. To this end, various approaches to facilitate flexibility analysis have been proposed in numerous studies published in the literature (Grossmann and Halemane, 1982; Halemane and Grossmann, 1983; Swaney and Grossmann, 1985a,b; Grossmann and Floudas, 1987; Varvarezos et al., 1995; Bansal et al., 2000, 2002; Floudas et al., 2001; Ostrovski et al., 2001; Ostrovski and Volin, 2002; Volin and Ostrovski 2002; Malcolm et al., 2007; Lima and Georgakis, 2008; Lima et al., 2010a,b). Similar flexibility analysis was also carried out in a series of subsequent studies to produce resilient grassroots and revamp designs (Chang et al., 2009; Riyanto and Chang, 2010). Since the steady-state material-and-energy balances are used as the equality constraints in the aforementioned MINLP model (Swaney and Grossmann, 1985a,b; Grossmann and Floudas, 1987; Varvarezos et al., 1995; Ostrovski et al., 2001; Ostrovski and Volin, 2002; Volin and Ostrovski, 2002), this original index can be viewed as a performance indicator of the *continuous* process under consideration (Pistikopoulos and Grossmann, 1988a,b, 1989a,b; Petracci, et al., 1996), and it is referred to as the *steady-state flexibility index* in this chapter.

10.1.1   MODEL FORMULATION

As mentioned previously, the steady-state flexibility index was defined by Swaney and Grossmann (1985a,b) as an overall measure of the allowable variations in all uncertain parameters. The basic framework of the flexibility index model (Biegler et al., 1997) for computing such an index is outlined in the sequel.

For illustration clarity, let us first introduce two label sets:

$$\mathbb{I} = \{i \mid i \text{ is the label of an equality constraint}\} \tag{10.1}$$

$$\mathbb{J} = \{j \mid j \text{ is the label of an inequality constraint}\} \tag{10.2}$$

The general design model can be expressed accordingly as

$$h_i(\mathbf{d}, \mathbf{z}, \mathbf{x}, \theta) = 0 \quad \forall i \in \mathbb{I} \tag{10.3}$$

$$g_i(\mathbf{d},\mathbf{z},\mathbf{x},\theta) \leq 0, \quad \forall j \in \mathbb{J} \tag{10.4}$$

where h_i is the ith equality constraint in the design model (e.g., the mass or energy balance equation for a processing unit); g_j is the jth inequality constraint (e.g., a capacity limit); $\mathbf{d}$ represents a vector in which all design variables are stored; $\mathbf{z}$ denotes the vector of adjustable controlling variables; $\mathbf{x}$ is the vector of state variables; θ denotes the vector of uncertain parameters.

The following mathematical program can be utilized to determine the so-called *feasibility function* $\psi(\mathbf{d},\theta)$, that is,

$$\psi(\mathbf{d},\theta) = \min_{\mathbf{x},\mathbf{z}} \max_{j \in \mathbb{J}} g_j(\mathbf{d},\mathbf{z},\mathbf{x},\theta) \tag{10.5}$$

subject to the equality constraints given in Equation 10.3. Notice that this formulation means that, for a *fixed* design defined by $\mathbf{d}$ and the *fixed* parameters given in θ, the largest g_j ($\forall j \in \mathbb{J}$) is minimized by adjusting the control variables $\mathbf{z}$ while keeping $h_i = 0$ ($\forall i \in \mathbb{I}$). If $\psi(\mathbf{d},\theta) \leq 0$, then the given system is operable (see Figure 10.2).

On the other hand, the above optimization problem can be posed alternatively by introducing an extra scalar variable u, that is,

$$\psi(\mathbf{d},\theta) = \min_{\mathbf{x},\mathbf{z},u} u \tag{10.6}$$

subject to Equation 10.3 and

$$g_j(\mathbf{d},\mathbf{z},\mathbf{x},\theta) \leq u \quad \forall j \in \mathbb{J} \tag{10.7}$$

Notice also that if $\psi(\mathbf{d},\theta) = 0$, then at least one of the inequality constraints should be active, that is, $g_j = 0$ ($\exists j \in \mathbb{J}$).

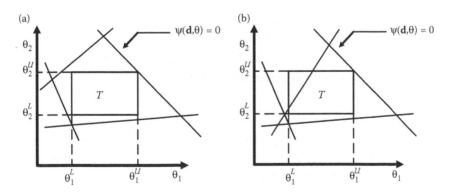

FIGURE 10.2 Feasible and infeasible designs in the parameter space. (Adapted from Biegler, L. T.; Grossmann I. E.; Westerberg, A. W. *Systematic Methods of Chemical Process Design.* Englewood Cliffs, New Jersey: Prentice-Hall, 690, 1997.)

Since the aforementioned test can be performed only on deterministic models with constant θ, it is still necessary to develop an improved feasibility criterion by considering all possible values of the uncertain parameters. To this end, let us first define a feasible region $\mathbf{T}$ in the parameter space, that is,

$$\mathbf{T} = \left\{ \theta | \theta^N - \Delta\theta^- \le \theta \le \theta^N + \Delta\theta^+ \right\} \tag{10.8}$$

where θ^N denotes a vector of given nominal parameter values and $\Delta\theta^+$ and $\Delta\theta^-$ represent vectors of the expected deviations in the positive and negative directions, respectively. Hence, an additional optimization problem can be formulated to facilitate this modified test

$$\chi(\mathbf{d}) = \max_{\theta \in \mathbf{T}} \psi(\mathbf{d}, \theta) \tag{10.9}$$

where $\chi(\mathbf{d})$ denotes the feasibility function of a *fixed* design defined by $\mathbf{d}$ over the entire feasible region $\mathbf{T}$. The given system should therefore be feasible if $\chi(\mathbf{d}) \le 0$, while infeasible if otherwise.

To develop a unified measure of the maximum tolerable range of variation in every uncertain parameter (Swaney and Grossmann, 1985a,b), the feasible region $\mathbf{T}$ is modified by introducing another scalar variable

$$\mathbf{T}(\delta) = \left\{ \theta | \theta^N - \delta\Delta\theta^- \le \theta \le \theta^N + \delta\Delta\theta^+ \right\} \tag{10.10}$$

where δ is a scalar variable to be determined by solving the flexibility index model given below

$$FI_s = \max \delta \tag{10.11}$$

subject to

$$\chi(\mathbf{d}) \le 0 \tag{10.12}$$

Note that the maximized objective value FI_s is the steady-state *flexibility index*, which represents the largest value of δ that guarantees $g_j \le 0$ ($\forall j \in \mathbb{J}$), that is, $\chi(\mathbf{d}) \le 0$, in the parameter space (see Figure 10.3).

10.1.2   SOLUTION STRATEGIES

Several effective strategies are available for solving the optimization problem defined by Equations 10.11 and 10.12. Two of them, that is, the active set method and the vertex method, are described in the sequel.

10.1.2.1   Active Set Method

Solving the flexibility index model is in general very difficult because Equations 10.11 and 10.12 represent a nonlinear, non-differentiable, multilevel optimization

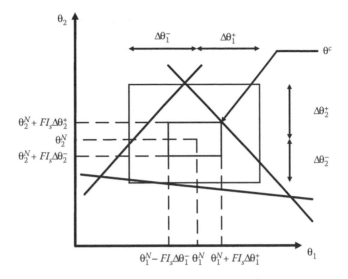

FIGURE 10.3 Geometrical interpretation of steady-state flexibility index. (Adapted from Biegler, L. T.; Grossmann I. E.; Westerberg, A. W. *Systematic Methods of Chemical Process Design*. Englewood Cliffs, New Jersey: *Prentice-Hall*, 690, 1997.)

problem. Grossmann and Floudas (1987) developed a solution strategy on the basis of the Karush–Kuhn–Tucker (KKT) conditions of the optimization problem for computing the function $\psi(\mathbf{d},\theta)$, that is, Equations 10.3, 10.6, and 10.7. To be able to apply these conditions, the aforementioned flexibility index model is first reformulated by imposing an extra equality constraint that keeps the feasibility function zero, that is,

$$FI_s = \min \delta \tag{10.13}$$

subject to

$$\psi(\mathbf{d},\theta) = 0 \tag{10.14}$$

Notice that the original maximization problem, that is, Equations 10.11 and 10.12, is now replaced with the present *minimization* problem. This is owing to the fact that if the chosen value of δ is not the smallest, at least one inequality constraint must be violated, that is, $g_j > 0 \ (\exists j \in \mathbb{J})$. Since Equations 10.3 and 10.4 are inherently satisfied in the optimization problem defined by Equations 10.3, 10.6, and 10.7, the corresponding KKT conditions should be applicable. Consequently, the flexibility index problem in Equations 10.13 and 10.14 can be written more explicitly as follows (Grossman and Floudas, 1987):

$$FI_s = \min_{\delta,\mu_i,\lambda_j,s_j,y_j,x_i,z_k} \delta \tag{10.15}$$

subject to the constraints in Equation 10.3 and also those presented below

$$g_i(\mathbf{d}, \mathbf{z}, \mathbf{x}, \theta) + s_j = 0, \quad j \in \mathbb{J} \tag{10.16}$$

$$\sum_{i \in \mathbb{I}} \mu_i \frac{\partial h_i}{\partial \mathbf{z}} + \sum_{j \in \mathbb{J}} \lambda_j \frac{\partial g_j}{\partial \mathbf{z}} = 0 \tag{10.17}$$

$$\sum_{i \in \mathbb{I}} \mu_i \frac{\partial h_i}{\partial \mathbf{x}} + \sum_{j \in \mathbb{J}} \lambda_j \frac{\partial g_j}{\partial \mathbf{x}} = 0 \tag{10.18}$$

$$\sum_{j \in \mathbb{J}} \lambda_j = 1 \tag{10.19}$$

$$\lambda_j - y_j \leq 0, \quad j \in \mathbb{J} \tag{10.20}$$

$$s_j - Q(1 - y_j) \leq 0, \quad j \in \mathbb{J} \tag{10.21}$$

$$\sum_{j \in \mathbb{J}} y_j = n_z + 1 \tag{10.22}$$

$$\theta^N - \delta \Delta \theta^- \leq \theta \leq \theta^N + \delta \Delta \theta^+ \tag{10.23}$$

$$y_j = \{0,1\}, \lambda_j \geq 0, s_j \geq 0, \quad j \in \mathbb{J} \tag{10.24}$$

$$\delta \geq 0 \tag{10.25}$$

where s_j is the slack variable for the jth inequality constraint; Q denotes a large enough positive number to be used as the upper bound of s_j; μ_i denotes the Lagrange multiplier of equality constraint h_i; λ_j is the Lagrange multiplier of inequality constraint g_j; y_j denotes the binary variable reflecting whether the corresponding inequality constraint is active, that is, $g_j = 0$ if $y_j = 1$, whereas $g_j < 0$ if $y_j < 0$; and n_z is the total number of independent controlling variables.

10.1.2.2 Vertex Method

The optimization procedure for Equations 10.11 and 10.12 can be greatly simplified if the optimal solution is always associated with one of the vertices of the feasible region $\mathbf{T}(\delta)$ (Halemane and Grossmann, 1983). Let $\Delta \theta^k$ ($\forall k \in V$) denote the kth vertex and V be the set of all vertices. Then the maximum value of δ along $\Delta \theta^k$ can be evaluated according to the following programming model:

$$\delta^k = \max_{\mathbf{x}, \mathbf{z}, \delta} \delta \tag{10.26}$$

subject to Equations 10.3, 10.4, and

$$\theta = \theta^N + \delta\Delta\theta^k \qquad (10.27)$$

Among the corresponding parameter polyhedrons, that is, $\mathbf{T}(\delta^k), \forall k \in V$, it is clear that only the smallest one can be totally inscribed within the feasible region. Hence,

$$FI_s = \min_{k \in V}\left\{\delta^k\right\} \qquad (10.28)$$

Thus, the following simple procedure applies:

Step 1: Solve the optimization problem defined by Equations 10.3, 10.4, 10.26, and 10.27 for each vertex $k \in V$.
Step 2: Select FI_s according to Equation 10.28.

Swaney and Grossmann (1985a,b) showed that, only under certain convexity conditions for the constraints in Equations 10.3 and 10.4, the optimal solution is guaranteed to be associated with one of the vertices. However, even when these conditions are not met, it can often be found that this approach is still applicable. Note also that the vertex method may be quite impractical in realistic case studies due to dimension explosion. For example, $2^{10} = 1024$ optimization runs that are needed for 10 uncertain parameters and, if the number of parameters is raised to 20, the computation load for the required $2^{20} = 1,048,576$ runs can be overwhelming.

The implementation steps of steady-state flexibility analysis in water network designs can be found in Liou (2006). Owing to the special model structure for water network designs, Li and Chang (2011) suggested that a simplified version of the vertex method could be applied by checking only a single corner of the parameter space. This critical point should be associated with the upper or lower limit of each uncertain parameter on the basis of physical insights (Chang et al., 2009). Specifically, they are located at

- The upper bounds of the mass loads of water using units and the pollutant concentrations at the primary and secondary sources.
- The lower bounds of the removal ratios of wastewater treatment units, the allowed maximum inlet and outlet pollutant concentrations of water using units, and the allowed maximum inlet pollutant concentration of wastewater treatment units.

The flexibility index of a water network can thus be determined on the basis of this most constrained point alone.

10.2 DYNAMIC FLEXIBILITY ANALYSIS

As suggested by Dimitriadis and Pistikopoulos (1995), the operational flexibility of a *dynamic* system should be evaluated differently. By adopting a system of differential algebraic equations (DAEs) as the model constraints, these authors developed a mathematical programming formulation for dynamic flexibility analysis. Clearly, this analysis is more rigorous than that based on the steady-state model since, even for a continuous process, the operational flexibility cannot be adequately characterized without accounting for the control dynamics. In an earlier study, Brengel and Seider (1992) advocated the need for design and control integration. The integration of flexibility and controllability in design was discussed extensively by several other groups (Chacon-Mondragon and Himmelblau, 1996; Mohideen et al., 1996; Bahri et al., 1997; Bansal et al., 1998; Georgiadis and Pistikopoulos, 1999; Aziz and Mujtaba, 2002; Malcolm et al., 2007). Soroush and Kravaris (1993a,b) addressed various issues concerning flexible operation for batch reactors. The effects of uncertain disturbances on the wastewater neutralization processes were also studied by Walsh and Perkins (1994). White et al. (1996) presented an evaluation method to assess the switchability of any given system, that is, its ability to perform satisfactorily when moving between different operating points. Dimitriadis et al. (1997) studied the feasibility problem from the safety verification point of view. Zhou et al. (2009) utilized a similar approach to assess the operational flexibility of batch systems.

10.2.1 Model Formulation

In the dynamic flexibility analysis, the equality constraints in Equation 10.3 are replaced with a system of differential-algebraic equations (Dimitriadis and Pistikopoulos, 1995), that is,

$$h_i\left(\mathbf{d},\mathbf{z}(t),\mathbf{x}(t),\dot{\mathbf{x}}(t),\theta(t)\right) = 0 \quad \forall i \in \mathbb{I} \tag{10.29}$$

where $t \in [0, H], i \in \mathbb{I}$, and $\mathbf{x}(0) = \mathbf{x}^0$. On the other hand, the inequality constraints in Equation 10.4 should also be time dependent, that is,

$$g_j\left(\mathbf{d},\mathbf{z}(t),\mathbf{x}(t),\theta(t)\right) \leq 0 \quad \forall j \in \mathbb{J} \tag{10.30}$$

Finally, the uncertain parameters and their upper and lower limits in this case should be functions of time and Equation 10.10 can be modified as

$$\theta^N(t) - \delta\Delta\theta^-(t) \leq \theta(t) \leq \theta^N(t) + \delta\Delta\theta^+(t) \tag{10.31}$$

Thus, the corresponding *dynamic flexibility index* FI_d can be computed with the following model:

$$FI_d = \max \delta \tag{10.32}$$

subject to Equation 10.29 and

$$\max_{\theta(t)} \min_{\mathbf{z}(t)} \max_{j,t} g_j(\mathbf{d},\mathbf{z}(t),\mathbf{x}(t),\theta(t)) \leq 0 \tag{10.33}$$

Note that this model is essentially the dynamic version of Equations 10.11 and 10.12.

10.2.2 Solution Strategies

Two alternative strategies are presented in this section for computing the dynamic flexibility index. First of all, Equations 10.29, 10.32, and 10.33 can obviously be transformed into a steady-state flexibility index model by approximating the embedded differential equations with a set of algebraic equations. Another viable option is to establish the KKT conditions for the minimum feasibility functional and then develop the dynamic version of the active set method. These two approaches are outlined below.

10.2.2.1 Transformation to Steady-State FI Model through Discretization

The simplest strategy to compute the dynamic flexibility index is to convert the DAEs in Equation 10.29 into a system of algebraic equations on the basis of a credible numerical discretization technique. Since the resulting optimization problems should be identical to those described in Section 10.1.1, they can be solved with any existing algorithm for the steady-state flexibility analysis (see Section 10.1.2).

Two popular discretization techniques have been utilized in the past, that is, the differential quadrature (DQ) method (Bellman et al., 1971, 1972) and the orthogonal collocation (OC) method (Biegler, 1984; Cuthrell and Biegler, 1987). Since they were adopted for essentially the same purpose and both yield satisfactory results, let us consider only the former for the sake of brevity. Notice that the accuracy of DQ approximation has been well documented in the literature (e.g., Civan and Sliepcevich, 1984; Quan and Chang, 1989a,b; Chang et al., 1993), and its implementation procedure is also very straightforward. As pointed out by Shu (2000), DQ is essentially equivalent to the finite difference scheme of a higher order. To improve the computation efficiency when a large number of grid points are required, a localized DQ scheme was introduced by Zong and Lam (2002). An extensive discussion of differential quadrature and its state-of-the-art developments can be found in Zong and Zhang (2009).

To illustrate the DQ discretization principle, let us consider the first-order derivative of the ith state variable ($i \in I$) as an example

$$\left. \frac{dx_i(t)}{dt} \right|_{t=t_m^e} \cong \sum_{n=1}^{N_{node}} w_{mn} x_i(t_n^e) \tag{10.34}$$

where $m = 1, 2, \ldots, N_{node}$; $e = 1, 2, \ldots, N_{element}$; t_m^e and t_n^e, respectively, denote the locations of the mth node and the nth node in time element e; w_{mn} denotes the weighting coefficient associated with the state value at t_n^e for the derivative at t_m^e, which is dependent only upon the *predetermined* node spacing. As a result, every differential equation in Equation 10.29 can be approximated with a set of algebraic equations. In addition, all inequality constraints in Equations 10.30 and 10.31 should be discretized at the node locations, that is,

$$g_j\left(\mathbf{d}, \mathbf{z}(t_m^e), \mathbf{x}(t_m^e), \theta(t_m^e)\right) \leq 0 \quad \forall j \in \mathbb{J} \tag{10.35}$$

$$\theta^N(t_m^e) - \delta\Delta\theta^-(t_m^e) \leq \theta(t_m^e) \leq \theta^N(t_m^e) + \delta\Delta\theta^+(t_m^e) \qquad (10.36)$$

Quan and Chang (1989a,b) suggested that, in most cases, it is beneficial to use the shifted zeros of a standard Chebyshev polynomial as the selected nodes. This node spacing in an arbitrary interval $t \in [a, b]$ yields the following formulas for calculating the weighting coefficients, that is,

$$w_{mn} = \frac{r_{N_{node}} - r_1}{b - a} \frac{(-1)^{(m-n)}}{r_m - r_n} \sqrt{\frac{1 - r_n^2}{1 - r_m^2}}, \quad m \neq n \qquad (10.37)$$

$$w_{mm} = \frac{1}{2} \frac{r_{N_{node}} - r_1}{b - a} \frac{r_m}{(1 - r_m^2)} \qquad (10.38)$$

where $m, n = 1, 2, \ldots, N_{node}$ and the locations of Chebyshev zeros in the standard interval $[-1, +1]$ are

$$r_m = \cos\frac{(2m - 1)\pi}{2N_{node}} \qquad (10.39)$$

where w_{mn} are the weighting coefficients for the first-order derivatives. With these formulas, all weighting coefficients can be easily calculated for any combination of element length $b - a$ and node number N_{node}. A typical example can be found in Table 10.1.

As mentioned previously, the time horizon H is supposed to be divided into $N_{element}$ elements. Continuity of every state variable at the border point of each pair of adjacent elements can be enforced with a boundary condition, that is, $x_k(t_{N_{node}}^e) = x_k(t_1^{e+1})$ and $e = 1, 2, \ldots, N_{element}$. The initial conditions of Equation 10.29 should be imposed at the left end of first element, that is, $x_k(t_1^1) = x_k^0$, whereas the states at right end of the last element are not constrained, that is, $x_k(t_{N_{node}}^{N_{element}}) = $ free. Finally, the element number and lengths should be allowed to be adjusted to achieve satisfactory accuracy.

10.2.2.2   Identification of Dynamic KKT Conditions

If one chooses *not* to discretize Equations 10.29 through 10.31 in the original formulation, then the corresponding feasibility functional can be defined in the same way as its steady-state counterpart in Equations 10.3, 10.6, and 10.7

TABLE 10.1

Weighting Coefficients for $b - a = 10$ and $N_{node} = 5$

n \\ m	1	2	3	4	5
1	−0.947214	1.370820	−0.647214	0.323607	−0.100000
2	−0.200000	−0.085410	0.400000	−0.161803	0.047214
3	0.061803	−0.261803	0.000000	0.261803	−0.061803
4	−0.047214	0.161803	−0.400000	0.085410	0.200000
5	0.100000	−0.323607	0.647214	−1.370820	0.947214

$$\psi\big(\mathbf{d},\theta(t)\big) = \min_{\mathbf{x}(t),\mathbf{z}(t),u(t)} u(t)\Big|_{t=H} \tag{10.40}$$

subject to the constraints in Equation 10.29 and

$$\dot{u}(t) = 0 \tag{10.41}$$

$$g_j\big(\mathbf{d},\mathbf{z}(t),\mathbf{x}(t),\theta(t)\big) \le u(t) \quad \forall\, j \in \mathbb{J} \tag{10.42}$$

To facilitate derivation of the KKT conditions for this functional optimization problem, let us rewrite Equation 10.29 more explicitly as

$$\dot{\mathbf{x}}(t) = \mathbf{f}_1\big(\mathbf{d},\mathbf{z}(t),\mathbf{x}(t),\theta(t)\big) \tag{10.43}$$

$$\mathbf{f}_2\big(\mathbf{d},\mathbf{z}(t),\mathbf{x}(t),\theta(t)\big) = \mathbf{0} \tag{10.44}$$

An aggregated objective functional can be constructed by introducing Lagrange multipliers to incorporate all constraints, that is,

$$L = u(H) + \int_0^H \Big\{ \mu_u(t)\big[0 - \dot{u}\big] + \mu(t)^T\big[\mathbf{f}_1 - \dot{\mathbf{x}}\big] + \nu(t)^T \mathbf{f}_2 + \lambda(t)^T\big[\mathbf{g} - u\mathbf{1}\big]\Big\}dt \tag{10.45}$$

where $\mathbf{g} = [g_1\ g_2\ g_3\ ...]^T$, $\mathbf{1} = [1\,1\,1\,...]^T$, and the multipliers $\mu_u(t)$, $\mu(t)$, and $\nu(t)$ are real numbers while $\lambda(t) \ge \mathbf{0}$. By taking the first variation of L and setting it to zero, the following four sets of necessary conditions can be obtained:

   i. $\mathbf{x}(0) = \mathbf{x}_0$; $\mu(H) = \mathbf{0}$; $\mu_u(0) = 0$; $\mu_u(H) = 1$

   ii. $\dot{\mu} = -\mu^T\left(\dfrac{\partial \mathbf{f}_1}{\partial \mathbf{x}}\right) - \nu^T\left(\dfrac{\partial \mathbf{f}_2}{\partial \mathbf{x}}\right) - \lambda^T\left(\dfrac{\partial \mathbf{g}}{\partial \mathbf{x}}\right)$; $\dot{\mu}_u = \lambda^T \mathbf{1}$

   iii. $\dot{\mathbf{x}} = \mathbf{f}_1$; $\mathbf{f}_2 = \mathbf{0}$; $\dot{u} = 0$; $\lambda^T\big(\mathbf{g} - u\mathbf{1}\big) = 0$; $\lambda \ge \mathbf{0}$

   iv. $\mu^T\left(\dfrac{\partial \mathbf{f}_1}{\partial \mathbf{z}}\right) + \nu^T\left(\dfrac{\partial \mathbf{f}_2}{\partial \mathbf{z}}\right) + \lambda^T\left(\dfrac{\partial \mathbf{g}}{\partial \mathbf{z}}\right) = \mathbf{0}^T$

By following the same rationale in developing the active set method for computing FI_s, that is, Equations 10.13 and 10.14, it is necessary to set $u(t) = 0$ and change conditions in (iii) to

$$(\text{iii})'\ \dot{\mathbf{x}} = \mathbf{f}_1; \quad \mathbf{f}_2 = \mathbf{0}; \quad u = 0; \quad \lambda^T\mathbf{g} = 0; \quad \lambda \ge \mathbf{0}$$

Therefore, the dynamic flexibility index FI_d can be determined by minimizing δ while subject to the constraints in (i), (ii), (iii)′, (iv), and Equation 10.31.

10.3 TEMPORAL FLEXIBILITY ANALYSIS

As indicated previously, the nominal values of uncertain parameters and their anticipated positive and negative deviations are assumed in the dynamic flexibility analysis
to be available in advance at every instance over the entire time horizon of operation
life. However, while an ill-designed system may become inoperable due to instantaneous variations in some process parameters at certain instances, the cumulative
effects of temporary disturbances in finite time intervals can also result in serious
consequences. Although the latter scenario is usually ignored in dynamic flexibility
analysis, it is in fact a more probable event in practical applications. To address this
important issue, a mathematical programming model has been developed by Adi and
Chang (2013) for computing the corresponding performance measure, which was
referred to as the *temporal flexibility index*.

10.3.1 MODEL FORMULATION AND SOLUTION STRATEGIES

Let us assume that the variations in uncertain parameters are possible only within
a finite time interval $[t_0, t_1] \subset [0, H]$. To characterize the cumulative effects, let us
integrate Equation 10.31 over this finite interval, that is,

$$-\delta \int_{t_0}^{t_1} \Delta\theta(\tau)^- \, d\tau \leq \int_{t_0}^{t_1} (\theta(\tau) - \theta(\tau)^N) d\tau \leq \delta \int_{t_0}^{t_1} \Delta\theta(\tau)^+ \, d\tau \tag{10.46}$$

Since the expected maximum deviations in uncertain parameters should be
regarded as given information, the expected *net* positive and negative cumulated
deviations over interval $[t_0, t_1]$ can also be computed in advance. Let us introduce the
following definitions to simplify notation:

$$\Delta\Theta^- = \int_{t_0}^{t_1} \Delta\theta(\tau)^- \, d\tau \tag{10.47}$$

$$\Delta\Theta^+ = \int_{t_0}^{t_1} \Delta\theta(\tau)^+ \, d\tau \tag{10.48}$$

and

$$\Theta(t) = \int_{t_0}^{t} \theta((\tau) - \theta_s(\tau)^N) d\tau \tag{10.49}$$

Additional constraints can be imposed upon the accumulated effects of the instantaneous variations in the uncertain parameters so as to ensure operational safety.
Specifically, Equation 10.46 can be expressed explicitly as follows:

$$-\delta\Delta\Theta^- \leq \Theta(t) \leq \delta\Delta\Theta^+ \tag{10.50}$$

Furthermore, since the time interval $[t_0, t_1]$ itself may be uncertain, Equation 10.49 can be rewritten as

$$\frac{d}{dt}\Theta(t) = \theta(t) - \theta^N(t) \tag{10.51}$$

where $\Theta(0) = 0$ and $t \in [0, H]$. Equations 10.29, 10.33, 10.50, and 10.51 can then be used as the constraints of a mathematical programming model to determine the temporal flexibility index FI_t by maximizing the scalar variable δ. Notice also that, since the only difference between the mathematical models for computing FI_d and FI_t lies in the inequality constraints that bound the uncertain parameters, the solution approaches described in Section 10.2.2 should be very similar to those adopted in the present case. The detailed descriptions of these strategies are thus omitted for the sake of brevity. The theoretical implications of temporal flexibility index can be summarized as follows:

If $FI_t < 1$, then the given batch process cannot withstand at least some of the temporary disturbances that satisfy Equations 10.50 and 10.51. If otherwise, then the operation should always be successful.

10.4   FLEXIBLE SMDDS DESIGNS

A realistic SMDDS design is expected to be fully functional in the presence of uncertain sunlight radiation and fluctuating freshwater demand. In this section, the *temporal flexibility index* (FI_t) is adopted as the evaluation criterion of SMDDS designs. All units embedded in the typical design (see Configuration I in Figure 10.1) must be properly sized to achieve a target FI_t. In addition, alternative thermal storage schemes should be evaluated accordingly. The stripped-down version of SMDDS design in Figure 10.4 (Configuration II) can certainly be analyzed and compared with Configuration I, while installation of an extra thermal storage tank (say, on an additional bypass from solar absorber to heat exchanger and/or vice versa) can also be considered.

10.4.1   Simplified Mathematical Models

The essential SMDDS units, that is, the solar absorber, the thermal storage tank, the counter-flow shell-and-tube heat exchanger, the AGMD modules, and the distillate tank, are interconnected in a typical system to form two separate processing routes for seawater desalination and solar energy conversion, respectively. For implementation convenience, the unit models given in Chang et al. (2010) have been simplified (Adi and Chang, 2013) and outlined below:

10.4.1.1   Solar Absorber

The solar energy is converted into heat using the SMDDS solar absorber. Two basic assumptions are adopted in formulating its mathematical model: (i) the fluid

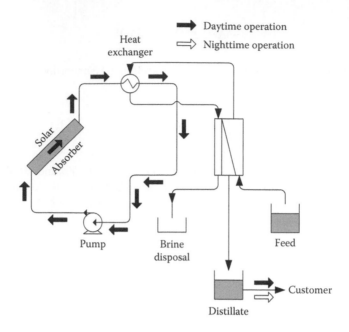

FIGURE 10.4   The stripped-down SMDDS design—Configuration II.

velocities in all absorber tubes are the same; (ii) the fluid temperature should be kept below 100°C; (iii) there is no water loss; and (iv) heat loss is negligible. The corresponding transient energy balance can be written as

$$\frac{dT_{f,SA_{\text{out}}}}{dt} = -L_{SA}\frac{\dot{m}_{f,SA}}{M_{f,SA}}\frac{T_{f,SA_{\text{out}}} - T_{f,SA_{\text{in}}}}{L_{SA}} + \frac{A_{SA}I(t)}{M_{f,SA}Cp_f^L} \tag{10.52}$$

$$T_{f,SA_{\text{out}}} \leq T_{f,SA_{\text{out}}}^{\text{max}} \tag{10.53}$$

where $T_{f,SA_{\text{in}}}$ and $T_{f,SA_{\text{out}}}$ denote the inlet and outlet temperatures (°C) of the solar absorber, respectively; $T_{f,SA_{\text{out}}}^{\text{max}}$ is the maximum allowable outlet temperature (100°C); $M_{f,SA}$ denotes the total mass of operating fluid in the solar absorber (kg); $\dot{m}_{f,SA}$ denotes the overall mass flow rate of operating fluid in solar absorber (kg/h); L_{SA} is the length of an absorber tube (m); A_{SA} is the exposed area of solar absorber (m²); Cp_f^L is the heat capacity of operating fluid (J/kg°C); $I(t)$ is the solar irradiation rate per unit area (W/m²).

10.4.1.2   Thermal Storage Tank

Based on the assumption that (i) the fluid within the thermal storage tank is well mixed, (ii) the inlet and outlet flow rates are identical, and (iii) the heat capacity of operating fluid is independent of temperature, the energy balance around thermal storage tank can be expressed as

$$M_{f,ST} \frac{dT_{f,ST_{out}}}{dt} = r_{f,ST} \dot{m}_{f,STL} \left(T_{f,ST_{in}} - T_{f,ST_{out}} \right) \tag{10.54}$$

$$r_{f,ST} = \frac{\dot{m}_{f,ST}}{\dot{m}_{f,STL}} \tag{10.55}$$

where $T_{f,ST_{in}}$ and $T_{f,ST_{out}}$ denote the inlet and outlet temperatures (°C), respectively; $M_{f,ST}$ represents the total mass of operating fluid in the thermal storage tank (kg); $\dot{m}_{f,STL}$ is the total mass flow rate driven by the pump in the thermal loop (kg/h); $\dot{m}_{f,ST}$ is the throughput of thermal storage tank (kg/h) which equals $r_{f,ST}\dot{m}_{f,STL}$.

In Configuration I, the thermal storage tank is employed during the daytime and nighttime according to the corresponding flow direction mode. In other words,

$$\dot{m}_{f,SA} = \begin{cases} m_{f,STL} & \text{if } I(t) > 0 \text{ (day time)} \\ 0 & \text{if } I(t) = 0 \text{ (night time)} \end{cases} \tag{10.56}$$

As a result, the flow ratio defined in Equation 10.55 can be treated as an adjustable control variable in daytime operation, that is, $0 \le r_{f,ST}(t) \le 1$, while $r_{f,ST}(= 1)$ is kept unchanged during nighttime.

Finally, in the case of the thermal storage tank that is not utilized, that is, Configuration II, there is really no need to distinguish the operation modes and thus one can simply fix $\dot{m}_{f,SA} = \dot{m}_{f,STL}$, and $r_{f,ST} = 0$.

10.4.1.3 Heat Exchanger

The hot fluid used in the counter-flow heat exchanger comes from the thermal storage tank and/or solar absorber, whereas the cold fluid is the seawater. The heat exchanger is assumed to be always in steady-state and there is no heat loss. Thus, the unit model of heat exchanger can be written as

$$\dot{m}_{f,MD} (T_{f,HX,CL_{out}} - T_{f,HX,CL_{in}}) = \dot{m}_{f,HX,HL} (T_{f,HX,HL_{in}} - T_{f,HX,HL_{out}}) \tag{10.57}$$

where $\dot{m}_{f,HX,HL}$ is the mass flow rate of hot fluid (kg/h); $T_{f,HX,HL_{in}}$ and $T_{f,HX,HL_{out}}$, respectively, denote the inlet and outlet temperatures of hot fluid (°C); $\dot{m}_{f,MD}$ is the mass flow rate of seawater in membrane distillation loop (kg/h); $T_{f,HX,CL_{in}}$ and $T_{f,HX,CL_{out}}$, respectively, denote the inlet and outlet temperatures of the cold fluid (°C). Note that the mass flow rate of hot fluid is essentially the same as that in the thermal loops in both Configurations I and II, that is,

$$\dot{m}_{f,HX,HL} = \dot{m}_{f,STL} \tag{10.58}$$

An energy balance around the valve V-2 yields

$$T_{f,HX,HL_{in}} = (1 - r_{f,ST})T_{f,SA_{out}} + r_{f,ST}T_{f,ST_{out}} \tag{10.59}$$

Again, this equation is also valid in the above two structures. Finally, let us consider the outlet temperature of hot fluid. According to the flow direction in Configuration I, the hot fluid leaving the heat exchanger is recycled back to the solar absorber in the daytime operation and to the thermal storage tank during nighttime, the following constraints should be imposed:

$$
T_{f,HX,HL_{\text{out}}} = \begin{cases} T_{f,SA_{\text{in}}} & \text{if } I(t) > 0 \text{ (day time)} \\ T_{f,ST_{\text{in}}} & \text{if } I(t) = 0 \text{ (night time)} \end{cases}
\tag{10.60}
$$

On the other hand, since Configuration II is not equipped with a thermal storage tank, only the first constraint in Equation 10.60 can be used in the corresponding model.

10.4.1.4   AGMD Module

A simplified model is adopted in this study for characterizing the AGMD unit. It is assumed that the mass flux of distillate across the membrane is a function of the rate of energy input. Specifically, this flux in a standard module can be expressed as

$$
N_{mem} \doteq \frac{m_{f,MD} Cp_f^L \left(T_{f,HX,CL_{\text{out}}} - T_{f,HX,CL_{\text{in}}}\right)}{STEC \cdot A_{MD} \cdot n_{AGMD}}
\tag{10.61}
$$

where N_{mem} denotes the distillate flux (kg/m² · h); A_{MD} is the fixed membrane area of a standard AGMD module (i.e., 10 m²); n_{AGMD} is the total number of standard modules; and STEC is the *specific thermal energy consumption* constant (kJ/kg), which can be considered as the ratio between energy supplied by the heat exchanger and mass of the distillate produced (Burgess and Lovegrove, 2005; Banat et al., 2007).

The mass flux through the AGMD membrane is driven primarily by the vapor pressure differential. However, to simplify calculations the corresponding flux is assumed here to be roughly proportional to the temperature difference. Since Equation 10.61 is used essentially as an empirical relation in this case, it should be only valid within a finite range of the seawater flow rate. Consequently, $\dot{m}_{f,MD}$ is treated in this work as an adjustable control variable, which is allowed to vary ±10% from its nominal value

$$
0.9 \dot{m}_{f,MD}^N \leq \dot{m}_{f,MD} \leq 1.1 \dot{m}_{f,MD}^N
\tag{10.62}
$$

Finally, note that the seawater entering the AGMD module should not be allowed to exceed a specified upper bound so as to avoid damaging the membrane, that is,

$$
T_{f,HX,CL_{\text{out}}} \leq T_{f,HX,CL_{\text{out}}}^{\max}
\tag{10.63}
$$

where $T_{f,HX,CL_{\text{out}}}^{\max}$ is the upper bound othe f cold stream temperature at the outlet of heat exchanger (90°C).

10.4.1.5 Distillate Tank

The distillate tank is acting as the buffer for the fluctuating water demand. The corresponding model can be written as

$$\rho_f^L A_{DT} \frac{dh_{DT}}{dt} = \dot{m}_{f,DT_{in}} - \dot{m}_{f,DT_{out}} \tag{10.64}$$

where ρ_f^L is the distillate density (kg/m^3); A_{DT} is the cross-sectional area of distillate tank (m^2); h_{DT} is the height of liquid in distillate tank (m); $\dot{m}_{f,DT_{in}}$ and $\dot{m}_{f,DT_{out}}$ denote the inlet and outlet flow rates, respectively, (kg/h). Note that the inlet flow is produced by the AGMD unit, that is,

$$\dot{m}_{f,DT_{in}} = n_{AGMD} N_{mem} A_{MD} \tag{10.65}$$

Finally, the liquid height in the distillate tank should be maintained within a specified range, that is,

$$h_{DT,low} \le h_{DT} \le h_{DT,high} \tag{10.66}$$

where $h_{DT,low}$ and $h_{DT,high}$, respectively, denote the given lower and upper bounds (m). Equations 10.52 through 10.66 are then incorporated into the temporal flexibility index model described in Section 10.3 and solved to determine the value of temporal flexibility index FI_t.

10.4.2 CASE STUDIES

The important role of temporal flexibility index in SMDDS design is demonstrated in the case studies presented below. Note that the specifications of a standard AGMD module are assumed to be the same as those given in Banat et al. (2007). The effluent of cold seawater flows into the shell side of a heat exchanger and then into the hot flow channel of the AGMD unit. Only water vapor can be transferred through the membrane pore because of the hydrophobic nature of porous membrane. Water vapor then is condensed on the wall surface of the cold seawater flow channel and then collected in a distillate tank for domestic consumption.

The AGMD desalination unit is driven by the thermal energy which is circulated and carried in the operating fluid (which is water in the present case studies) as depicted in Figures 10.1 and 10.4. In the daytime operation, the heat generated by the solar absorber can be consumed entirely in either Configuration I or II if the irradiation level is low. In the case of strong sunlight, a portion of the absorbed energy can be kept in the thermal storage tank of Configuration I and then used later to enable an extended period of desalination operation after sunset. Since Configuration II is not equipped with any thermal storage facility, it is therefore necessary to utilize a relatively small absorber to ensure complete consumption of solar energy in daytime operation and satisfy the freshwater demand during the night with a large enough distillate tank.

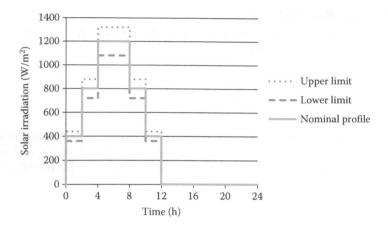

FIGURE 10.5   Solar irradiation profile.

There are two uncertain parameters considered in this case study. First is the solar irradiation rate $I(t)$ and the corresponding nominal profile $I^N(t)$, the expected upper and lower bounds are all depicted in Figure 10.5. Note that the expected positive and negative deviations at any time are both set at 10% of the nominal level. The other uncertainty parameter is the water demand rate $\dot{m}_{f,DT_{out}}(t)$ with the nominal value set at 18 kg/h × $wdf(t)$. $wdf(t)$ is the ratio between the demand rate at time t and a reference value, that is, 18 kg/h. The expected deviations in $m_{f,DT_{out}}$ are also selected to be 10% of its nominal value. The nominal level of $wdf(t)$ and also the corresponding upper and lower limits are sketched in Figure 10.6 where the time-dependent household water consumption rate can be closely characterized according to the nominal profile of $wdf(t)$. Better designs may be acquired accurately when more realistic solar irradiation and water demand profiles are available and they can be easily incorporated in the temporal flexibility analysis formulation.

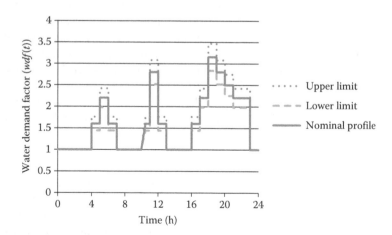

FIGURE 10.6   Normalized water demand profile.

On the basis of Equations 10.61 and 10.65, the production rate of each AGMD module at $T_{f,HX,CL_{out}} = 74°C$ is estimated to be 16.54 kg/h (Banat et al., 2007) (assuming that the feed temperature is $T_{f,HX,CL_{in}}$ 25°C). The nominal mass flow rate of seawater in membrane distillation loop $\dot{m}_{f,MD}^{N}$ is set to be 1125 kg/h per AGMD module according to Banat et al. (2007). Moreover, a maximum daily demand of 750.42 kg/day can be computed according to Figure 10.6. By adopting an average online period of 12 h/day, the approximate number of parallel AGMD modules can be calculated: $n_{AGMD} = (750.42/16.54 \times 12) = 3.78 \approx 4$, thus the total membrane area is 40 m². In the solar absorber, the total mass of operating fluid per unit area, that is, $M_{f,SA}/A_{SA}$, is set to be 15 kg/m² (Chang et al., 2010). The flow rate in the solar thermal loop ($\dot{m}_{f,STL}$) is set to be 36,000 kg/h, which is eight times the total nominal flow rate of seawater in the membrane distillation loop $\left(\dot{m}_{f,MD}^{N} = 1125 \times 4 = 4500 \text{ kg/h}\right)$ to ensure quick temperature response in the desalination loop. The volume of the distillate tank in each configuration is assumed to be ($A_{DT} = 0.35 \text{ m}^2; h_{DT,low} = 0 \text{ m}; h_{DT,high} = 2.143 \text{ m}$), whereas a 10 m³ thermal storage tank ($M_{f,ST} = 10,000$ kg) is utilized in Configuration I. Finally, it is assumed that the heat capacity of operating fluid Cp_{f}^{L} is held constant at 4200 J/kg°C and its density ρ_{f}^{L} is also assumed to be constant at 1000 kg/m³.

To facilitate a proper decision, the asymptotic energy utilization ratio between the solar absorber and the AGMD capacity is formulated as follows:

$$\phi_{util} = \frac{\text{Maximum supply rate of solar energy}}{\text{Maximum consumption rate of thermal energy}}$$

$$= \frac{A_{SA}I(t)^{\max}}{\dot{m}_{f,MD}^{\max} Cp_{f}^{L}(T_{f,HX,CL_{out}}^{\max} - T_{f,HX,CL_{in}}^{\min})} \tag{10.67}$$

Hence, the energy collected by the solar absorber can be fully utilized by the AGMD module if $\phi_{util} \leq 1$ and the excess heat need to be stored when $\phi_{util} > 1$. From Figure 10.5, it can be observed that $I(t)^{\max} = 1320 \text{ W/m}^2$. On the basis of Equation 10.62, one could deduce $\dot{m}_{f,MD}^{\max} = 1.1\dot{m}_{f,MD}^{N} = 1237.5 \text{ kg/h}$. Moreover, from the model description given in Section 10.4.1, it is reasonable to assume that $T_{f,HX,CL_{out}}^{\max} = 90°C$ and $T_{f,HX,CL_{in}}^{\min} = 25°C$. For a given ϕ_{util}, the calculation for the solar absorber size is simply straightforward according to Equation 10.67. For example, the absorber area for $\phi_{util} = 1$ should be $A_{SA} = (1237.5 \times 4200 \times (90 - 25)/1320 \times 3600) = 71.09 \text{ m}^2$. For the sake of completeness, all model parameters and variables used in the case studies are also listed in Table 10.2.

Given an AGMD module size, the solar absorber can be sized on the basis of Equation 10.67. By adopting the aforementioned thermal storage tank and distillate tank, the temporal flexibility indices of Configurations I and II can be computed for different utilization ratios. Table 10.3 summarizes the corresponding optimization results. It can be seen that when $\phi_{util} \leq 1$ the same flexibility indices can be obtained with both configurations. This is because the absorbed solar energy is consumed almost immediately and completely, therefore, the thermal storage tank in Configuration I is not needed at all, that is, $r_{f,ST} = 0$. On the other hand, one can see that $r_{f,ST} = 1$ if $\phi_{util} > 1$, which implies that the thermal storage tank is fully utilized for storing the excess solar energy acquired during daytime operation in Configuration I.

TABLE 10.2

Model Parameters and Variables

Symbol	Definition	Value	Classification
$T_{f,SA_{out}}^{max}$	Maximum allowable outlet temperature of solar absorber	100°C	d
$M_{f,SA}$	Total mass of operating fluid in solar absorber	–	d
L_{SA}	Length of absorber tube	–	d
A_{SA}	Exposed area in solar absorber	–	d
Cp_f^L	Heat capacity of operating fluid	4200 J/kg°C	d
$M_{f,ST}$	Total mass of operating fluid in thermal storage tank	10,000 kg	d
$\dot{m}_{f,STL}$	Mass flow rate in thermal loop	36,000 kg/h	d
$T_{f,HX,CL_{in}}$	Cold fluid inlet temperature of heat exchanger	25°C	d
A_{MD}	Membrane area of standard AGMD module	10 m² (Banat et al., 2007)	d
n_{AGMD}	Total number of standard AGMD modules	–	d
STEC	Specific Thermal Energy Consumption	14,000 kJ/kg (Banat et al., 2007)	d
$T_{f,HX,CL_{out}}^{max}$	Maximum cold fluid outlet temperature of heat exchanger	90°C	d
ρ_f^L	Distillate density	1000 kg/m³	d
A_{DT}	Cross-sectional area of distillate tank	0.35 m²	d
$h_{DT,low}$	Lower bound of liquid height in distillate tank	0 m	d
$h_{DT,high}$	Upper bound of liquid height in distillate tank	2.14 m	d
ϕ_{util}	Energy utilization ratio	To be selected	d
$I(t)^{max}$	Maximum solar irradiation rate per unit area	1320 W/m²	d
$\dot{m}_{f,MD}^{max}$	Maximum mass flow rate in membrane distillation loop	1237.5 kg/h	d
$T_{f,HX,CL_{in}}^{min}$	Minimum cold fluid inlet temperature of heat exchanger	25°C	d
$T_{f,SA_{in}}$	Inlet temperature of solar absorber	–	x
$T_{f,SA_{out}}$	Outlet temperature of solar absorber	–	x
$m_{f,SA}$	Mass flow rate of operating fluid in solar absorber	–	x
$T_{f,ST_{in}}$	Inlet temperature of thermal storage tank	–	x
$T_{f,ST_{out}}$	Outlet temperature of thermal storage tank	–	x
$\dot{m}_{f,ST}$	Throughput of thermal storage tank	–	x
$r_{f,ST}$	Flow ratio for thermal storage tank	–	x
$\dot{m}_{f,HX,HL}$	Mass flow rate of hot fluid in heat exchanger	–	x
$T_{f,HX,HL_{in}}$	Hot fluid inlet temperature of heat exchanger	–	x
$T_{f,HX,HL_{out}}$	Hot fluid outlet temperature of heat exchanger	–	x
$T_{f,HX,CL_{out}}$	Cold fluid outlet temperature of heat exchanger	–	x
h_{DT}	Liquid height in distillate tank	–	x
$\dot{m}_{f,DT_{in}}$	Inlet flow rate of distillate tank	–	x
N_{mem}	Distillate flux through AGMD membrane	–	x
$\dot{m}_{f,MD}$	Mass flow rate in membrane distillation loop	4500 kg/h (nominal)	z
I	Solar irradiation rate per unit area	–	θ
$\dot{m}_{f,DT_{out}}$	Outlet flow rate of distillate tank	–	θ

TABLE 10.3

Optimization Results

	Case No.	1	2	3	4	5	6	7
Configuration	ϕ_{util}	0.785	0.955	1	1.16	1.28	1.385	1.576
	FI_t	0	1	1.24	1.65	1.65	1	0
I	$r_{f,ST}$ (day)	0	0	0	1	1	1	1
	$g_j = 0$	$h_{DT,low}$	$h_{DT,low}$	$h_{DT,low}$	$h_{DT,low}$	$h_{DT,high}$	$h_{DT,high}$	$h_{DT,high}$
	$\dot{m}_{f,MD}$ factor	1.1	1.1	1.1	1.1	0.9	0.9	0.9
	FI_t	0	1	1.24	1	0	Inf	Inf
II	$r_{f,ST}$ (day)	0	0	0	0	0	N/A	N/A
	$g_j = 0$	$h_{DT,low}$	$h_{DT,low}$	$h_{DT,low}$	$T^{max}_{f,SA_{out}}$	$T^{max}_{f,SA_{out}}$	N/A	N/A
	$\dot{m}_{f,MD}$ factor	1.1	1.1	1.1	1.1	1.1	N/A	N/A

Note that, although $r_{f,ST}$ is allowed to assume a real value between 0 and 1 in this situation, this optimal operating policy is adopted mainly to avoid violating the temperature upper bounds in Equations 10.53 and 10.63.

Notice that the information of the active constraint in each optimum solution can also be found in Table 10.3. When $\phi_{util} < 1$, the consumed energy may not be enough to meet the demand; hence, the distillate tank is expected to be empty sometimes. The optimization results of the corresponding two cases are analyzed below:

1. Let us first consider Case 1 when $\phi_{util} = 0.785$. Note that $FI_t = 0$ in this case, that is, no deviations from the nominal parameters are allowed for both configurations. This is due to the fact that the nominal absorption rate of solar energy is just enough to meet the nominal demand by maximizing the control variable $\dot{m}_{f,MD}$ at all times.

2. Let us next consider Case 2 when $\phi_{util} = 0.955$. Note that $FI_t = 1$ in this case, that is, the expected deviations from the nominal parameters can be exactly accommodated in both configurations. To validate this prediction, the worst-case scenario (which is corresponding to the lower bound of solar irradiation rate and the upper bound of water demand rate) has been numerically simulated with Simulink® (The Mathworks, Inc., 2012). The Simulink program was built according to Equations 10.52 through 10.66 and also the parameter values listed in Table 10.2. To facilitate the simulation run, three time profiles were adopted as inputs, that is, (1) the lower limit of the solar irradiation profile in Figure 10.5, (2) the upper limit of the water demand profile in Figure 10.6, and (3) the control variable $\dot{m}_{f,MD}(t)$ obtained by solving the temporal flexibility index model. The simulated temperature of operating fluid at the exit of the solar absorber (i.e., $T_{f,SA_{out}}$) and also that of seawater at the exit of the heat exchanger (i.e., $T_{f,HX,CL_{out}}$) are both plotted in Figure 10.7. It can be clearly observed that, both temperatures are always well below their respective upper bounds. The corresponding water level in the distillate tank can also be found in Figure 10.8. Note that the tank is

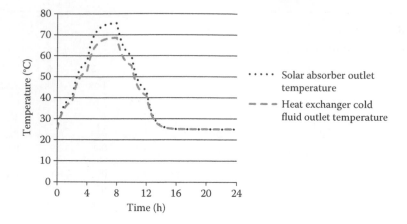

FIGURE 7 The time profiles of $T_{f,SA_{out}}$ and $T_{f,HX,CL_{out}}$ for both configurations in the worst-case scenario ($\phi_{util} = 0.955$).

just emptied at the end of 24 h. This observation essentially confirms the optimization result of $FI_t = 1$ for both configurations. Thus, if the desired value of temporal flexibility index is one, Configuration II should be chosen since the equipment cost of thermal storage facility can be saved.

3. In Case 3 when $\phi_{util} = 1$, notice that the optimization results are still the same for both configurations, that is, $r_{f,ST} = 0$ in daytime operations and $FI_t = 1.24$. If the target value of temporal flexibility index is 1, then Configurations I and II in this case are both slightly overdesigned since $FI_t > 1$. However, if a higher operational flexibility is called for in the design, then there is an incentive to consider additional cases where the solar absorbers are larger, that is, $\phi_{util} > 1$. Following are the corresponding case studies:

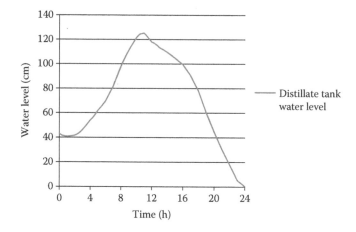

FIGURE 10.8 The time profiles of h_{DT} for both configurations in the worst-case scenario ($\phi_{util} = 0.955$).

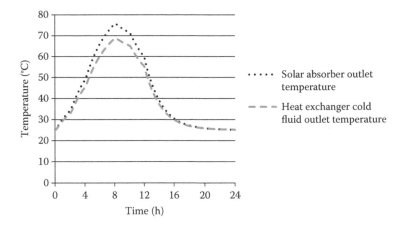

FIGURE 10.9 The time profiles of $T_{f,SA_{out}}$ and $T_{f,HX,CL_{out}}$ for Configuration I in the worst-case scenario ($\phi_{util} = 1.16$).

4. In Case 4 ($\phi_{util} = 1.16$), Configuration I can be made more flexible ($FI_t = 1.65$) by operating the thermal storage tank, that is, $r_{f,ST} = 1$. The corresponding worst-case scenario can be simulated and the time profiles of three critical variables, that is, $T_{f,SA_{out}}$, $T_{f,HX,CL_{out}}$, and h_{DT}, can be found in Figures 10.9 and 10.10. On the other hand, note that the temporal flexibility index of Configuration II equals one. This is because, since there is no thermal storage capacity, the exit temperature of the solar absorber ($T_{f,SA_{out}}$) reaches its upper limit at a certain instance during daytime operation. The corresponding system behavior can be characterized with Figures 10.11 and 10.12.

5. In Case 5 ($\phi_{util} = 1.28$), the temporal flexibility index of Configuration I can be raised to 1.65 and the active constraint is now associated with the upper

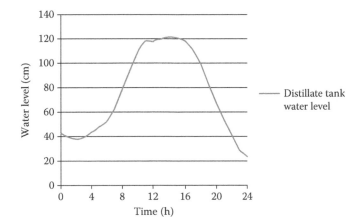

FIGURE 10.10 The time profile of h_{DT} for Configuration I in the worst-case scenario ($\phi_{util} = 1.16$).

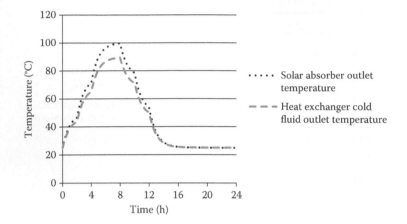

FIGURE 10.11   The time profiles of $T_{f,SA_{out}}$ and $T_{f,HX,CL_{out}}$ for Configuration II in the worst-case scenario ($\phi_{util} = 1.16$).

bound of water level in the distillate tank, that is, $h_{DT} \leq h_{DT,high}$. This is obviously due to the fact that the solar energy is introduced at a rate that is much faster than the consumption rate of thermal energy. On the other hand, note that $FI_t = 0$ for Configuration II. This drastic reduction of flexibility can also be attributed to the high intake rate of solar energy. Since there is no thermal storage tank, it is very difficult to keep the exit temperature of the solar absorber ($T_{f,SA_{out}}$) below 100°C.

6. In Case 6 ($\phi_{util} = 1.385$) and Case 7 ($\phi_{util} = 1.576$), the selected solar absorbers are larger than those used in the other cases. Since more water is produced in Configuration I while the size of distillate tank remains the same in either Case 6 or 7, the resulting flexibility index becomes much

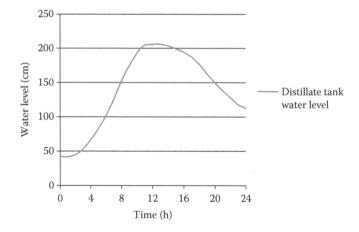

FIGURE 10.12   The time profile of h_{DT} for Configuration II in the worst-case scenario ($\phi_{util} = 1.16$).

lower than that achieved in Case 5. On the other hand, note that $FI_t = 0$ for Configuration II in Case 5. Thus, any further increase in the utilization ratio inevitably renders Configuration II infeasible.

If there is a need to make the SMDDS system even more flexible ($FI_t > 1.65$), one can deduce from Case 5 that this goal can be reached by relaxing the active constraint, that is, by enlarging the thermal storage tank ($M_{f,ST} \geq 10,000$ kg) and also the distillate tank ($h_{DT,high} \geq 2.143$ m). Finally, it should be noted that the temporal flexibility may be further enhanced by introducing additional structural modifications, for example, by operating more than one thermal storage tank in parallel. The merits of these new configurations can be easily assessed on the basis of the proposed temporal flexibility analysis.

10.5 CONCLUDING REMARKS AND UNSETTLED ISSUES

Three different types of flexibility analyses are discussed in this chapter with emphases on their model formulations and solution strategies. Specifically,

1. For the steady-state flexibility analysis, the available basic formulation (Swaney and Grossmann, 1985a,b) is presented for quantifying the ability of any continuous process to maintain feasible operation under uncertainty influences. Two main strategies are described for solving the corresponding multilevel optimization problems. On the basis of KKT conditions for the lower levels, a single-level formulation can be derived to compute the steady-state flexibility index FI_s. Another approach to achieve the same purpose is the so-called vertex method, which is valid if the solution lies on one of the vertices of the feasible region $\mathbf{T}(\delta)$ in the parameter space (Halemane and Grossmann, 1983). This method can also be further simplified when physical insights (Chang et al., 2009) are available for eliminating the unlikely candidates (Li and Chang, 2011).
2. The conventional formulation for the dynamic flexibility analysis (Dimitriadis and Pistikopoulos, 1995) is also described briefly in this chapter. Since the system dynamics are characterized with DAEs in the model constraints, the required solution strategies must be devised accordingly. In the first approach, a discretization method, for example, the DQ (Bellman et al., 1971, 1972), is adopted to convert the DAEs into a system of algebraic equations and solve the resulting model with existing strategies for the steady-state flexibility analysis. On the other hand, the KKT conditions can also be derived via calculus of variation to produce the dynamic version of the active set method.
3. For the temporal flexibility analysis, a novel concept of *temporal flexibility* has been developed to address the practical issues caused by short-term disturbances that may occur in operating the batch chemical plants. A generic mathematical programming model is formulated accordingly for characterizing the corresponding performance measure. The solution strategies in this case are essentially the same as those for the dynamic flexibility analysis.

The usefulness of the above analyses is demonstrated with a realistic application, that is, the SMDDS design problem where several design alternatives are evaluated according to their temporal flexibility indices. Although satisfactory results are obtained in this work, there are still several unsettled issues which require further attention. Specifically,

1. In the case of the steady-state flexibility analysis, the uncertain parameter space is characterized according to Equation 10.8, that is, it is a hypercube bound by the following inequality constraints:

$$\theta^N - \Delta\theta^- \leq \theta \leq \theta^N + \Delta\theta^+$$

   However, this space may be generalized to any region bound by

$$\psi(\theta) \leq 0$$

   where $\psi(\theta)$ is a vector of linear or nonlinear functions of θ.

2. Lai and Hui (2008) have developed a promising flexibility metric which was referred to as the volumetric flexibility (FI_v). It is defined according to the hypervolume ratio of the feasible region and the region containing all possible combinations of expected uncertain parameters. In certain applications, this index is believed to be capable of providing more reliable assessments of the system performance with relatively mild computation requirements. Since it was developed only for the steady-state processes, there seems to be a need for extending this approach for the dynamic problems.

3. The discretization strategy for solving dynamic flexibility index problems needs further improvement so as to enhance computational efficiency. The conventional DQ method (Bellman et al., 1971, 1972; Quan and Chang, 1989a, b) has been adopted with arbitrarily chosen grid points and elements. The computation loads were quite heavy, which is considered necessary to develop a systematic strategy to determine the minimum numbers of grid points and elements, the proper grid spacing, and the optimal element length(s) for solving the dynamic and temporal flexibility index models.

REFERENCES

Adi, V. S. K.; Chang, C. T. A mathematical programming formulation for temporal flexibility analysis. *Comput. Chem. Eng.*, *57*, 151, 2013.

Adi, V. S. K.; Chang, C. T. SMDDS design based on temporal flexibility analysis. *Desalination*, *320*, 96, 2013.

Aziz, N.; Mujtaba, I. M. Optimal operation policies in batch reactors. *Chem. Eng. J.*, *85*, 313, 2002.

Bahri, P. A.; Bandoni, J. A.; Romagnoli, J. A. Integrated flexibility and controllability analysis in design of chemical processes. *AIChE J.*, *43*, 997, 1997.

Banat, F.; Jwaied, N.; Rommel, M.; Koschikowski, J.; Wieghaus, M. Performance evaluation of the "Large SMADES" autonomous desalination solar-driven membrane distillation plant in Aqaba, Jordan. *Desalination*, *217*, 17, 2007.

Bansal, V.; Perkins, J. D.; Pistikopoulos, E. N. Flexibility analysis and design of dynamic processes with stochastic parameters. *Comput. Chem. Eng.*, 22, S817, 1998.

Bansal, V.; Perkins, J. D.; Pistikopoulos, E. N. Flexibility analysis and design of linear systems by parametric programming. *AIChE J.*, *46*, 335, 2000.

Bansal, V.; Perkins, J. D.; Pistikopoulos, E. N. Flexibility analysis and design using a parametric programming framework. *AIChE J.*, *48*, 2851, 2002.

Bellman, R. E.; Casti, J. Differential quadrature and long-term integration. *J. Math. Anal. Appl.*, *34*, 235, 1971.

Bellman, R. E.; Kashef, B. G.; Casti, J. Differential quadrature: A technique for the rapid solution of nonlinear partial differential equations. *J. Comput. Phys.*, *10*, 40, 1972.

Ben Bacha, H.; Dammak, T.; Ben Abdalah, A. A.; Maalej, A. Y.; Ben Dhia, H. Desalination unit coupled with solar collectors and storage tank: Modeling and simulation. *Desalination*, *206*, 341, 2007.

Biegler, L. T. Solution of dynamic optimization problems by successive quadratic programming and orthogonal collocation. *Comput. Chem. Eng.*, *8*, 243, 1984.

Biegler, L. T.; Grossmann I. E.; Westerberg, A. W. *Systematic Methods of Chemical Process Design*. Englewood Cliffs, New Jersey: Prentice-Hall, 690, 1997.

Brengel, D. D.; Seider, W. D. Coordinated design and control optimization of nonlinear processes. *Comput. Chem. Eng.*, *16*, 861, 1992.

Bui, V. A.; Vu, L. T. T.; Nguyen, M. H. Simulation and optimization of direct contact membrane distillation for energy efficiency. *Desalination*, *259*, 29, 2010.

Burgess, G.; Lovegrove, K. Solar thermal powered desalination: Membrane versus distillation technologies, Centre for Sustainable Energy Systems, Department of Engineering, Australian National University, Canberra, Australia. 2005. http://people.eng.unimelb.edu.au/lua/039.pdf (Accessed November 2014).

Cabassud C.; Wirth, D. Membrane distillation for water desalination: How to choose an appropriate membrane? *Desalination*, *157*, 307, 2003.

Chacon-Mondragon, O. L.; Himmelblau, D. M. Integration of flexibility and control in process design. *Comput. Chem. Eng.*, *20*, 447, 1996.

Chang, C. T.; Li, B. H.; Liou, C. W. Development of a generalized mixed integer nonlinear programming model for assessing and improving the operational flexibility of water network designs. *Ind. Eng. Chem. Res.*, *48*, 3496, 2009.

Chang, C. T.; Tsai, C. S.; Lin, T. T. Modified differential quadratures and their applications. *Chem. Eng. Commun.*, *123*, 135, 1993.

Chang, H.; Lyu, S. G.; Tsai, C. M.; Chen, Y. H.; Cheng, T. W.; Chou, Y. H. Experimental and simulation study of a solar thermal driven membrane distillation desalination process. *Desalination*, *286*, 400, 2012.

Chang, H.; Wang, G. B.; Chen, Y. H.; Li, C. C.; Chang, C. L. Modeling and optimization of a solar driven membrane distillation desalination System. *Renew. Energ.*, *35*, 2714, 2010.

Civan F.; Sliepcevich, C. M. Differential quadrature for multidimensional problems. *J. Math. Anal. Appl.*, *101*, 423, 1984.

Cuthrell, J. E.; Biegler, L. T. On the optimization of differential algebraic process systems. *AIChE J.*, 33, 1257, 1987.

Dimitriadis, V. D.; Pistikopoulos, E. N. Flexibility analysis of dynamic systems. *Ind. Eng. Chem. Res.*, *34*, 4451, 1995.

Dimitriadis, V. D.; Shah, N.; Pantelides, C. C. Modeling and safety verification of discrete/continuous processing systems. *AIChE J.*, *43*, 1041, 1997.

Floudas, C. A.; Gunus, Z. H.; Ierapetritou, M. G. Global optimization in design under uncertainty: Feasibility test and flexibility ondex problems. *Ind. Eng. Chem. Res.*, *40*, 4267, 2001.

Gálvez, J. B.; García-Rodríguez, L.; Martín-Mateos, I. Seawater desalination by an innovative solar-powered membrane distillation system: The MEDESOL project. *Desalination*, *246*, 567, 2009.

Georgiadis, M. C.; Pistikopoulos, E. N. An integrated framework for robust and flexible process systems. *Ind. Eng. Chem. Res.*, *38*, 133, 1999.

Grossmann, I. E.; Floudas, C. A. Active constraint strategy for flexibility analysis in chemical process. *Comput. Chem. Eng.*, *11*, 675, 1987.

Grossmann, I. E.; Halemane, K. P. Decomposition strategy for designing flexible chemical plants. *AIChE J.*, *28*, 686, 1982.

Guillen-Burrieza, E.; Blanco, J.; Zaragoza, G.; Alarcon, D. C.; Palenzuela, P.; Ibarra, M.; Gernjak, W. Experimental analysis of an air gap membrane distillation solar desalination pilot system. *J. Membr. Sci.*, *379*, 386, 2011.

Halemane, K. P.; Grossmann, I. E. Optimal process design under uncertainty. *AIChE J.*, *29*, 425, 1983.

Koschikowski, J.; Wieghaus, M.; Rommel, M. Solar thermal-driven desalination plants based on membrane distillation. *Desalination*, *156*, 295, 2003.

Lai, S. M.; Hui, C.-W. Process flexibility for multivariable systems. *Ind. Eng. Chem. Res.*, *47*, 4170, 2008.

Li, B. H.; Chang, C. T., 2011, Efficient flexibility assessment procedure for water network designs. *Ind. Eng. Chem. Res.*, *50*, 3763, 2011.

Lima F. V.; Georgakis C. Design of output constraints for model-based non-square controllers using interval operability. *J. Process Control*, 18, 610, 2008.

Lima, F. V.; Georgakis C.; Smith, J. F.; Schnelle, P. D.; Vinson D. R. Operability-based determination of feasible control constraints for several high-dimensional nonsquare industrial processes. *AIChE J.*, *56*, 1249, 2010a.

Lima, F. V.; Jia, Z.; Ierapetritou, M.; Georgakis, C. Similarities and differences between the concepts of operability and flexibility: The steady-state case, *AIChE J.*, *56*, 702, 2010b.

Liou, C. W. *The impacts of mixers and buffer tanks on the operational flexibility of water usage and treatment networks.* MS Thesis, Chemical Engineering Department, National Cheng Kung University, Taiwan, 2006.

Malcolm, A.; Polan, J.; Zhang, L.; Ogunnaike, B. A.; Linninger, A. A. Integrating systems design and control using dynamic flexibility analysis, *AIChE J.*, *53*, 2048, 2007.

Meindersma, G. W.; Guijt, C. M.; de Haan, A. B. Desalination and water recycling by air gap membrane distillation. *Desalination.*, *187*, 291, 2006.

Mohideen, M. J.; Perkins, J. D.; Pistikopoulos, E. N. Optimal design dynamic systems under uncertainty. *AIChE J.*, *42*, 2251, 1996.

Ostrovski, G. M.; Achenie, L. E. K.; Wang, Y. P.; Volin, Y. M. A new algorithm for computing process flexibility. *Comput. Chem. Eng.*, *39*, 2368, 2001.

Ostrovski, G. M.; Volin, Y. M. Flexibility analysis of chemical process: Selected global optimization sub-problems. *Optim. Eng.*, *3*, 31, 2002.

Petracci, N. C.; Hoch, P. M.; Cliche, A. M. Flexibility analysis of an ethylene plant. *Comput. Chem. Eng.*, *20*, S443, 1996.

Pistikopoulos, E. N.; Grossmann, I. E. Evaluation and redesign for improving flexibility in linear systems with infeasible nominal conditions. *Comput. Chem. Eng.*, *12*, 841, 1988a.

Pistikopoulos, E. N.; Grossmann, I. E. Optimal retrofit design for improving process flexibility in linear systems. *Comput. Chem. Eng.*, *12*, 719, 1988b.

Pistikopoulos, E. N.; Grossmann, I. E. Optimal retrofit design for improving process flexibility in nonlinear systems—I: Fixed degree of flexibility. *Comput. Chem. Eng.*, *13*, 1003, 1989a.

Pistikopoulos, E. N.; Grossmann, I. E. Optimal retrofit design for improving process flexibility in nonlinear systems—II: Optimal level of flexibility. *Comput. Chem. Eng.*, *13*, 1087, 1989b.

Quan, J. R.; Chang, C. T. New insights in solving distributed system equations by the quadrature methods— I. *Comput. Chem. Eng.*, *13*, 779, 1989a.

Quan, J. R.; Chang, C. T. New insights in solving distributed system equations by the quadrature methods— II. *Comput. Chem. Eng.*, *13*, 1017, 1989b.

Riyanto, E.; Chang, C. T. A Heuristic revamp strategy to improve operational flexibility of water networks based on active constraints, *Chem. Eng. Sci.*, 65, 2758, 2010.

Shu, C. *Differential Quadrature and Its Applications in Engineering*. London: Springer-Verlag, 340, 2000.

Soroush, M.; Kravaris, C. Optimal design and operation of batch reactors. 1. Theoretical framework. *Ind. Eng. Chem. Res.*, *32*, 866, 1993a.

Soroush, M.; Kravaris, C. Optimal design and operation of batch reactors. 2. A case study. *Ind. Eng. Chem. Res.*, *32*, 882, 1993b.

Swaney, R. E.; Grossmann, I. E. An index for operational flexibility in chemical process design. Part I: Formulation and theory. *AIChE J.*, *31*, 621, 1985a.

Swaney, R. E.; Grossmann, I. E. An index for operational flexibility in chemical process design. Part II: Computational algorithms. *AIChE J.*, *31*, 631, 1985b.

Varvarezos, D. K.; Grossmann, I. E.; Biegler, L. T. A sensitivity based approach for flexibility analysis and design of linear process systems. *Comput. Chem. Eng.*, *19*, 1301, 1995.

Volin, Y. M.; Ostrovski, G. M. Flexibility analysis of complex technical systems under uncertainty. *Automat. Rem. Control*, *63*, 1123, 2002.

Walsh, S.; Perkins, J. D. Application of integrated process and control system design to waste water neutralization. *Comput. Chem. Eng.*, *18*, S183, 1994.

White, V.; Perkins, J. D.; Espie, D. M. Switchability analysis. *Comput. Chem. Eng.*, *20*, 469, 1996.

Zhou, H.; Li, X. X.; Qian,Y.; Chen, Y.; Kraslawski, A. Optimizing the initial conditions to improve the dynamic flexibility of batch processes. *Ind. Eng. Chem. Res.*, *48*, 6321, 2009.

Zong, Z.; Lam, K. Y. A Localized differential quadrature method and its application to the 2D wave equation. *Comput. Mech.*, *29*, 382, 2002.

Zong, Z.; Zhang, Y. Y. Advanced *Differential Quadrature Methods*. Boca Raton, FL: Chapman & Hall/CRC, 2009.

11 Automated Targeting Model for Batch Process Integration

Dominic Chwan Yee Foo

CONTENTS

11.1 INTRODUCTION

In recent years, the mismanagement of natural resources (such as the overextraction of water resources), population growth, industrial pollution problems, and climate change are pressing problems to the global community (Sandia National Laboratories, 2005; Rockstrom et al., 2009). For instance, it is expected that approximately 40% of the global population will live in areas of severe water stress by 2050 (WWAP, 2014). The efficient use of industrial resources has become an important element to ensure both environmental and business sustainability. Many problems pertaining to efficient use of resources in industrial process plants share common features, which allow them to be handled using a common family of systematic techniques known collectively as *process integration*. One of the widely accepted definitions for process integration is given as *a holistic approach to process design, retrofitting and operation which emphasises the unity of the process* (El-Halwagi, 2006, p. 15).

In the past three decades, process integration techniques such as *pinch analysis* and *mathematical optimization* have been developed to address various resource conservation problems, such as energy and material recovery. To date, many of the established techniques are documented in various textbooks (Linnhoff et al., 1982; Smith, 1995, 2005; El-Halwagi, 1997, 2006; Klemeš et al., 2010; Foo, 2012) and review papers (Gundersen and Naess, 1988; Linnhoff, 1993; Furman and Sahinidis, 2002; Foo, 2009; Klemeš and Kravanja, 2013). For batch process integration,

265

different techniques have also been developed for energy and material recovery. Readers may refer to the textbook (Majozi, 2010) or review papers (Gouws et al., 2010; Fernández et al., 2012) for more details of these works.

One important subproblem for batch process integration is water minimization for batch processes, or *batch water network* (BWN) synthesis in short, for which different variants of pinch analysis (Foo et al., 2005; Majozi et al., 2006; Chen and Lee, 2008) and mathematical optimization techniques (Almató et al., 1999; Kim and Smith, 2004; Majozi, 2005, 2006; Li and Chang, 2006; Cheng and Chang, 2007; Chen et al., 2008, 2009, 2010; Gouws and Majozi, 2008; Majozi and Gouws, 2009) have been developed. Readers may refer to the review paper (Gouws et al., 2010) to understand the strength and weakness of the various process integration techniques.

In this chapter, an optimization model known as the *automated targeting model* (ATM) is illustrated for BWN synthesis. In the following section, conceptual understanding of process integration for material resource conservation is given first. Next, the problem statement for a BWN is given. Next, the ATM is illustrated, followed by a case study on water recovery in a polyvinyl chloride (PVC) plant.

11.2 CONCEPTUAL UNDERSTANDING

In the past two decades, generic process integration tools were developed for various material *resource conservation networks* (RCNs) including *water minimization, gas recovery,* and *property integration* (Foo, 2012). Different strategies for RCNs are formally defined from the perspective of process integration, that is, *direct reuse/ recycle* and *regeneration–reuse/recycle*. Direct reuse refers to the scheme where a process effluent is sent to other processes and does not re-enter its original process. On the other hand, direct recycle refers to the recovery scheme where the process effluent re-enters its original process. A process effluent may be partially purified in an *interception unit* to improve its quality prior to reuse/recycle; these are known as regeneration–reuse and regeneration–recycle. Figure 11.1 shows the recovery schemes for an RCN. In most cases, the priority is given to the direct reuse/recycle scheme (focus of this chapter), as it involves the lowest investment cost and ease of implementation.

To better understand an RCN problem, it is important to understand the concept of *process sink* and *source*. Source refers to a process stream (normally the outlet) that can be reused/recycled for material recovery. On the other hand, sink refers to a process unit where a resource (typically a fresh material) is needed (see Figure 11.2).

For a water minimization problem, the water sinks and sources exist in all water-using processes. Note that the latter can exist in different configurations. Figure 11.3 shows two typical batch operation modes. For a *truly batch* operation in Figure 11.3a, Process 1 takes place between t_2 and t_3. Before the commencement of the operation, a water sink exists within the interval $(t_1 - t_2)$, where freshwater is charged to the operation. Upon the completion of the operation, a water source emerges from the process in the interval $(t_3 - t_4)$. On the other hand, another type of batch process is the *semicontinuous* operation shown in Figure 11.3b. In this case, the water sink and source

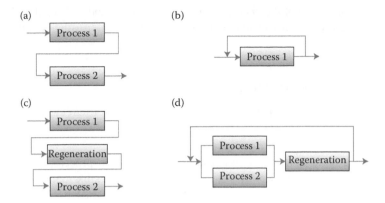

FIGURE 11.1   Strategies for an RCN: (a) direct reuse, (b) direct recycle, (c) regeneration–reuse, (d) regeneration–recycling. (Adapted from Wang, Y. P. and Smith, R., 1994. *Chemical Engineering Science*, 49: 981–1006; Foo, D. C. Y., 2012. *Process Integration for Resource Conservation*. CRC Press: Boca Raton, Florida, US.)

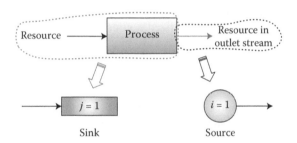

FIGURE 11.2   Conceptual understanding of process source and sink. (From Foo, D. C. Y., 2012. *Process Integration for Resource Conservation*. CRC Press: Boca Raton, Florida, US.)

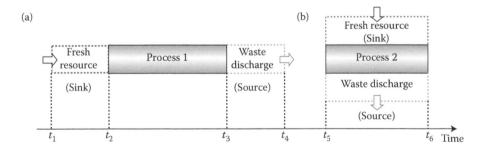

FIGURE 11.3   Types of batch resource-consumption units: (a) truly batch; and (b) semi-continuous operations. (Adapted from Gouws, J. et al., 2010. *Industrial & Engineering Chemistry Research*, 49(19): 8877–8893; Foo, D. C. Y., 2012. *Process Integration for Resource Conservation*. CRC Press: Boca Raton, Florida, US.)

exist during the course of the operation, that is, between t_5 and t_6. Note that there are also cases where water sink or source exists only for the water-using operations (see discussion in Foo, 2012).

11.3   PROBLEM STATEMENT

The formal problem statement for a BWN with direct reuse/recycle is given as follows:

- Given a number of resource-consuming units in the process, designated as *process sink*, or SK = {j = 1, 2, 3, ..., N_{SK}}. Each sink requires a fixed water flow (F_{SKj}) with a targeted quality index (q_{SKj}). The latter varies from one case to another, which may take the form of impurity concentration of a concentration-based RCN (e.g., ppm, mass fraction, percentage, etc.), or property operator values for a property-based problem (see detailed discussion in Foo, 2012).
- Given a number of resource-generating units/streams, designated as *process source*, or SR = {i = 1, 2, 3, ..., N_{SR}}. Each source has a given water flow (F_{SRi}) and a quality index (q_{SRi}). Each source can be sent for reuse/recycle to the sinks. The unutilized source(s) are sent for environmental discharge (treated as a sink).
- The process sinks and sources exist in different time intervals. Storage tanks may be used to recover process sources across different time intervals.
- When the process source is insufficient for use in the sinks, external freshwater may be purchased to satisfy sink requirements.

The objective of the problem may be set to minimize the total freshwater flows (which then lead to minimum wastewater), or minimum cost. The superstructural representation for a BMN is shown in Figure 11.4.

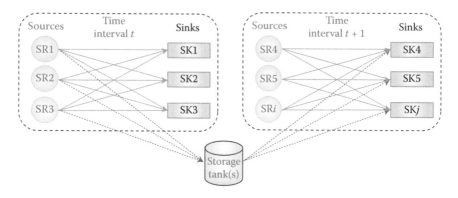

FIGURE 11.4   Superstructural model for BWN. (Adapted from Foo, D. C. Y., 2012. *Process Integration for Resource Conservation.* CRC Press: Boca Raton, Florida, US.)

11.4 ATM FOR DIRECT WATER REUSE/RECYCLE

The ATM was developed by Foo and co-workers (Ng et al., 2009a,b,c, 2010; Chew and Foo, 2009) for resource conservation problems for continuous processes. It was then extended to batch process integration by Foo (2010). Note that the ATM is a mathematical optimization model that is built on the concept of pinch analysis, which enables the minimum resource/cost targets to be identified prior to the detailed network design.

The ATM framework for the direct reuse/recycle scheme for a BWN is shown in Figure 11.5. Note that each flow term has a subscript t that denotes the time interval it belongs to. The flow terms for water stored from earlier ($F_{ST, t-1, t}$) and later time intervals ($F_{ST, t, t+1}$) allow the transfer of water across different time intervals. Note also that an inherent assumption made here is that water storage is potentially available at each quality level. Their actual existence is to be decided by the ATM.

The ATM has the following constraints that are imposed for all time intervals:

$$\delta_{k,t} = \delta_{k-1,t} + (\Sigma_i F_{SRi} - \Sigma_j F_{SKj})_{k,t} + (F_{ST, t-1,t} - F_{ST,t,t+1})_k \quad k = 1, 2, \ldots, n-1; \forall t \quad (11.1)$$

$$\delta_{k-1,t} \geq 0 \quad\quad\quad\quad k = 1, n; \forall t \quad (11.2)$$

$$\varepsilon_{k,t} = \begin{cases} 0 & k = 1; \forall t \\ \varepsilon_{k-1,t} + \delta_{k,t}(q_{k+1,t} - q_{k,t}) & k = 2, 3, \ldots, n; \forall t \end{cases} \quad (11.3)$$

$$\varepsilon_{k,t} \geq 0 \quad\quad\quad\quad k = 2, 3, \ldots, n; \forall t \quad (11.4)$$

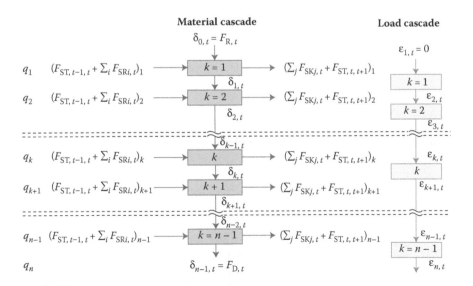

FIGURE 11.5 ATM framework for BMNs for direct reuse/recycle with mass storage.

Equation 11.1 describes the *water cascades* across all quality levels for all time intervals. The *net material flow* of level k of interval t ($\delta_{k,t}$) is a result of summation of the net material flowrate cascaded from the previous quality level ($\delta_{k-1,t}$) with the flow balance at quality level k ($\Sigma_i F_{SRi} - \Sigma_j F_{SKj})_{k,t}$, as well as the net stored material flow at each level ($F_{ST, t-1,t} - F_{ST,t,t+1})_k$. In Equation 11.2, it is stated that the net water flowrates entering the first ($\delta_{0,t}$) and last levels ($\delta_{n-1,t}$) of each interval t should take nonnegative values. These variables correspond to the freshwater ($F_{R,t}$) and wastewater flows ($F_{D,t}$) of interval t. Next, Equation 11.3 states that the *residual impurity/property load* ($\varepsilon_{k,t}$) at the first quality level of each time interval t has zero value, that is, $\varepsilon_{1,t} = 0$, while those in other quality levels are contributed by the residual load cascaded from the previous level ($\varepsilon_{k-1,t}$), as well as the net material flowrate within each quality interval. The latter is given by the product of the net material flowrate from level k ($\delta_{k,t}$) and the difference between two adjacent quality levels ($q_{k+1,t} - q_{k,t}$). All residual loads take nonnegative values following Equation 11.4, except at level 1 (which is set to zero following Equation 11.3).

In addition, constraints given in Equations 11.5 through 11.8 are useful in sizing the water storage system, so that water may be recovered across different time intervals. Equation 11.5 defines the cumulative content for water storage at quality level k for time interval t ($F_{MS,t,k}$), which is contributed by the stored mass flows from the earlier ($F_{ST,t-1,t}$) and to the later ($F_{ST,t,t+1}$) time intervals. The size of the storage at each quality level k ($F_{STG, k}$) is then determined by the maximum cumulative content across all time intervals, as given in Equation 11.6. Equations 11.7 and 11.8 indicate that both the stored mass flow and the cumulative content of mass storage should take positive values

$$F_{MS,t+1,k} = (F_{MS,t} - F_{ST,t-1,t} + F_{ST,t,t+1})_k \quad \forall k, \; \forall t \tag{11.5}$$

$$F_{STG,k} \geq F_{MS,t,k} \quad \forall k, \; \forall t \tag{11.6}$$

$$F_{ST,t,t+1,k} \geq 0 \quad \forall k, \; \forall t \tag{11.7}$$

$$F_{MS,t+1,k} \geq 0 \quad \forall k \tag{11.8}$$

The optimization objective of this problem may be set to minimize the minimum freshwater and wastewater flows, with a second objective to minimize the water storage capacity (see detailed discussion in Foo, 2012). Other alternatives are to minimize various cost elements of the BWN, which may take the form of minimum operating cost (*OC*), or minimum total annualized cost (*TAC*), as given in Equations 11.9 and 11.10. The operating cost of a BWN takes into account freshwater (with unit cost of CT_{FW}) and wastewater treatment costs (with unit cost of CT_{WW}). To calculate the *TAC*, the capital cost (*CC*) of the BWN (contributed mainly by storage tanks, Equations 11.10c and 11.10d) should be included and annualized before being added to the annual operating cost (product of *OC* and the annual number of batches—*BATCH*). As shown in Equation 11.10b, the second term of the equation gives the annual operating cost, while the third term gives the annualized capital cost of the BWN (IR = interest rate; YR = number of years)

$$\text{minimize } OC \tag{11.9a}$$

where

$$OC = \Sigma_t F_{\text{FW},t}\, CT_{\text{FW}} + \Sigma_t F_{\text{WW},t} CT_{\text{WW}} \tag{11.9b}$$

$$\text{minimize } TAC \tag{11.10a}$$

where

$$TAC = OC(BATCH) + CC \left[\frac{IR(1 + IR)^{YR}}{(1 + IR)^{YR} - 1} \right] \tag{11.10b}$$

where

$$CC = \Sigma_k\, SC_{\text{STG},k} \tag{11.10c}$$

$$SC_{\text{STG},k} = (235\, F_{\text{STG},k} + 20{,}300\, B_{\text{STG},k}) \quad k = 1, \ldots, n - 1 \tag{11.10d}$$

Equation 11.10d is used in conjunction with Equation 11.11 to examine the presence/absence of a water storage tank at level k

$$\frac{F_{\text{STG},k}}{M} \leq B_{\text{STG},k} \quad k = 1, \ldots, n - 1 \tag{11.11}$$

where M is an arbitrary large number.

Owing to the existence of the binary variable, the above problem is a *mixed integer linear program* (MILP). The model may be solved using any commercial optimization software to achieve a global optimum solution.

11.5   CASE STUDY: WATER RECOVERY FOR A PVC PLANT

Figure 11.6 shows the process flow diagram for a PVC plant (Chan et al., 2008; Foo, 2012). The process is operated in a mixed-batch and continuous mode. Fresh vinyl chloride monomer (VCM) is mixed with the recycled VCM to be fed to a series of batch polymerization reactors. A huge amount of water is used as a reaction carrier in the reactor. Toward the end of the polymerization process, water is added to the reactor in order to maintain its temperature. At the end of the polymerization reaction, the reactor effluent which is in slurry form (containing PVC resin and water)

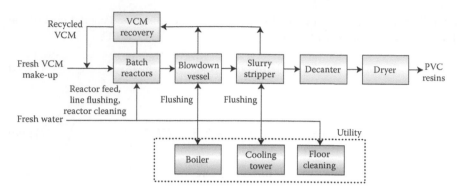

FIGURE 11.6   Process flow diagram for PVC plant.

is sent to a blowdown vessel; the latter serves as a buffer for the downstream purification processes that are operated continuously. The reactor effluent is sent to a steam stripper where the unconverted VCM is recovered from the reactor effluent. The stripped VCM, along with the VCM that emits from the reactors and blowdown vessel, is sent to the VCM recovery system. The stripper effluent next enters the decanter and the product dryer where moisture is removed before the PVC product is sent to the storage tank.

A huge amount of water is consumed in the plant as process and utility water. The latter includes general plant cleaning, boiler, and cooling tower make-up. Process water includes those for reactors feed, line flushing, blowdown vessel flushing, reactor cleaning, as well as for the stripper. Water recovery is to be conducted with direct reuse/recycle for overall water reduction. Suspended solid content is the main impurity in the concern for water recovery. Since the polymerization reactors are operated in batch mode, the associated water sinks only exist in certain periods of time. Their limiting water flows (F_{SKj}), concentrations (C_{SKj}), start (T^{STT}) and end (T^{END}) time are given in Table 11.1. Note that Table 11.1 includes data for three operational batches ($N-1$, N and $N+1$) that exist within the duration of interest, that is, the cycle time (0–14.78 h). In contrast, the decanter effluent is the only water source of the process, and is continuously produced, with its limiting flowrate and concentration (C_{SRi}) given in Table 11.2.

It is assumed that the PVC plant has an annual operating time of 7982 h. With the cycle time of 14.78 h, the plant has an annual production (*BATCH*) of 540 batches. The unit costs of freshwater (CT_{FW}) and wastewater treatment (CT_{WW}) are both given as US$1/t. It is further assumed that the piping cost may be ignored. Note that the storage cost is annualized to a period of 5 years (*YR*), with an interest rate (*IR*) of 5%.

Solving the optimization objective in Equation 11.10, subject to the constraints in Equations 11.1 through 11.8, 11.9b, and 11.11, the *TAC* of the BWN is determined as US$159,925,700. The MILP model is solved using LINGO v13. Other important results of the case study are summarized in Table 11.3. It is interesting to note that the freshwater and wastewater flows are identical to the minimum flow solution reported in the original work (Chan et al., 2008), where a pinch analysis technique was used. This is because, in this case study the freshwater and wastewater costs are dominating.

TABLE 11.1
Limiting Data for Water Sinks

Batch	SK_j	Operation	F_{SKj} (kg)	C_{SKj} (ppm)	T^{STT} (h)	T^{END} (h)
	2a	Line flushing	12,500	10	0	4
$N-1$	5a	Reactor cleaning	5400	20	0	4.5
	6a	Cooling tower	28,350	50	0	4.5
	7a	Flushing	8000	200	0	4
	1b	Reactor feed	105,000	10	0	3.33
	2b	Line flushing	15,625	10	9.28	14.28
	3b	Boiler make-up 1	3300	10	0	0.25
	4b	Boiler make-up 2	3300	10	8.00	8.25
N	5b	Reactor cleaning	6000	20	9.78	14.78
	6b	Cooling tower	93,114	50	0	14.78
	7b	Flushing	10,000	200	9.28	14.28
	8b	Floor cleaning 1	1500	500	0	0.25
	9b	Floor cleaning 2	1500	500	8.00	8.25
	1c	Reactor feed	105,000	10	10.28	13.61
$N+1$	3c	Boiler make-up 1	3300	10	10.28	10.53
	6c	Cooling tower	28,350	50	10.28	14.78
	8c	Floor cleaning 1	1500	500	10.28	10.53

TABLE 11.2
Limiting Data for Water Source

SR_i	Operation	Flowrate (kg/h)	C_{SRi} (ppm)
1	Decanter	9244	50

TABLE 11.3
Important Results for BWN Model in PVC Plant

Parameters (Unit)	Values
Freshwater (t/batch)	295,114
Wastewater (t/batch)	0
Total water cost ($/batch)	295,114
Number of storage	1
Storage size (t)	10,303
Capital (storage) cost ($)	2,441,658

11.6 CONCLUSION

This chapter presents an ATM for BWN synthesis, with an illustration on water recovery for a PVC case study. Note that the ATM may be extended to other water minimization problems, such as regeneration networks and property integration, as well as other batch process integration problems such as heat and mass exchange networks. Readers may refer to Foo (2010, 2012) for more details.

REFERENCES

Almató, M., Espuña, A., and Puigjaner, L., 1999. Optimisation of water use in batch process industries. *Computers and Chemical Engineering*, 23: 1427–1437.

Chan, J. H., Foo, D. C. Y., Kumaresan, S., Aziz, R. A., and Hassan, M. A. A., 2008. An integrated approach for water minimisation in a PVC manufacturing process. *Clean Technologies and Environmental Policy*, 10(1): 67–79.

Chen, C.-L.; Chang, C.-Y., and Lee, J.-Y., 2008. Continuous-time formulation for the synthesis of water-using networks in batch plants. *Industrial & Engineering Chemistry Research*, 47: 7818–7832.

Chen, C. L. and Lee, J. Y., 2008. A graphical technique for the design of water-using networks in batch processes. *Chemical Engineering Science*, 63: 3740–3754.

Chen, C.-L., Lee, J.-Y., Ng, D. K. S., and Foo, D. C. Y., 2010. A unified model of property integration for batch and continuous processes. *AIChE J*, 56: 1845–1858.

Chen, C.-L., Lee, J.-Y., Tang, J.-W., and Ciou, Y.-J., 2009. Synthesis of water-using network with central reusable storage in batch processes. *Computers and Chemical Engineering*, 33: 267–276.

Cheng, K. F. and Chang, C. T., 2007. Integrated water network designs for batch processes. *Industrial & Engineering Chemistry Research*, 46: 1241–1253.

Chew, I. M. L. and Foo, D. C. Y., 2009. Automated targeting for inter-plant water integration. *Chemical Engineering Journal*, 153(1–3): 23–36.

El-Halwagi, M. M., 1997. *Pollution Prevention through Process Integration: Systematic Design Tools*. Academic Press: San Diego, CA, USA.

El-Halwagi, M. M., 2006. *Process Integration*. Elsevier: Amsterdam.

Foo, D. C. Y., 2009. A state-of-the-art review of pinch analysis techniques for water network synthesis. *Industrial and Engineering Chemistry Research*, 48(11): 5125–5159.

Foo, D. C. Y., 2010. Automated targeting technique for batch process integration. *Industrial & Engineering Chemistry Research*, 49(20), 9899–9916.

Foo, D. C. Y., 2012. *Process Integration for Resource Conservation*. CRC Press: Boca Raton, FL, US.

Foo, D. C. Y., Manan, Z. A., and Tan, Y. L., 2005. Synthesis of maximum water recovery network for batch process systems. *Journal of Cleaner Production*, 13(15): 1381–1394.

Fernández, I., Renedo, C. J., Pérez, S. F., Ortiz, A., and Mañana, M., 2012. A review: Energy recovery in batch processes. *Renewable and Sustainable Energy Reviews*, 16: 2260–2277.

Furman, K. C. and Sahinidis, N. V., 2002. A critical review and annotated bibliography for heat exchanger network synthesis in the 20th century. *Industrial & Engineering Chemistry Research*, 41(10): 2335–2370.

Gouws, J. F. and Majozi, T., 2008. Impact of multiple storage in wastewater minimisation for multi-contaminant batch plants: Towards zero effluent. *Industrial & Engineering Chemistry Research*, 47: 369–379.

Gouws, J., Majozi, T., Foo, D. C. Y., Chen, C. L., and Lee, J.-Y., 2010. Water minimisation techniques for batch processes. *Industrial & Engineering Chemistry Research*, 49(19): 8877–8893.

Gundersen, T. and Naess, L., 1988. The synthesis of cost optimal heat exchange networks—An industrial review of the state of the art. *Computers and Chemical Engineering*, 6: 503–530.

Kim, J.-K. and Smith, R., 2004. Automated design of discontinuous water systems. *Process Safety and Environmental Protection*, 82(B3): 238–248.

Klemeš, J., Friedler, F., Bulatov, I., and Varbanov, P., 2010. *Sustainability in the Process Industry: Integration and Optimization*. New York: McGraw-Hill.

Klemeš, J. J. and Kravanja, Z., 2013. Recent developments in process integration. *Current Opinion in Chemical Engineering*, 2: 461–474.

Li, B. H. and Chang, C. T., 2006. A mathematical programming model for discontinuous water-reuse system design. *Industrial & Engineering Chemistry Research*, 45: 5027–5036.

Linnhoff, B., 1993. Pinch analysis: A state-of-art overview. *Transactions of IChemE (Part A)*, 71: 503–522.

Linnhoff, B., Townsend, D. W., Boland, D., Hewitt, G. F., Thomas, B. E. A., Guy, A. R., and Marshall, R. H., 1982. Last edition 1994. *A User Guide on Process Integration for the Efficient Use of Energy*. IChemE: Rugby.

Majozi, T., 2005. Wastewater minimization using central reusable storage in batch plants. *Computers and Chemical Engineering*, 29: 1631–1646.

Majozi, T., 2006. Storage design for maximum wastewater reuse in multipurpose batch plants. *Industrial & Engineering Chemistry Research*, 45: 5936–5943.

Majozi, T., 2010. *Batch Chemical Process Integration Analysis, Synthesis and Optimization*. Springer: Springer, London.

Majozi, T., Brouckaert, C. J., and Buckley, C. A., 2006. A graphical technique for wastewater minimization in batch processes. *Journal of Environmental Management*, 78: 317–329.

Majozi, T. and Gouws, J., 2009. A mathematical optimization approach for wastewater minimization in multiple contaminant batch plants. *Computers and Chemical Engineering*, 33: 1826–1840.

Ng, D. K. S., Foo, D. C. Y., and Tan, R. R., 2009a. Automated targeting technique for single-component resource conservation networks—Part 1: Direct reuse/recycle. *Industrial & Engineering Chemistry Research*, 48(16): 7637–7646.

Ng, D. K. S., Foo, D. C. Y., and Tan, R. R., 2009b. Automated targeting technique for single-component resource conservation networks—Part 2: Single pass and partitioning waste interception systems. *Industrial & Engineering Chemistry Research*, 48(16), 7647–7661.

Ng, D. K. S., Foo, D. C. Y., Tan, R. R., and El-Halwagi, M. M., 2010. Automated targeting technique for concentration- and property-based total resource conservation network. *Computers and Chemical Engineering*, 34(5): 825–845 (most cited articles 2010–2012).

Ng, D. K. S., Foo, D. C. Y., Tan, R. R., Pau, C. H., and Tan, Y. L., 2009c. Automated targeting for conventional and bilateral property-based resource conservation network. *Chemical Engineering Journal*, 149: 87–101.

Rockstrom, J., Steffen, W., Noone, K., Persson A., Chapin, F. S., Lambin, E. F., Lenton, T. M. et al., 2009. A safe operating space for humanity. *Nature* 461: 472–475.

Sandia National Laboratories 2005. Global water futures, p. 4. Retrieved February 1, 2010 from: www.sandia.gov.

Smith, R. 1995. *Chemical Process Design*. McGraw-Hill: New York.

Smith, R. 2005. *Chemical Process Design and Integration*. John Wiley & Sons: New York.

United Nations World Water Assessment Programme (WWAP, 2014). *The United Nations World Water Development Report 2014: Water and Energy*. UNESCO, Paris.

Wang, Y. P. and Smith, R., 1994. Wastewater minimisation. *Chemical Engineering Science*, 49: 981–1006.

FURTHER READING

Chen, C.-L. and Ciou, Y.-J., 2006. Superstructure-based MINLP formulation for synthesis of semicontinuous mass exchanger networks. *Industrial & Engineering Chemistry Research*, 45: 6728–6739.

Chen, C.-L. and Ciou, Y.-J., 2007. Synthesis of a continuously operated mass-exchanger network for a semiconsecutive process. *Industrial & Engineering Chemistry Research*, 46: 7136–7151.

Foo, C. Y., Manan, Z. A., Yunus, R. M., and Aziz, R. A., 2004. Synthesis of mass exchange network for batch processes—Part I: Utility targeting. *Chemical Engineering Science*, 59(5): 1009–1026.

Foo, C. Y., Manan, Z. A., Yunus, R. M., and Aziz, R. A., 2005. Synthesis of mass exchange network for batch processes—Part II: minimum units target and batch network design. *Chemical Engineering Science*, 60(5): 1349–1362.

12 Integration of Batch Process Schedules and Water Allocation Network

Xiong Zou, Yi Zhang, and Hong-Guang Dong

CONTENTS

12.1 INTRODUCTION

Water is one of the most important resources for process industry. For instance, a large amount of water is used for cooling and heating operations, absorption and extraction processes, equipment cleaning procedures, and so on. In turn, the generated wastewater in these processes requires treatment before releasing it into the environment. The increasing water consumed leads to not only higher freshwater purchase cost but also more investment on effluent treatment units. The water network (WN) is a collection of water using, regeneration, and treatment processes. Owing to the increasing cost of fresh water and more stringent environment regulations, the designing of more efficient process WNs has been a major topic of process system engineering since the 1990s. The WN consists of mass-transfer type water-using operations that could be regarded as a special mass exchanger network (MEN) and are closely related to heat exchanger networks (HENs) (Savulescu et al. 2005a,b; Dong et al. 2008). Inevitably, the techniques proposed in the HENs and MENs field extended to WNs and evolved into a more generalized methodology. There are noticeable achievements on this topic and it could be further classified into various categories based on the basic elements involved, which represent different provisions and complexities of the problem. The classification of WNs is summarized below:

- Units: water-using network (fixed load and/or fixed flow rate; mass transfer and/or non-mass transfer), regeneration and treatment network, total WN
- Substances: single-contaminant WN, multi-contaminant WN
- Operation model: continuous WN, semicontinuous WN, batch WN, mixed batch, and continuous WN
- Operation region: single plant WN, interplant WN, enterprise-wide WN
- Objective function: minimum freshwater consumption WN; minimum total annual cost (TAC) WN

In literature, most of the early research on WNs focused on continuous processes and the optimization of batch WNs has been ignored in the long past. The main reason is that the profit obtained by implementing a continuous industrial water recovery project is much higher than batch process due to the scale effects. Besides, the batch WN problems embedded with a time dimension are generally more complex than those for continuous processes. However, when batch production of low-volume and high-value-added fine chemicals and pharmaceuticals in the market rose, academic and industrial practitioners steadily gained a driving force to develop systematic design techniques to retrofit existing poorly designed batch WNs and an optimal design of a new efficient batch WN.

In this chapter, the one-step optimization methodology based on the state–time–space (STS) superstructure is presented as a mixed integer nonlinear programming (MINLP) formulation and aims for the optimal design batch schedules, water-reuse subsystems, and wastewater-treatment subsystems simultaneously. In Section 12.2, we will briefly review the milestone contributions in continuous WN designs and batch water-allocation networks with a fixed schedule, which build a theoretical basis for further research in integrated methodology. This is followed by a comprehensive comparison study of integrated design approaches in Section 12.3, which will focus on the state of the art and the limitations of current models and solution strategy. In Section 12.4, we will introduce the basic ideas of the STS superstructure and the resulting MINLP mathematical model; its solution strategy is also described. Next, a benchmark example is proposed in Section 12.5 to illustrate the applicability of the proposed approaches and the discussion on the complexity of networks is also provided. Conclusions are summarized in Section 12.6.

12.2 BACKGROUND

12.2.1 Classical Methods for Continuous WN Design

The problem of a water-allocation network was first defined by Takama et al. (1980) as the seminal paper in this area. The authors made an important contribution by converting the problem of maximizing water reuse into a problem of optimizing water allocation into a total system consisting of water-using units and wastewater-treating units. In the years that followed, the problem has received a lot of attention from the academic community and industrial practitioners. Various approaches have been presented and can roughly be divided into water pinch analysis and mathematical programming techniques.

Similar to pinch analysis technique for heat integration, water pinch technology relies on graphic representations and completes the synthesis tasks in two stages: freshwater flow rate targeting and network design. In most cases, network design techniques are actually dependent on the minimum flow rates established in the targeting stage. Professor Robin Smith and his co-workers initiated the water pinch analysis technique in the mid-1990s (Wang and Smith, 1994a,b; Kuo and Smith, 1997, 1998). Later this approach was extended and improved by a number of researchers (Dhole et al. 1996; Sorin and Bédard, 1999). A detailed description and summary of these insight-based techniques and case studies developed in the twentieth century are provided in a dedicated textbook (Mann and Liu, 1999). Approaches developed in 1994 through 2000 mainly focused on fixed load processes, that is, mass-transfer-based water-using units, such as solvent extraction, gas absorption, and so on; the main concern is to remove the contaminant load from the rich stream. Foo (2009) outlined various targeting and network design techniques developed for water reuse/recycle, generation, and wastewater-treatment network in the new century and to address fixed flow rate problems from the water sink and source perspective becomes a new trend (Hallale, 2002; Manan et al. 2004; Prakash and Shenoy, 2005; Bandyopadhyay et al. 2006; Alva-Argaez et al. 2007; Ng et al. 2007a,b).

For process engineers, the water pinch analysis technique is easy to use as a strong interactive graphical tool. The insights are important in practice because a better understanding of the WN allows the engineers to solve the problem efficiently with the consideration of many operating factors. However, the success of the insight-based approach has been mainly reported for single impurity systems and the method proved to have limitations as it failed to identify optimal solutions for multi-contaminant systems. Furthermore, in more complex situations, such as TAC optimization, simultaneous heat and water recovery, water integration across plants and integration of batch process schedules and water-allocation network, a more general methodology is required. As a consequence, in recent years the methods based on mathematical programming are becoming the mainstream approach to addressing the limitations of the "old" conceptual techniques.

Mathematical programming techniques often apply superstructure optimization with the ability to capture nonlinearities and all feasible structures. Hence, targeting and network design may be performed simultaneously. The superstructures for WNs have been shown in many papers, the mixer–unit–splitter (MUS) superstructure by Li and Chang (2011), the generalized superstructure by Ahmetović and Grossmann (2011) and the complete WN superstructure by Faria and Bagajewicz (2010) are three kinds of classical representations. Owing to the fixed number of water-using processes, the superstructures for WNs are always simpler than previous HEN superstructures, which makes it more attractive. In addition, the ease of modification is another important advantage of the superstructure optimization method. For example, the superstructures presented for mass-transfer water-using can be easily extended to non-mass-transfer units. Generally, the superstructure optimization model is performed with complex MINLP formulation, which consists of nonlinear material conservation equations and binary variables to account for fixed charges on equipment capital costs and for topological constraints. It is difficult to solve, particularly for global optimization of large-scale cases. The proposed representative solution strategies are listed in Table 12.1. The central claim made in this section is that synergistic combination of both approaches is the most effective alternative to date. The conceptual insight is very helpful in formulating better mathematical programming models by generating good initial points and pursuing effective solution procedures. To have a more comprehensive view of these approaches on continuous processes, the reader is referred to the paper written by Bagajewicz (2000) and Jezowski (2010).

TABLE 12.1
Solution Strategies for MINLP Model

Solution Strategies	Representative Literatures
Direct linearization	Bagajewicz et al. (2000)
	Bagajewicz and Savelski (2001)
Good initiation	Li and Chang (2007)
	Teles et al. (2008); Chew et al. (2009)
Sequential solution procedure	Takama et al. (1980)
	Hernandez-Suarez et al. (2004)
	Gunaratnam et al. (2005)
Meta-heuristic (stochastic) optimization	Tsai and Chang (2001)
	Lavric et al. (2005)
Global (deterministic) optimization	Zamora and Grossmann (1998)
	Floudas et al. (2005)

12.2.2   SYNTHESIS OF BATCH WATER-ALLOCATION NETWORK WITH FIXED SCHEDULE

In a continuous process, the main concern is the impurity concentration. However, for the batch processes, the time dimension becomes the major issue for water reuse, in addition to the concentration constraints. The discreteness of tasks in batch processes means water using and wastewater generation are not always active during the time horizon of interest. Hence, the availability of unit and water is another important factor that needs to be simultaneously addressed in a batch WN. Batch WNs are time dependent not only with respect to the water usage and wastewater generation but also to the network architecture. This underlines the importance of process scheduling in the optimization of batch WN. The time constraints add a new dimension to the WN problem, and it renders batch WN intrinsically multidimensional, even for single-contaminant processes. In addition, the operating condition of batch water-using processes also has important effects on the superstructure and formulations. For the truly batch operation, water is charged before the operation and discharged at a later duration. On the other hand, freshwater charging and wastewater discharging are normally carried out within the course of the operation for a semicontinuous water-using unit. For a detailed introduction, classifications, and remarks on water minimization techniques for discontinuous processes, we refer the reader to the review written by Gouws et al. (2010).

Like its continuous counterpart, the techniques to synthesis batch WNs can further be classified as insight-based and mathematical-based optimization techniques. Wang and Smith (1995) developed the first pinch-based approach for the fixed load batch WN problem. The major idea is to calculate targets for successive concentration intervals and for time intervals at each concentration interval. However, the technique is limited to semicontinuous single-contaminant processes with uniform inlet and outlet flows. Later, Majozi et al. (2006) further extended this targeting method to the truly batch processes based on the water demand profile, which consists of the horizontal segments in all time intervals. An inherent storage availability diagram is also presented to reduce the required storage by using idle-processing vessels for reusable water storage. The first work on the fixed flow problem is proposed by Foo et al. (2005), where a two-stage targeting and network design approach based on a time-dependent water cascade analysis (WCA) technique was developed. Subsequently, another algebraic targeting technique called the time-dependent concentration interval analysis (CIA) was developed by Liu et al. (2007). Chen and Lee (2008a) presented a quantity–time diagram graphical technique for the design of a batch fixed flow WN, which could capture the minimum storage capacity and the simplified network. However, all insight-based approaches fail to handle multiple-contaminant problems efficiently. Furthermore, these insight-based approaches could be very complex when the capital cost of network, many concentration and/or time intervals are considered. Hence, mathematical optimization techniques have been developed to handle these more complex cases.

Almost all mathematical-based techniques for batch WNs are based on a superstructure, which considers all of the water-using processes and possible interconnections between them. These superstructures always stem from their earlier work done in continuous processes and could be regarded as a variant of MUS superstructure or source–tank–sink representation. Similar to insight-based techniques, most of the mathematical optimization techniques determine the optimal batch WN with a predefined schedule (Almato et al., 1999; Kim and Smith, 2004; Majozi, 2005a; Li and Chang, 2006; Chen et al., 2008b, 2009, 2010; Ng et al., 2008; Shoaib et al., 2008; Tokos and Pintaric, 2009; Lee et al., 2013). Therefore, the inherent time dependence of batch processes is actually avoided rather than addressed. An interesting perspective on the batch WN with a pre-established schedule by Dogaru and Lavric (2011) should be mentioned. The authors regard a batch WN as a dynamic structure, which changes its topology at fixed time intervals delimited by events. During each time interval, the WN could be abstracted into an oriented graph. In these aforementioned methods, time is treated as a parameter and the minimum wastewater or TAC achieved is only a minimum for the specified schedule.

12.3   COMPREHENSIVE COMPARISON OF INTEGRATED DESIGN METHODOLOGIES

It is really an oversimplification to assume that the production schedule of an overall plant is known *a priori*. Thus, it is necessary to incorporate the scheduling framework into previous procedures. Integration of batch process schedules and water-allocation network means that both scheduling and water minimization comprise the same modeling framework, which allows time to be treated as a variable rather than a parameter. Thus, the global minimum water flow and/or cost can be determined for a variety of operational requirements. Majozi (2005b) first proposed a method for batch WN synthesis by combining the state sequence network (SSN) with an established underlying scheduling framework developed by Majozi and Zhu (2001), in which starting and finishing times become optimization variables. Later, this approach was extended to a multiple-contaminant system with multiple storage vessels (Gouws and Majozi, 2008; Majozi and Gouws, 2009). Gouws and Majozi (2009) exploited the inherent storage possibilities in idle-processing units to minimize single-contaminant wastewater. New water-reuse opportunities in the plant by a wastewater regenerator were included in the framework proposed by Adekola and Majozi (2011). However, the common drawback of these works is that the overall batch production scheme was not fully incorporated. The tasks of optimizing batch schedules, water reuse, and wastewater-treatment subsystems were performed individually in the past. An effective procedure to incorporate the three subsystems into a single comprehensive model is reported by Cheng and Chang (2007). The source–mixer–unit–splitter–sink superstructure is similar to that of Li and Chang (2006), in which a number of sinks, sources, water-using units, buffer tanks, and tasks are incorporated. Water reuse/recycle is only allowed through buffer tanks, and hence there are no direct connections between the units. However, in their study, not all possible network configurations can be generated by their superstructure and this superstructure fails to reflect the essential relationship between units and corresponding operations. Furthermore, to define one operation in a scheduling module, three binary variables are required in their discrete representation of the time horizon. The explosive binary dimension that could be encountered in more complex problems leads to the increase of computational time. Hence, a comprehensive superstructure with more efficient scheduling framework is the key to address the interactions between batch production and water allocation network (WAN). Zhou et al. (2009) integrated batch production and multiple-contaminants WAN into a simultaneous optimization model based on the unit-specific event-based continuous time representation and the modified state task network (STN) and state equipment network (SEN). However, the relationship between operations and units was not fully demonstrated in batch production. More importantly, assigning a fixed mixer and splitter to each unit during the whole time horizon inevitability leads to the preclusion of a class of optimal network structures, where the best cost-optimal scheme may actually lie. To circumvent these problems, Li et al. (2010) proposed a completely integrated and interconnected framework for modeling, optimizing, and illustrating the batch WAN in both time and space dimensions based on two STS superstructures. Chen et al. (2011) introduced a mathematical model, which consists of a short-term scheduling module and a water reuse and recycle module, for simultaneous scheduling and WN synthesis for batch processes. The scheduling framework is based on the resource-task network (RTN) (Chen and Chang, 2009) and the WN superstructure is similar to those by Chen et al. (2008b). Most recently, Chaturvedi and Bandyopadhyay (2014) extended the integrated optimization methodology to minimize the overall operating cost by utilizing multiple freshwater resources. To understand more of these approaches, a theoretical analysis and comparison of seven representative papers is provided in the following paragraphs. These seven papers are Majozi (2005b), Adekola and Majozi (2011), Cheng and Chang (2007), Zhou et al. (2009), Li et al. (2010), Chen et al. (2011), and Chaturvedi and Bandyopadhyay (2014). For the sake of convenience, we will use the first author to represent each paper in the following comparison. A summary of the comparison is presented in Table 12.2.

12.3.1 PROBLEM STATEMENT

A problem statement is a concise description of the issues that need to be addressed by the authors. First, the general conditions and parameters given in these seven papers are as follows:

1. Time horizon of interest
2. A specific process production flow sheet or production recipe for each product (units and tasks)
3. The mean processing time of each operation
4. The constant and variable term of processing time of each operation
5. The capacity of processing units and the maximum storage capacity for each material
6. Supply of each raw material and demand of each product
7. Freshwater resources specification data
8. Contaminant mass load in each operation of each contaminant (proportional coefficient correlating mass load/flow rate and amount proceed of each operation)

TABLE 12.2

Summary of Various Integrated Methodologies

	Majozi (2005b)	Adekola (2011)	Cheng (2007)	Zhou (2009)	Li et al. (2010)	Chen (2011)	Chaturvedi (2014)
Objective function	Min. freshwater max. profit	Max. profit min. effluent	Max. profit	Max. profit	Max. profit	Max. profit	Min. operating cost
Capital cost	No	No	Yes	Yes	Yes	No	No
Batch type	Truly batch	Truly batch	Semicontinuous	Semicontinuous	Semicontinuous	Truly batch	Truly batch
Production schedule	SSN	SSN	STN	STN	STS	RTN	SSN
Time representation	Continuous global event	Continuous global event	Discrete	Continuous unit-specific event	Discrete–continuous	Continuous global event	Continuous global event
Processing time	Constant	Constant	Constant	Variable	Variable	Variable	Variable
Processing cost	No	No	No	No	No	Yes	No
WN elements	Unit/tank	Unit/tank/regenerator	Unit/tank/treatment	Unit/tank/treatment/mixer/splitter	Unit/treatment/tank/junction/pipeline	Unit/tank	Unit (water using) multiple freshwater resources
Superstructure (WN)	M-(U/T)-S	M-(U/T-R)-S	M-U-S	STN/SEN	STS	M-U-S	Source-sink
Contaminant	Single	Multiple	Single	Multiple	Multiple	Single	Multiple
Flow types	Fixed load	Fixed load	Fixed load/flow	Fixed load/flow	Fixed load/flow	Fixed load	Fixed load
Water loss/gain	No	No	Yes	Yes	Yes	Yes	No
Formulation	MINLP/MILP	MINLP	MINLP	MINLP	MINLP	MINLP	MILP
Nonlinearity	Bilinear	Bilinear	Bilinear	Bilinear	Bilinear	Bilinear	Linear
Solution strategy	Direct	Relaxation	Direct	Two-stage hybrid deterministic and GA	Two-stage hybrid deterministic and GA	Reformulation linearization	Direct (sequential)
Solver	B&B	DICOPT	DICOPT	DICOPT-GA	DICOPT-GA	CPLEX/BARON	XPRESS

9. Water requirement of each unit (including the maximum inlet and outlet concentrations and flow rate for operations in the units)
10. The maximum reusable water storage of buffer tank
11. The performance and cost coefficients of wastewater-treatment units
12. The performance and cost coefficients of regeneration units
13. The amount and concentration of water gain/loss (if any) in a operation
14. The cost of raw material
15. The prices of products
16. The cost of freshwater
17. The cost coefficients of buffer tanks
18. The cost coefficients of junctions (splitters and mixers)
19. The cost coefficients of pipelines
20. The cost coefficients of handling material in production operations

Conditions (1) through (6) represent the production scheduling data, including product recipes and resource specification, equipment and storages types and capacities, durations of tasks, and time horizon of interest. The mean processing time of each operation as condition (3) is adopted by Majozi, Adekola and Cheng, on the contrary, Cheng, Zhou, Li, and Chaturvedi regard processing time as a function of batch size given in condition (4). Conditions (7) through (12) show the water-allocation network data and the cost coefficients of resources and equipment are given in conditions (13) through (20). Obviously, the given parameters and conditions depend on the definition of the research subject (types of equipment, operating mode, contaminant) and the objective functions.

12.3.2   OBJECTIVE FUNCTION

One of the major advantages of mathematical programming techniques is the flexibility and adaptability of the performance index, that is, the objective function. The simplest objective is to minimize the amount of freshwater required, which is similar to the minimization of wastewater generation or the maximization of profit with emphasis on operating costs of freshwater and/or effluent treatment. Another commonly used objective function is the maximization of profit. The choice of the objective function always depends on whether the production is given or not. In the situation where the production is given, the objective function is the minimization of effluent. When the production is not given, it is the maximization of profit, which equals the net income of production minus the operating cost of the WN. The minimization of freshwater operating costs is used by Chaturvedi. The maximization of profit is favored by all the rest authors. Besides, only Chen introduced the processing cost of material into the objective function. Furthermore, the investment costs are always assumed to be negligible. The capital cost of buffer tanks was first taken into account by Cheng. Later, Zhou and Li took into account the production profit and costs associated with freshwater supply, wastewater treatment, buffer tanks, as well as piping and junctions to link the entire network.

12.3.3 PRODUCTION SCHEDULE AND TIME REPRESENTATION

The main challenge of production scheduling is to specify the appropriate time interval and allocate the suitable unit(s) to perform every task of the batch process. First, we need a processing network to represent the batch production process. In the STN process representation used by Cheng and Zhou, raw materials, intermediates, and final products are represented as states denoted by circles, and an operation is represented as a task denoted by a rectangle box. The STN is concerned with short-term scheduling of a large number of batch processes that are specified by recipes. Chen proposed the schedule model based on RTN, which can be interpreted as an advanced form of the STN. The RTN regards all processing and storage equipment and utilities as resources and provides a unified framework for the description and solution of a variety of process scheduling problems. Majozi, Adekola, and Chaturvedi use SSN as a superstructure when dealing with batch WN design. In the representation of SSN, states are expressed explicitly and tasks/units are implicitly incorporated. Only states are considered in this network, thereby eliminating the need for task and unit binary variables required by the STN representation. Li developed STS as a representation of batch schedule by incorporating the concept of STN and Gantt chart into a combined SS structure. The STS integrates state, task, unit, and time axis in one graph which contains completed connection possibilities between units and states.

Time representation is the key criterion to formulate scheduling mathematical models, which can be classified into two typical categories: discreet-time models and continuous-time models. The discrete-time approach divided the entire time horizon into a finite number of identical time intervals and events that only occur at the beginning and end of the intervals (Kondili et al. 1993). To precisely approximate the original problem, the model contains a great number of time intervals to result in an enormous combinatorial problem of intractable scale for real-world problems. Owing to these limitations, only Chen adopted the discrete-time model. Other researchers prefer to use continuous-time models, in which the starting and ending time can happen at any point in the time horizon, leading to a smaller size and requiring less computational effort. Continuous-time models can be further divided into three categories: slot-based, global event-based, and unit-specific event-based formulations. Slot-based formulation is first presented to reduce the binary variable in which time duration is presented as ordered block of unknown slots or intervals (Pinto and Grossman, 1994; Karimi and Mcdonald, 1997). In the global event-based model (Castro et al. 2001; Majozi and Zhu, 2001; Maravelias and Grossmann, 2003), events are used for all the units and the task in any unit starts and ends at the event points. This continuous-time mode was established by Majozi, Adekola, Chen, and Chaturvedi, respectively. The scheduling model of Zhou is based on the continuous-time formulation for short-term scheduling of batch processes proposed by Ierapetritou and Floudas (1998). This formulation introduces unit-event points which may vary from different units, decoupling the task from the unit. Later, Li adapted this representation by combining both discrete- and continuous-time formulation to adjust the relationship between the operating time and the amount processed to reflect the scenarios of different plants. Shaik and Floudas (2008) presented a comparative study of all existing approaches in the scheduling of batch processes and found that unit-specific event-based models require a less number of events, have the least problem statistics, and have a faster computational performance in most scenarios. It is worth noting that no matter what kind of continuous model is chosen, the reason why the continuous model is superior to the conventional discrete model is the fewer number of time points. In this connection, the continuous characteristics would raise the complexity of the model, but the number of binary variables can be reduced remarkably.

12.3.4 ELEMENTS OF BATCH WN

The purpose of a superstructure for a batch WN is to represent all the water-related units in a unified manner while considering all possible interconnections between them. Most of researches are based on a variant of the MUS superstructure. Chaturvedi developed a methodology, where all water reuse/recycle between multiple freshwater sources and sinks are obtained. As the starting and finishing times of batch water-using tasks are dependent on the production schedule as the inherent time dependence in batch processes, buffer tanks are equipped by Majozi and Chen for the temporary storage of reusable water to partially bypass the time limitation. There are two types of regenerators: the one with a fixed outlet concentration and the other with constant removal ratio. The potential for water reuse by the latter wastewater regenerator was introduced by Adekola. The author assumes that when the regenerator is active, it operates in a continuous mode with steady inlet and outlet streams, until the sink process is satisfied and the wastewater to the regenerator only comes from the storage tank. The tasks of optimizing batch schedules, water-reuse subsystems, and wastewater-treatment subsystems simultaneously were performed by Cheng, Zhou, and Li. More detailed discussion on the superstructures is provided in Section 12.4.

12.3.5 MATHEMATICAL FORMULATION AND SOLUTION STRATEGY

The truly batch processes and fixed load problems were dealt by Majozi, Adekola, Chen, and Chaturvedi, while Cheng, Zhou, and Li focused both on fixed load and fixed flow problems for the semicontinuous processes. The difference mainly lies in the operation types and scheduling constraints. The general mathematical constraints used in different models can be categorized under two groups: scheduling constraints and WAN-related constraints. Constraints related to allocation, capacity, duration, and sequence are included in scheduling constraints (12.1) through (12.6) whereas the constraints related to flow and concentration requirements of units are included in WAN-related constraints (12.7) through (12.12):

1. Material balance for processing units
2. Allocation constraints for processing units
3. Capacity constraints of each operation
4. Storage constraints of units and tanks
5. Duration constraints
6. Demand and purchase constraints
7. Overall/component material balance of water sources

8. Overall/component material balance of water sinks
9. Overall/component material balance of tanks
10. Overall/component material balance of water treatment units
11. Water treatment/regeneration/reuse constraints
12. Structure constraints

Most of the models finally form an MINLP problem, except the MILP formulation proposed by Chaturvedi to minimize the operating cost of batch process while keeping the overall production at its maximum. Generally, the binary variable is introduced to express the existence of a state at a given time point. Another set of binary variables to express the connection between storage vessels to water-consuming units was proposed by Majozi. Besides, Cheng introduced a set of binary variables to signify the existence of a pipeline. Similarly, Zhou and Li proposed binary variables to judge the existence of junctions (splitters and mixers). The component material balances that contain the product between the concentration and flow rate may yield the bilinear constraints and lead to a nonconvex problem. The product between the continuous variable of flow rate and binary variable of operation existence is another source of bi-linearity. There are three kinds of solution strategies in these seven papers to solve the resulting MINLP problem: Relaxation, linearization and hybrid deterministic and stochastic method.

12.4 SIMULTANEOUS OPTIMIZATION APPROACH FOR BATCH WATER-ALLOCATION NETWORK DESIGN

12.4.1 The Evolution of STS Superstructure

Similar to any other mathematical optimization study in process synthesis, superstructure is one of the core topics in the synthesis of batch WN with a variable schedule. Oftentimes the structure of the problem is not only as important as the solution, but often determines the solution. As mentioned in Section 12.3, it is necessary to build the superstructure not only for WNs but also for process scheduling. And the interaction between both of them should also be taken into consideration. Most of the previous research focused on the superstructure for WN and put less value on representation for process scheduling, let alone the interaction and integration of both. Li et al. (2010) attempted to address this problem based on the STS superstructure. In the following section, we will explain the idea of the STS superstructure and its evolution from previous approaches. The discussion on its advantages and limitations is provided to inspire new research into this topic.

For batch process synthesis, the basic elements of superstructures are identified as states, tasks, equipment, and the basic dimensions of space, time, and contaminants. The standard of superstructure is the exact representation of system elements and dimensions. There was no general framework that clearly illustrates these basic elements and dimensions simultaneously. In STS methodology, it still required two superstructures to capture the characteristics of batch WNs, one for batch production schedule and the other for the WN. It should be noted that the interactions between schedules and network are addressed in this framework.

The basic idea of the STS superstructure is to incorporate the concepts of STN, SEN, and Gantt chart into the modified state space (SS) notion. The predominant superiority of this superstructure over previous ones is that all batch processes can be projected onto the time and space dimension, thereby facilitating simultaneous optimization of batch WANs in both dimensions. The STS superstructures are derived from SS superstructure in the traditional process synthesis field, which constitutes a significant departure from the previous superstructure-based model in that it does not contain simplifying assumptions and completed connections are achieved. Of course, the completed network commonly includes more redundant structures that lead to more complex problems.

12.4.1.1 STS Superstructure for Batch Production Schedule

Processing networks such as STN, RTN, and recipe diagrams are always used to represent batch and continuous complex production recipes and flowsheets in which different products follow different production routes. These representations focus on the processing sequences and do not consider the time dimension and the relation between operations and units. The Gantt chart is a very useful tool to illustrate the production schedule results with collecting information on time, equipment, and tasks in one graph. The main drawback is that the Gantt chart fails to display the connections between units and equipment. By combining the processing network and Gantt chart in the framework of the SS superstructure, the STS superstructure for the batch schedule is proposed and shown in Figure 12.1. The left-hand side

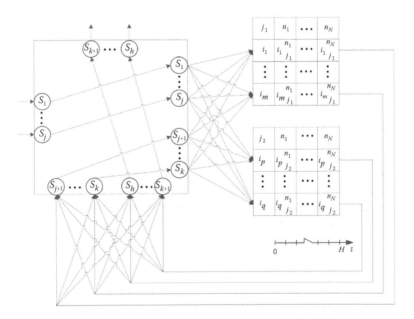

FIGURE 12.1   STS superstructure for batch schedule. (Reprinted with permission from Li, L.-J., R.-J. Zhou, and H.-G. Dong. State-time-space superstructure-based MINLP formulation for batch water-allocation network design. *Ind. Eng. Chem. Res.* 49:236–251. Copyright 2010, American Chemical Society.)

block is the so-called distribution network (DN) which contains three groups of materials—raw material (on the left side), intermediate product (on the right and bottom side), and final product (on the upper side). Each group of material contains two states—the state to be consumed (sold) and the state to be produced (brought). Moreover, in the process operator (OP) block, which contains equipment sub-blocks on the upper side and the time axis on the bottom side, each unit is illustrated as a block with many sub-blocks that correspond to the operations performing at that event point. To be more specific, supposing there are a total of M operations performing in each unit and N event points to be partitioned, the unit can be viewed as a set of $M \times N$ subgrids, which also could be regarded as a small Gantt chart for one unit. Material states and units should be connected with streams to reflect the real situation. Finally, a time axis is set to indicate the starting and ending point of each time interval.

12.4.1.2   STS Superstructure for Batch Water Networks

Most of superstructures in the literature for batch WNs are very similar to their continuous counterparts and could be also regarded as variants of MUS or source–tank–sink superstructure. The major differences between these superstructures are the number and types of units involved and the connections between them. Two distinctive characteristic of batch WN—multipurpose and time dependent—are often overlooked by most previous batch WN superstructures. A multipurpose time-dependent process means more than one operation can be specified to certain equipment in a time event. Thus, the optimal number of processing operations and the corresponding assignment of operations to equipment in a production cycle are not fixed *a priori*. The continuous-based superstructure may fail to wholly reflect the relationship between equipment and corresponding operations simultaneously and can often involve ambiguities when applied to batch processes.

Zhou et al. (2009) first introduced STN/SEN-based representation to deal with multipurpose features in the SS superstructure framework, which is viewed as a system of two interconnected blocks—OP and DN. Time dimension has also been incorporated into this structure to represent the tasks and operations running on different event point. In the STN-based representation, the modified OP can be further divided into three sub-blocks: equipment which corresponds to units in batch production, buffer tanks, and distributed wastewater-treatment units. Each equipment in the OP block is illustrated as a set of operations that can be performed in the equipment itself. To be more specific, suppose there are N event points and each equipment can be viewed as N potential operations. Correspondingly, the potential splitters and mixers for all equipment attached on DN block are also divided into a set of N "fictitious" splitting and mixing nodes, which correspond to operations performed at different event points in certain equipment. In this way, time dimension is integrated in the superstructure by dividing equipment, splitters, and mixers into operations, splitting, and mixing nodes, respectively. Moreover, operations performed in all time intervals can be clearly illustrated.

The difference between SEN-based and STN-based SS superstructures lies in the first sub-block of the OP block. If we view the operation as a set of tasks that can be performed in the equipment itself correspondingly, an STN-based SS superstructure is adopted. On the other hand, if we intend to capture the actual pipeline network and network complexity of the resulting system design, a SEN-based SS superstructure is suggested. Moreover, it should be noted that an STN-based formulation can always be translated into the equivalent SEN-based representation with Equations 12.1 and 12.2. It should also be noted that only one kind of operation can be performed in exclusive equipment, which contains the buffer tanks, wastewater-treatment units, water sinks, and freshwater sources during the time horizon.

$$f_{ua,t}^{in} = \max_{i} \left\{ f_{i,t}^{in} \right\} \quad \forall ua \in UA, \ i \in I, \ t \in T \tag{12.1}$$

$$f_{ua,t}^{out} = \max_{i} \left\{ f_{i,t}^{out} \right\} \quad \forall ua \in UA, \ i \in I, \ t \in T \tag{12.2}$$

However, there is a one-to-one correspondence between mixers/splitters and water users in DN, and the possibilities of streams bypassing water users without mass exchange were completely eliminated. As a result, streams can only mix once before entering the water user and in the same way they are only allowed to split once after leaving the water user. To overcome these deficiencies, a unified superstructure should also be introduced into the model to capture more reasonable structures systemically and visually. Figure 12.2 depicts the STS superstructures on which the batch WAN mathematical model is based.

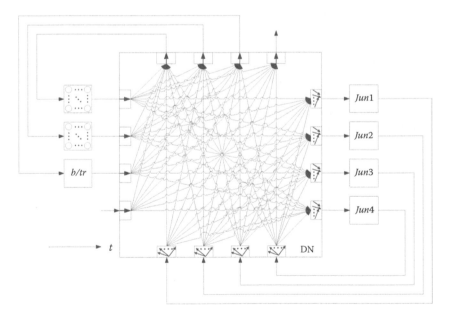

FIGURE 12.2   STS superstructure for batch WAN. (Reprinted with permission from Li, L.-J., R.-J. Zhou and H.-G. Dong. State-time-space superstructure-based MINLP formulation for batch water-allocation network design. *Ind. Eng. Chem. Res.* 49:236–251. Copyright 2010, American Chemical Society.)

Specifically, the overall STS framework for the WAN design is viewed as a system of two interconnected blocks—DN block and OP block. Moreover, the OP block consists of two sub-blocks, OP_UNIT and OP_JUN. In this superstructure, all water users, potential wastewater-treatment units, and buffer tanks are placed in the OP_UNIT and all junctions are set in the OP_JUN. Moreover, at the bottom of the OP_UNIT, a time axis is set to indicate the starting and ending interval of each operation.

In the DN block, the input streams, which contain the internal inputs (recycle and reuse streams) and external inputs (freshwater streams), are considered. In each time interval, input streams can be connected to all exits which include an OP block or the environment. On the basis of the characteristics of the structure, only internal input/output from OP_JUN are allowed to be split/mixed. Splitting/mixing from the unit in OP_UNIT is strictly forbidden in this structure.

Every single water-consuming unit should be posed in the OP_UNIT based on the process requirements, whereas water treatment units and buffer tanks are viewed as off-line candidate equipment, which are held in a possible installation. The operations in a water-using unit can be divided into four categories: operations generating wastewaters, operations with nonidentical charging and discharging, operations with identical charging and discharging, and operations consuming wastewaters. Similar to the superstructure in batch production, each unit operations perform at certain event points.

Before introducing the OP_JUN sub-block, the concept of multistage mixing/splitting is discussed. Suppose unit 1, unit 2, and unit 3 are three water users operating in the same time interval and their inlet flow conditions are shown in Figure 12.3. The inlet flow configuration can be revamped by allowing the stream to mix and split on a sequential basis when multistage mixing/splitting is introduced. Moreover such policy, which has never been revealed before, extends the possibilities of interconnection. According to such analysis, junctions are introduced into the superstructure to obtain all the possible and reasonable matches from OP_JUN to DN blocks including multistage splitting and mixing policies.

All junctions can be equally shared by different units during the time horizon, so we can obtain a better balanced and higher integrated WAN. Finally, it is imperative

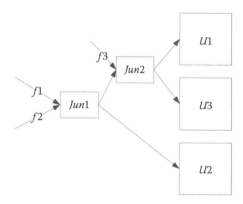

FIGURE 12.3 Multistage mixing/splitting.

to place sufficient junctions to provide enough opportunities to match all streams with different concentration grades. So the initial number of junctions is determined heuristically as the total number of the units.

It should be noted that the OP block in batch production schedule could be translated into the OP_UNIT sub-block in the WN by considering the water-using characteristics of production unit. So time dimension is shared by two STS superstructures. This framework displays all the information of batch schedule and the WN clearly and visually. On the contrary, some elements and dimensions are implicit in other methodologies.

12.4.2 MODEL DESCRIPTION

The overall integrated mathematical model is made up of two modules. One of the modules focuses on batch schedules and the other on the water-allocation network. For the sake of clearness and simplicity, these two parts will be presented separately in the following paragraphs, where the indices, sets, parameters, and variables used are annotated in the Appendix.

12.4.2.1 Objective Function

In this scheme, the objective function is to maximize the overall profit in a production cycle, while taking into account the profit of production and cost associated with freshwater supply, wastewater treatment, buffer tanks, as well as pipelines and junctions linking to the entire network. The objective function is decoupled into several terms to facilitate a clear-cut description.

1. Net income in production:

$$\text{Net profit in production} = \sum_{s_p \in S_p} \sum_{n \in N} PR_{s_p} \cdot d_{s_p,n} - \sum_{s_m \in S_m} \sum_{n \in N} PR_{s_m} \cdot r_{s_m,n} \quad (12.3)$$

where PR_{s_p}, PR_{s_m} represent, respectively, the cost coefficient of raw materials (s_m) and final products (s_p); $d_{s_p,n}$ denotes the amount of product s_p sold at the event point n, and $r_{s_m,n}$ denotes the amount of raw material s_m purchased at the event point n.

2. Overall cost of network:

$$\text{Overall cost of network} = \text{Cost of fresh water} + \text{Cost of treatment unit}$$
$$+ \text{Cost of buffer tanks} + \text{Cost of junctions and pipelines}$$

$$(12.4)$$

Terms on the right-hand side of Equation 12.4 are estimated as follows:

$$\text{Cost of freshwaters} = \sum_{sa} PR_{sa} \sum_{t \in T} f_{sa,t}^{\text{in}} \cdot \Delta t \quad (12.5)$$

$$\text{Cost of wastewater treatment} = \sum_{tr \in Tr} PR_{tr} \sum_{t \in T} f_{tr,t}^{in} \cdot \Delta t \qquad (12.6)$$

$$\text{Cost of buffer tanks} = \sum_{b \in B} [PR_{b1} \cdot nb(b) + PR_{b2} \cdot v_b^{max} \cdot \exp(PR_{b3})] \qquad (12.7)$$

$$\text{Cost of junctions and pipelines} = \sum_{jun \in JUN} PR_{jun} \cdot n(jun)$$
$$+ \sum_{eq \in EQ} \sum_{eq' \in EQ'} PR_{pipe} \cdot ne(eq,eq') \qquad (12.8)$$

where PR_{sa}, PR_{tr} are the unit cost of freshwater and wastewater treatment over the time horizon; PR_{b1}, PR_{b2}, and PR_{b3} are the cost coefficients concerned with the installation of buffer tanks; PR_{jun} and PR_{pipe} represent, respectively, the cost of each junction and pipeline, respectively.
3. Objective function:
    The overall objective function is:

$$f(\text{obj.}) = \text{net profit in production} - \text{overall cost of network} \qquad (12.9)$$

12.4.2.2 Formulations for a Batch Process Schedule

The batch process schedule model is discussed in this section. As aforementioned, unit-specific event-based time representation is efficient for most of schedule problems, but it is unable to mark the intervals in which all operating status are kept constant and the optimization strategy has to be executed interactively between different components. Combining both discrete- and continuous-time formulation, the new schedule model allows the starting and ending point of operations lying in the interval boundaries. In this model, the operating time of batch operations is expressed in terms of the amount of material processed. To be more specific, the discrete features are used to identify the time intervals in which the water-consuming conditions of all operations are kept identical, while the event points are adopted to specify the operations performed in the units.

1. Distribution network:
    a. Material balances: The constraints (12.10) and (12.11) are the material balance of the state s from the event point $n-1$ to n. The constraint (12.12) indicates that the amount of each state should be kept identical at each starting and ending point of time in an operation cycle.
    b. Storage constraints: Constraints (12.13) and (12.14) represent that the storage capacity of any state (including initial state) should be less than the amount of maximum storage.
    c. Demand and purchase constraints: Constraints (12.15) and (12.16) indicate the bound of the actual quantities of state s purchased and sold at

event point n. The actual amount of state s purchased and sold at every event point n should follow the constraints (12.15) and (12.16).

2. Process operators:

   a. Allocation constraints: Constraint (12.17) indicates that only one of the tasks can be performed at each unit j and at each event point n. $yv(j,n)$ should be equal to 1 if any one of the $wv(i,n)$ is activated, and only one $wv(i,n)$ can be activated at the same time due to the characteristic of the process operation.

   b. Capacity constraints: Constraint (12.18) limits the range of actual amount of B_{ij^n} if $wv(i, n)$ is equal to 1.

   c. Duration constraints: Constraints (12.19) through (12.23) give the duration restriction. The constraint (12.19) indicates that the processing time is dependent on the amount of the state being processed. The constraint (12.20) provides the upper and lower bound of variable b_{ij^n}. The constraints (12.21) and (12.22) guarantee that the starting and ending time of any operations are only allowed at interval boundaries. Moreover, constraint (12.23) provides the method to estimate $\beta_{i,j}$.

   d. Time sequence constraints for the same task in the same units: Constraints (12.24) through (12.26) indicate that the task i starting at the event point $n + 1$ should start only after the end of the same task performed in the same unit j, which has already performed at event point n.

   e. Time sequence constraints for different tasks in the same units: Constraint (12.27) deals with the conflict that the task i should await until the end of the task i' which has been already performed at the event point n, when tasks i and i' are performed at the same unit.

   f. Time sequence constraints for the completion of previous tasks: Constraint (12.28) restricts that the task i should only start after the completion of all tasks performed at the previous event points at the same unit j.

   g. Time horizon constrains: Constraints (12.29) and (12.30) indicate that each task i should start and end within the range of time horizon H.

$$ST_{s,n} = ST_s^{in} - \sum_{i \in I_s} \rho_{s,i}^c \sum_{j \in J_i} B_{ij^n} + r_{s,n} - d_{s,n} \quad \forall s \in S, n \in N(n = n_1) \quad (12.10)$$

$$ST_{s,n} = ST_{s,n-1} - \sum_{i \in I_s} \rho_{s,i}^c \sum_{j \in J_i} B_{ij^n} + \sum_{i \in I_s} \rho_{s,i}^p \sum_{j \in J_i} B_{ij^{n-1}} + r_{s,n} - d_{s,n} \quad \forall s \in S, n \in N(n > n_1)$$

$$(12.11)$$

$$ST_{s,N} = ST_s^{in} \quad \forall s \in S \quad (12.12)$$

$$ST_{s,n} \leq ST_s^{max} \quad \forall s \in S, n \in N \quad (12.13)$$

$$ST_s^{in} \leq ST_s^{max} \quad \forall s \in S \quad (12.14)$$

$$r_{s,n} \geq R_{s,n}^{\max} \quad \forall s \in S, n \in N \tag{12.15}$$

$$d_{s,n} \geq D_{s,n}^{\min} \quad \forall s \in S, n \in N \tag{12.16}$$

$$\sum_{i \in I_j} wv(i,n) = yv(j,n) \quad \forall j \in J, n \in N \tag{12.17}$$

$$BC_{i,j}^{\min} \cdot wv(i,n) \leq B_{i_j^n} \leq BC_{i,j}^{\max} \cdot wv(i,n) \quad \forall i \in I, j \in J_i, n \in N \tag{12.18}$$

$$T_{i_j^n}^f = T_{i_j^n}^s + wv(i,n) \cdot \alpha_{i,j} + b_{i_j^n} \cdot \beta_{i,j} \quad \forall i \in I, j \in J_i, n \in N \tag{12.19}$$

$$\frac{B_{i_j^n}}{BC_{i,j}} \leq b_{i_j^n} < \frac{B_{i_j^n} + BC_{i,j}}{BC_{i,j}} \quad \forall i \in I, j \in J_i, n \in N \tag{12.20}$$

$$T_{i_j^n}^s = \chi_{i_j^n} \cdot \Delta t \quad \forall i \in I, j \in J_i, n \in N \tag{12.21}$$

$$T_{i_j^n}^f = \varphi_{i_j^n} \cdot \Delta t \quad \forall i \in I, j \in J_i, n \in N \tag{12.22}$$

$$\beta_{i,j} = \frac{t_{i,j}^{\max} - t_{i,j}^{\min}}{BC_{i,j}^{\max} - BC_{i,j}^{\min}} \cdot BC_{i,j} \quad \forall i \in I, j \in J_i \tag{12.23}$$

$$T_{i_j^{n+1}}^s \geq T_{i_j^n}^f - T \cdot [2 - wv(i,n) - yv(j,n)] \quad \forall i \in I, j \in J_i, n \in N(n \neq N) \tag{12.24}$$

$$T_{i_j^{n+1}}^s \geq T_{i_j^n}^s \quad \forall i \in I, j \in J_i, n \in N(n \neq N) \tag{12.25}$$

$$T_{i_j^{n+1}}^f \geq T_{i_j^n}^f \quad \forall i \in I, j \in J_i, n \in N(n \neq N) \tag{12.26}$$

$$T_{i_j^{n+1}}^s \geq T_{i'_j^n}^f - T \cdot [2 - wv(i',n) - yv(j,n)]$$
$$\forall j \in J, i \in I_j, i' \in I_j, i \neq i', n \in N(n \neq N) \tag{12.27}$$

$$T_{i_j^{n+1}}^s \geq \sum_{n' \in N, n' \leq n} \sum_{i' \in I_j} \left(T_{i'_j^{n'}}^f - T_{i'_j^{n'}}^s \right) \quad \forall j \in J, i \in I_j, n \in N, n' \in N, n \neq N \tag{12.28}$$

$$T_{i_j^n}^s \leq T \quad \forall i \in I_j, j \in J, n \in N \tag{12.29}$$

$$T_{i_j^n}^f \leq T \quad \forall i \in I_j, j \in J, n \in N \tag{12.30}$$

To integrate the batch schedule and batch WAN in one step, the status of all batch operations in each interval (Δt) should be specified. Therefore, we proposed the following two groups of constraints.

i.   Operations with identical charging and discharging:
     ε is a sufficient small positive, and constraint (12.31) enforces $w(i_j^n, t)$ to be 1 if t locates between the starting and ending point of the interval. Constraint (12.32) enforces $w(i_j^n, t)$ to be 0 if t locates at the starting point. In the same way, constraint (12.33) enforces $w(i_j^n, t)$ to be 1 if t locates at the ending point.

$$\frac{1}{H^2} \cdot \left(\Delta t \cdot t - T_{i_j^n}^s - \varepsilon \right) \cdot \left(T_{i_j^n}^f - \Delta t \cdot t \right) < w\left(i_j^n, t \right) \quad i \in SB \cup UB \cup OB, j \in J_i, n \in N, t \in T$$

$$\tag{12.31}$$

$$w\left(i_j^n, t \right) < \frac{1}{H} \left(\Delta t \cdot t - T_{i_j^n}^s \right) + 1 \quad i \in SB \cup UB \cup OB, j \in J_i, n \in N, t \in T \tag{12.32}$$

$$w\left(i_j^n, t \right) \leq \frac{1}{H} \left(T_{i_j^n}^f - \Delta t \cdot t \right) + 1 \quad i \in SB \cup UB \cup OB, j \in J_i, n \in N, t \in T \tag{12.33}$$

ii. Operations with nonidentical charging and discharging:
     For operations with nonidentical charging and discharging (such as operation ua), variable $T_{i,j,n}^m$ is introduced to specify the time boundary between charging and discharging. The boundary points can be determined with the following constraints:

$$T_{i_j^n}^M = \gamma_{i_j^n} \cdot \Delta t \quad \forall i \in I, j \in J_i, n \in N \tag{12.34}$$

$$\frac{1}{2} \left(T_{i_j^n}^s + T_{i_j^n}^f \right) \leq T_{i_j^n}^M < \frac{1}{2} \left(T_{i_j^n}^s + T_{i_j^n}^f + \Delta t \right) \quad \forall i \in I, j \in J_i, n \in N \tag{12.35}$$

$$\frac{1}{H^2} \cdot \left(\Delta t \cdot t - T_{i_j^n}^s - \varepsilon \right) \cdot \left(T_{i_j^n}^M - \Delta t \cdot t \right) < w^{in}\left(i_j^n, t \right) \quad i \in UA, j \in J_i, n \in N, t \in T \tag{12.36}$$

$$w^{in}\left(i_j^n, t \right) < \frac{1}{H} \left(\Delta t \cdot t - T_{i_j^n}^s \right) + 1 \quad i \in UA, j \in J_i, n \in N, t \in T \tag{12.37}$$

$$w^{in}\left(i_j^n, t\right) \le \frac{1}{H}\left(T_{i_j^n}^M - \Delta t \cdot t\right) + 1 \quad i \in UA, j \in J_i, n \in N, t \in T \qquad (12.38)$$

$$\frac{1}{H^2} \cdot \left(\Delta t \cdot t - T_{i_j^n}^M - \varepsilon\right) \cdot \left(T_{i_j^n}^f - \Delta t \cdot t\right) < w^{out}\left(i_j^n, t\right) \quad i \in UA, j \in J_i, n \in N, t \in T \quad (12.39)$$

$$w^{out}\left(i_j^n, t\right) < \frac{1}{H}\left(\Delta t \cdot t - T_{i_j^n}^M\right) + 1 \quad i \in UA, j \in J_i, n \in N, t \in T \qquad (12.40)$$

$$w^{out}\left(i_j^n, t\right) \le \frac{1}{H}\left(T_{i_j^n}^f - \Delta t \cdot t\right) + 1 \quad i \in UA, j \in J_i, n \in N, t \in T \qquad (12.41)$$

12.4.2.3 Formulations for Water-Allocation Network

The superstructure-based mathematical models for batch WAN are formulated and discussed in this section.

1. OP:
   a. Wastewater-treatment units: Constraints (12.42) through (12.46) give mass balance and boundary of the flow rate and concentration of the wastewater-treatment unit.
   b. Buffer tanks: Since the batch operations can be viewed as the periodic operations, the initial state of every single buffer tank should be kept identical with its final process conditions. Constraints (12.47) through (12.52) indicate the existence/nonexistence and volume capacity of buffer tanks.
   c. Operations in batch production: The water generating/consuming rate is proportional to the processing volume if the operations run without material income or discharge (such as SB and OB). On the other hand, for other operations (such as UA and UB), the mass exchange process contains a contaminant load which is determined by the production and transferred to a stream. In this section, five categories of constrains are introduced to identify the characteristics between schedules and network.
      i. Operations generating wastewaters:
         Constraint (12.53) denotes that the water-generating rate of sb_j^n is proportional to the process throughput of the corresponding task in each time unit and λ_{sb} is the relevant proportion coefficient. Moreover, constraints (12.54) through (12.56) ensure that once sb_j^n is performed in interval t, wastewater with fixed concentration grade will be released accordingly.
      ii. Operations with nonidentical charging and discharging:
         The constraint (12.57) indicates that the accumulated mass load of pollutant k in ua_j^n ($M_{ua_j^n,k}$) is proportional to the process throughput

in one time internal. Constraints (12.58) and (12.59) give the mass balance by assuming a fixed amount of water loss (DV_{ua}^{loss}) and relevant pollutant concentration ($C_{ua,k}^{loss}$). Constraints (12.60) through (12.71) introduce the restrictions to fix the actual flow rate and pollutant concentration to optimal level when ua_j^n is implemented.

iii. Operations with identical charging and discharging:
Constraints (12.72) through (12.74) indicate that the instantaneous mass load of the pollutant k in ub_j^n during the interval t ($m_{ub_j^n,k,t}$) is proportional to the production throughput in one time interval. Constraints (12.77) through (12.79) give the limitation of the range of the actual flow rate and concentration when ub_j^n is performed.

iv. Operations consuming wastewaters:
Constraints (12.81) through (12.83) limit the range of water-consuming rate and concentration to the desirable levels when ob_j^n is being performed.

v. Relationship between operations and units:
We assume that only one operation can be assigned to each unit in each time interval. Constraints (12.84) through (12.87) indicate that the water consuming and/or generating rate of each operation are actually the inlet and/or outlet flow rates of the corresponding unit.

vi. Junctions:
Constraints (12.88) through (12.90) are about the relevant mass balance for all junctions. In addition, to identify the existence or nonexistence of a junction, the inequalities (12.91) and (12.92) are introduced. The constraint (12.91) is introduced to identify if *jun* is occupied in the interval t, while the constraint (12.92) ensures that *jun* exists if *jun* is occupied in more than one time interval.

2. Distribution network:

a. System external inputs (freshwater sources): Constraints (12.93) and (12.94) restrict the outlet flow rate of sa and the concentration of pollutant k in waters generated from *sa*.

b. System outputs to the environment (water sinks): Constraints (12.95) and (12.96) ensure that wastewater discharged to the environment must satisfy the relevant flow rate and concentration regulations.

c. Mass balance at the inlet and outlet of each unit and junction: Constraints (12.97) through (12.99) give the mass balances of inlet and outlet of all units and junctions.

d. Structural constraints: On the basis of the superstructure we proposed above, only one stream can be allowed to enter/release from each unit at a certain time interval and such constraints can be written as Equations 12.100 and 12.101. Furthermore, uneconomical amounts of water should not be allowed in the optimal operating policy of the network and such negligible matches can be eliminated by the addition of constraint (12.102).

e. Number of pipelines:
Constraint (12.103) guarantees that such pipeline exits if water has been transferred from e to *Jun* during the time horizon. To reflect the

overall cost and complexity of the network, the cost of pipelines should be incorporated. Constraints (12.104) through (12.106) identified the existence of all other pipelines.

$$f_{tr,t}^{in} = f_{tr,t}^{out} + f_{tr,t}^{loss} \quad \forall tr \in Tr, t \in T \tag{12.42}$$

$$f_{tr,t}^{in} \cdot c_{tr,k,t}^{in} \cdot (1 - R_{tr,k}) = f_{tr,t}^{out} \cdot c_{tr,k,t}^{out} + f_{tr,t}^{loss} \cdot c_{tr,k,t}^{loss} \quad \forall tr \in Tr, k \in K, t \in T \tag{12.43}$$

$$f_{tr,t}^{in} \leq F_{tr}^{in,max} \cdot n(tr) \quad \forall tr \in Tr, t \in T \tag{12.44}$$

$$c_{tr,k,t}^{in} \leq C_{tr,k}^{in,max} \quad \forall tr \in Tr, k \in K, t \in T \tag{12.45}$$

$$c_{tr,k,t}^{out} \leq C_{tr,k}^{out,max} \quad \forall tr \in Tr, k \in K, t \in T \tag{12.46}$$

$$v_{b,t} \cdot c_{b,k,t}^{out} = v_{b,t-1} \cdot c_{b,k,t-1}^{out} + \Delta t \cdot \left(f_{b,t}^{in} \cdot c_{b,k,t}^{in} - f_{b,t}^{out} \cdot c_{b,k,t}^{out} \right) \quad \forall b \in B, k \in K, t \in T \tag{12.47}$$

$$v_{b,t} = v_{b,t-1} + \Delta t \cdot \left(f_{b,t}^{in} - f_{b,t}^{out} \right) \quad \forall b \in B, t \in T \tag{12.48}$$

$$c_{b,k,0}^{out} = c_{b,k,T}^{out} \quad \forall b \in B, k \in K \tag{12.49}$$

$$v_{b,0} = v_{b,T} \quad \forall b \in B \tag{12.50}$$

$$0 \leq v_{b,t} \leq v_b^{max} \quad \forall b \in B, t \in T \tag{12.51}$$

$$V_b^{min} \cdot n(b) \leq v_{b,t} \leq V_b^{max} \cdot n(b) \quad \forall b \in B \tag{12.52}$$

$$F_{sb_j^n}^{out} = \lambda_{sb} \cdot \frac{B_{sb_j^n}}{T_{sb_j^n}^f - T_{sb_j^n}^s} \quad \forall sb \in SB, j \in J_{sb}, n \in N \tag{12.53}$$

$$F_{sb_j^n}^{out} - F_{sb}^{out,max} \cdot \left[1 - w\left(sb_j^n,t\right) \right] \leq f_{sb_j^n,t}^{out} \leq F_{sb_j^n}^{out} - F_{sb}^{out,min} \cdot \left[1 - w\left(sb_j^n,t\right) \right]$$
$$\forall sb \in SB, j \in J_{sb}, n \in N, t \in T \tag{12.54}$$

$$f_{sb_j^n,t}^{out} \leq F_{sb}^{out,max} \cdot w\left(sb_j^n,t\right) \quad \forall sb \in SB, j \in J_{sb}, n \in N, t \in T \tag{12.55}$$

$$c_{sb_j^n,k,t}^{out} = C_{sb,k} \cdot w\left(sb_j^n,t\right) \quad \forall sb \in SB, j \in J_{sb}, n \in N, t \in T, k \in K \tag{12.56}$$

$$M_{ua_j^n,k} = \lambda_{ua,k} \cdot B_{ua_j^n} \quad \forall ua \in UA, j \in J_{ua}, n \in N, k \in K \tag{12.57}$$

$$\sum_{t \in T} f_{ua_j^n,t}^{in} \cdot \Delta t = \sum_{t \in T} f_{ua_j^n,t}^{out} \cdot \Delta t + DV_{ua}^{loss} \cdot wv(ua,n) \quad \forall ua \in UA, j \in J_{ua}, n \in N \tag{12.58}$$

$$\sum_{t \in T} f_{ua_j^n,t}^{in} \cdot c_{ua_j^n,k,t}^{in} \cdot \Delta t + M_{ua_j^n,k} = \sum_{t \in T} f_{ua_j^n,t}^{out} \cdot c_{ua_j^n,k,t}^{out} \cdot \Delta t + DV_{ua}^{loss} \cdot C_{ua,k}^{loss} \cdot wv(ua,n)$$
$$\forall ua \in UA, j \in J_{ua}, n \in N, k \in K \tag{12.59}$$

$$F_{ua_j^n}^{in} - F_{ua}^{in,max} \cdot \left[1 - w^{in}\left(ua_j^n,t\right)\right] \le f_{ua_j^n,t}^{in} \le F_{ua_j^n}^{in} - F_{ua}^{in,min} \cdot \left[1 - w^{in}\left(ua_j^n,t\right)\right]$$
$$\forall ua \in UA, j \in J_{ua}, n \in N, t \in T \tag{12.60}$$

$$f_{ua_j^n,t}^{in} \le F_{ua}^{in,max} \cdot w^{in}\left(ua_j^n,t\right) \quad \forall ua \in UA, j \in J_{ua}, n \in N, t \in T \tag{12.61}$$

$$F_{ua}^{in,min} \cdot wv(ua,n) \le F_{ua_j^n}^{in} \le F_{ua}^{in,max} \cdot wv(ua,n) \quad \forall ua \in UA, j \in J_{ua}, n \in N \tag{12.62}$$

$$F_{ua_j^n}^{out} - F_{ua}^{out,max} \cdot \left[1 - w^{out}\left(ua_j^n,t\right)\right] \le f_{ua_j^n,t}^{out} \le F_{ua_j^n}^{out} - F_{ua}^{out,min} \cdot \left[1 - w^{out}\left(ua_j^n,t\right)\right]$$
$$\forall ua \in UA, j \in J_{ua}, n \in N, t \in T \tag{12.63}$$

$$f_{ua_j^n,t}^{out} \le F_{ua}^{out,max} \cdot w^{out}\left(ua_j^n,t\right) \quad \forall ua \in UA, j \in J_{ua}, n \in N, t \in T \tag{12.64}$$

$$F_{ua}^{out,min} \cdot wv(ua,n) \le F_{ua_j^n}^{out} \le F_{ua}^{out,max} \cdot wv(ua,n) \quad \forall ua \in UA, j \in J_{ua}, n \in N \tag{12.65}$$

$$C_{ua_j^n,k}^{in} - C_{ua,k}^{in,max} \cdot \left[1 - w^{in}\left(ua_j^n,t\right)\right] \le c_{ua_j^n,k,t}^{in} \le C_{ua_j^n,k}^{in} - C_{ua,k}^{in,min} \cdot \left[1 - w^{in}\left(ua_j^n,t\right)\right]$$
$$\forall ua \in UA, j \in J_{ua}, n \in N, k \in K, t \in T \tag{12.66}$$

$$c_{ua_j^n,k,t}^{in} \le C_{ua,k}^{in,max} \cdot w^{in}\left(ua_j^n,t\right) \quad \forall ua \in UA, j \in J_{ua}, n \in N, k \in K, t \in T \tag{12.67}$$

$$C_{ua,k}^{in,min} \cdot wv(ua,n) \le C_{ua_j^n,k}^{in} \le C_{ua,k}^{in,max} \cdot wv(ua,n) \quad \forall ua \in UA, j \in J_{ua}, n \in N, k \in K$$
$$\tag{12.68}$$

$$C_{ua_j^n,k}^{\text{out}} - C_{ua,k}^{\text{out,max}} \cdot \left[1 - w^{\text{out}}\left(ua_j^n, t\right) \right] \le c_{ua_j^n,k,t}^{\text{out}} \le C_{ua_j^n}^{\text{out}} - C_{ua}^{\text{out,min}} \cdot \left[1 - w^{\text{out}}\left(ua_j^n, t\right) \right]$$
$$\forall ua \in UA, j \in J_{ua}, n \in N, k \in K, t \in T \tag{12.69}$$

$$c_{ua_j^n,t}^{\text{out}} \le C_{ua_j^n}^{\text{out,max}} \cdot w^{\text{out}}\left(ua_j^n, t\right) \quad \forall ua \in UA, j \in J_{ua}, n \in N, k \in K, t \in T \tag{12.70}$$

$$C_{ua,k}^{\text{out,min}} \cdot wv(ua,n) \le C_{ua_j^n,k}^{\text{out}} \le C_{ua,k}^{\text{out,max}} \cdot wv(ua,n) \quad \forall ua \in UA, j \in J_{ua}, n \in N, k \in K \tag{12.71}$$

$$\mu_{ub_j^n,k} = \lambda_{ub,k} \cdot \frac{B_{ub_j^n}}{T_{ub_j^n}^f - T_{ub_j^n}^s} \quad \forall ub \in UB, j \in J_{ub}, n \in N, k \in K \tag{12.72}$$

$$\mu_{ub_j^n,k} - \mu_{ub,k}^{\text{max}} \cdot \left[1 - w\left(ub_j^n, t\right) \right] \le m_{ub_j^n,k,t} \le \mu_{ub_j^n,k} - \mu_{ub,k}^{\text{min}} \cdot \left[1 - w\left(ub_j^n, t\right) \right]$$
$$\forall ub \in UB, j \in J_{ub}, k \in K, n \in N, t \in T \tag{12.73}$$

$$m_{ub_j^n,k,t} \le \mu_{ub,k}^{\text{max}} \cdot w\left(ub_j^n, t\right) \quad \forall ub \in UB, j \in J_{ub}, k \in K, n \in N, t \in T \tag{12.74}$$

$$f_{ub_j^n,t}^{\text{in}} = f_{ub_j^n,t}^{\text{out}} + dv_{ub}^{\text{loss}} \cdot w\left(ub_j^n, t\right) \quad \forall ub \in UB, j \in J_{ub}, n \in N, t \in T \tag{12.75}$$

$$f_{ub_j^n,t}^{\text{in}} \cdot c_{ub_j^n,k,t}^{\text{in}} + m_{ub_j^n,k,t} = f_{ub_j^n,t}^{\text{out}} \cdot c_{ub_j^n,k,t}^{\text{out}} + dv_{ub}^{\text{loss}} \cdot C_{ub,k}^{\text{loss}}$$
$$\forall ub \in UB, j \in J_{ub}, k \in K, n \in N, t \in T \tag{12.76}$$

$$F_{ub}^{\text{in,min}} \cdot w\left(ub_j^n, t\right) \le f_{ub_j^n,t}^{\text{in}} \le F_{ub}^{\text{in,max}} \cdot w\left(ub_j^n, t\right) \quad \forall ub \in UB, j \in J_{ub}, n \in N, t \in T \tag{12.77}$$

$$C_{ub,k}^{\text{in,min}} \cdot w\left(ub_j^n, t\right) \le c_{ub_j^n,k,t}^{\text{in}} \le C_{ub,k}^{\text{in,max}} \cdot w\left(ub_j^n, t\right) \quad \forall ub \in UB, j \in J_{ub}, k \in K, n \in N, t \in T \tag{12.78}$$

$$C_{ub,k}^{\text{out,min}} \cdot w\left(ub_j^n, t\right) \le c_{ub_j^n,k,t}^{\text{out}} \le C_{ub,k}^{\text{out,max}} \cdot w\left(ub_j^n, t\right) \quad \forall ub \in UB, j \in J_{ub}, k \in K, n \in N, t \in T \tag{12.79}$$

$$F_{ob_j^n}^{\text{in}} = \lambda_{ob} \cdot \frac{B_{ob_j^n}}{T_{ob_j^n}^f - T_{ob_j^n}^s} \quad \forall ob \in OB, j \in J_{ob}, n \in N \tag{12.80}$$

$$F_{ob_j^n}^{\text{in}} - F_{ob}^{\text{in,max}} \cdot \left[1 - w\left(ob_j^n, t\right)\right] \le f_{ob_j^n, t}^{\text{in}} \le F_{ob_j^n}^{\text{in}} - F_{ob}^{\text{in,min}} \cdot \left[1 - w\left(ob_j^n, t\right)\right]$$
$$\forall ob \in OB, j \in J_{ob}, n \in N, t \in T \tag{12.81}$$

$$f_{ob_j^n, t}^{\text{in}} \le F_{ob}^{\text{in,max}} \cdot w\left(ob_j^n, t\right) \quad \forall ob \in OB, j \in J_{ob}, n \in N, t \in T \tag{12.82}$$

$$C_{ob,k}^{\text{in,min}} \cdot w\left(ob_j^n, t\right) \le c_{ob_j^n, k, t}^{\text{in}} \le C_{ob,k}^{\text{in,max}} \cdot w\left(ob_j^n, t\right)$$
$$\forall ob \in OB, j \in J_{ob}, n \in N, t \in T, k \in K \tag{12.83}$$

$$f_{j,t}^{\text{in}} = \sum_{i_j^n \in I_j^n} f_{i_j^n, t}^{\text{in}} \quad \forall j \in J, t \in T \tag{12.84}$$

$$f_{j,t}^{\text{out}} = \sum_{i_j^n \in I_j^n} f_{i_j^n, t}^{\text{out}} \quad \forall j \in J, t \in T \tag{12.85}$$

$$c_{j,k,t}^{\text{in}} = \sum_{i_j^n \in I_j^n} c_{i_j^n, k, t}^{\text{in}} \quad \forall j \in J, k \in K, t \in T \tag{12.86}$$

$$c_{j,k,t}^{\text{out}} = \sum_{i_j^n \in I_j^n} c_{i_j^n, k, t}^{\text{out}} \quad \forall j \in J, k \in K, t \in T \tag{12.87}$$

$$f_{jun,t}^{\text{in}} = f_{jun,t}^{\text{out}} \quad \forall jun \in JUN, t \in T \tag{12.88}$$

$$c_{jun,k,t}^{\text{in}} = c_{jun,k,t}^{\text{out}} \quad \forall jun \in JUN, k \in K, t \in T \tag{12.89}$$

$$F_{jun}^{\text{in,min}} \cdot w(jun, t) \le f_{jun,t}^{\text{in}} \le F_{jun}^{\text{in,max}} \cdot w(jun, t) \quad \forall jun \in JUN, t \in T \tag{12.90}$$

$$N_{jun}^{\text{min}} \cdot w(jun, t) \le \sum_{eq \in EQ} nfs(eq, jun, t) + \sum_{eq \in EQ} nfs(jun, eq, t) < N_{jun}^{\text{max}} \cdot w(jun, t)$$
$$\forall jun \in JUN, t \in T \tag{12.91}$$

$$w(jun, t) \le n(jun) \le \sum_{t \in T} w(jun, t) \quad \forall jun \in JUN, t \in T \tag{12.92}$$

$$f_{sa,t}^{\text{out}} \le F_{sa}^{\text{out,max}} \quad \forall sa \in SA, t \in T \tag{12.93}$$

$$c_{sa,k,t}^{\text{out}} = C_{sa,k}^{\text{out}} \quad \forall sa \in SA, k \in K, t \in T \tag{12.94}$$

$$f_{oa,t}^{\text{in}} \leq F_{oa}^{\text{in,max}} \quad \forall oa \in OA, t \in T \tag{12.95}$$

$$c_{oa,k,t}^{\text{in}} \leq C_{oa,k}^{\text{in,max}} \quad \forall oa \in OA, k \in K, t \in T \tag{12.96}$$

$$f_{eq,t}^{\text{in}} = \sum_{eq' \in EQ} fs_{eq',eq,t} \quad \forall eq \in EQ, t \in T \tag{12.97}$$

$$f_{eq,t}^{\text{in}} \cdot c_{eq,k,t}^{\text{in}} = \sum_{eq' \in EQ} fs_{eq',eq,t} \cdot c_{eq',k,t}^{\text{out}} \quad \forall eq \in EQ, k \in K, t \in T \tag{12.98}$$

$$f_{eq,t}^{\text{out}} = \sum_{eq' \in EQ} fs_{eq,eq',t} \quad \forall eq \in EQ, t \in T \tag{12.99}$$

$$\frac{f_{e,t}^{\text{in}}}{Fr_e^{\text{in,max}}} \leq \sum_{eq \in EQ} nfs(eq,e,t) < \frac{f_{e,t}^{\text{in}}}{Fr_e^{\text{in,max}}} + 1 \quad \forall e \in E, t \in T \tag{12.100}$$

$$\frac{f_{e,t}^{\text{out}}}{Fr_e^{\text{out,max}}} \leq \sum_{eq \in EQ} nfs(eq,e,t) < \frac{f_{e,t}^{\text{out}}}{Fr_e^{\text{out,max}}} + 1 \quad \forall e \in E, t \in T \tag{12.101}$$

$$Fs^{\min} \cdot nfs(eq,eq',t) \leq fs_{eq,eq',t} \leq Fs^{\max} \cdot nfs(eq,eq',t)$$
$$\forall eq \in EQ, eq' \in EQ, t \in T \tag{12.102}$$

$$ne(e,jun) \leq \sum_{t \in T} nfs(e,jun,t) \leq M \cdot ne(e,jun) \quad \forall e \in EQ, jun \in JUN, t \in T \tag{12.103}$$

$$ne(jun,e) \leq \sum_{t \in T} nfs(jun,e,t) \leq M \cdot ne(jun,e) \quad \forall e \in EQ, jun \in JUN, t \in T \tag{12.104}$$

$$ne(e,e') \leq \sum_{t \in T} nfs(e,e',t) \leq M \cdot ne(e,e') \quad \forall e,e' \in E, jun \in JUN, t \in T \tag{12.105}$$

$$ne(jun',jun) \leq \sum_{t \in T} nfs(jun',jun,t) \leq M \cdot ne(jun',jun) \quad jun, jun' \in JUN, t \in T$$

$$\tag{12.106}$$

12.4.3   SOLUTION PROCEDURE

The optimization model we proposed above belongs to a nonconvex MINLP problem. It is difficult to solve this problem in one step as it has already overwhelmed the capability of any solvers. Therefore, an interactive hybrid solution strategy, which incorporates both deterministic and stochastic algorithm, is proposed to obtain a high-quality feasible solution. Specifically, the interactive procedure can be broadly divided into two stages. In the first stage, the original MINLP is divided into a set of MILP–MINLP formulations, which can be solved in a sequential manner to provide feasible solutions as the initial starting points for the next stage. Moreover, in the second stage, genetic algorithm is adopted to jump out of the local optimal point to obtain a better result.

12.5   ILLUSTRATIVE EXAMPLE

All case studies in aforementioned papers have already demonstrated the benefits that can be achieved by integrating the scheduling model and the WAN design model for batch processes. A benchmark example includes five cases that are illustrated to show the evolution and advantages of the aforementioned models. First, a common background of the example is given to provide the model parameters and process details. In a practical case, more than one water-using operation may be carried out with the same equipment. However, the superstructure used in this study is only a fictitious process configuration. Additional structural constraints must therefore be incorporated in the mathematical programming model to automatically translate the optimal solution into the actual pipeline network of the resulting system design. Notice that if a branch in the superstructure is not used to facilitate the operation of a water user, it can be regarded as a physical pipeline.

To simplify the network structure, let us consider the STN shown in Figure 12.4. In this process, feed A is heated to produce intermediate hot A, while 50% of feed B and 50% of feed C are mixed and then reacted to form intermediate BC. Forty

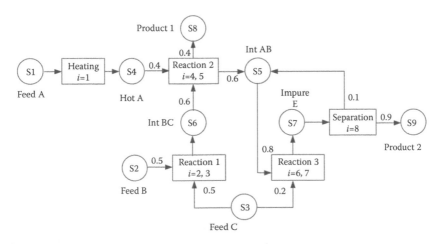

FIGURE 12.4   Production flow sheet.

percent of Hot A and 60% of BC are then mixed and reacted to form product 1 (40%) and intermediate AB (60%). On the other hand, 20% of feed C is reacted with 80% of intermediate AB to form impure E. Finally, the impure E is sent to a distillation column to separate product 2 (90%) and intermediate AB (10%). The available units, storage capacities, and processing times of this process are given in Tables 12.3 through 12.5. It is also assumed that the maximum amount of every feed supply is 1000 kg.

TABLE 12.3

Process Data (1) of the Example

Units	Capacity (kg)	Suitability	Average Processing Time (h)
Heater(j1)	100	Heating(i1)	1.0
Reactor1(j2)	50	Reaction1,2,3(i2,i4,i6)	2.0, 2.0, 1.0
Reactor2(j3)	80	Reaction1,2,3(i3,i5,i7)	2.0, 2.0, 1.0
Separator(j4)	200	Separation(i8)	2.0

TABLE 12.4

Process Data (2) of the Example

States	Storage Capacity in Example 1 (kg)	Price Per Unit
Feed A,B,C(s1,s2,s3)	Unlimited	10
Hot A(s4), Impure E(s7)	1000	–
Int AB(s5)	500	–
Int BC(s6)	0	–
Product1,2(s8,s9)	Unlimited	20

TABLE 12.5

Process Data (3) of the Example

Equipments	Operations	Inlet/Outlet Flowrate (m³/h)	Inlet/Outlet Concentration (ppm)
j1(Heater)	Ub	0–80	0/0–5
j2(Reactor 1)	ua1	0–80	0–6/0–14
	sb1	0–60	–/10
	ob1	0–60	0–7/–
j3(Reactor 2)	ua2	0–80	0–6/0–14
	sb2	0–60	–/10
	ob2	0–60	0–7/–
j4(Separator)	ua3	0–80	0–10/0–15
Sa	Sa	0–200	–/0
b1	b1	0–200	0–15/0–15
b2	b2	0–200	0–15/0–15
tr1	tr1	0–200	7–15/0–6
tr2	tr2	0–200	5–10/3–5
Oa	Oa	0–200	0–2.5/–

The minimum market demand for products and the maximum supply for materials at any event point are designed to be 0 and 120 kg, respectively; the cost of fresh water is set to 1 unit/m³; the installation fee for a conventional buffer tank is taken as the form of $4.8 + 6(v_b^{max})^{0.6}$ units; the costs of each junction and pipeline are assumed to be 5 and 3 units, respectively; the maximum and minimum of the throughput of all junctions are set to be 0 and 200 m³/h, respectively, the maximum and minimum number of streams connected to junctions are assigned to be 3 and 20, respectively; and the proportional parameters λ_{sb} and λ_{ob} are assigned 1.

Suppose that the reaction 1, separation, and heating operation are water users. Moreover, reaction 2 is a wastewater generating operation without consuming any usable water; while reaction 3 is a water-consuming operation without generating any wastewater. It is assumed that the amounts of consumed and/or generated waters of each batch operation are directly proportional to the processing volume.

Case 1:
Case 1 is studied to illustrate the advantage of simultaneous optimization of the WN and batch schedule. This case comes from Example 6 of Cheng and Chang (2007). The objective function of case 1 is to maximize the overall profit per time horizon. The time horizon of interest is 4 h and time interval is set to 1 h with the discrete-time representation. The optimal objective value is 987.815 and two buffer tanks, three splitters and three mixers, and 16 pipelines are required. The optimal schedule and the WN configuration are shown in Figures 12.5 and 12.6, respectively. Compared to the WN design under a predetermined batch schedule, the results achieved by integrated design approach represent a 5% improvement. Case 1 shows that the M–U–S framework proposed by Cheng and Chang (2007) can capture interaction between WN and production schedule in a reasonable way.

Case 2:
Let us consider case 2, which is intended to compare the SEN-based approach proposed by Zhou et al. (2009) with original computational results in case 1. The cost of the network is optimized to be 310.32 units, which decreases 18.1% due to the amelioration of the network configuration. In Figure 12.7a, the optimal WN is assembled

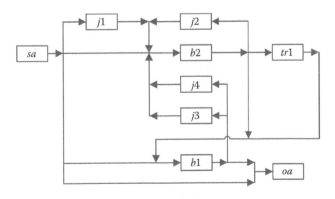

FIGURE 12.5 Optimal WN configuration of case 1.

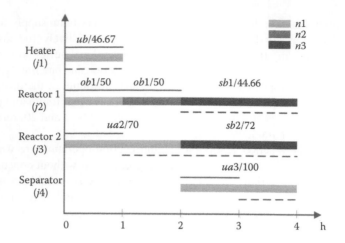

FIGURE 12.6 Optimal batch schedule scheme of case 1.

with 1 buffer tank, 5 splitters, 5 mixers, and 16 pipelines. Water reuse opportunities are improved by incorporating all the possible flow configurations and removing the limit of the number of pipelines attached to splitters and mixers adopted by Cheng and Chang (2007). Thus, more reasonable interconnections between units are pursued in SEN-based approach, which cannot be converted into M–U–S equivalently. It should also be noted that continuous-time representation is adopted in this case to reduce the expense of calculation. The optimal schedule scheme is almost the same as case 1 (see Figure 12.7). This is due to the fact that trade-offs between schedule

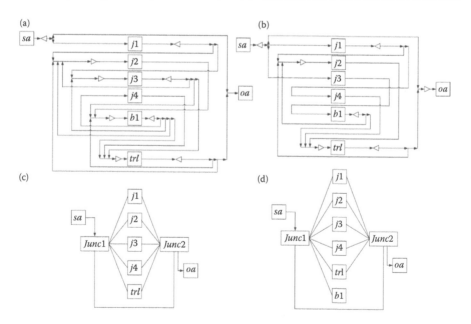

FIGURE 12.7 Optimal pipeline configuration for WN of cases 2 through 5.

and syntheses are properly balanced. In case 2, only 1 buffer tank is required, but 4 extra splitters and mixers are added. So the assumption that splitters and mixers are free of cost is impractical and leads to a more complicated network.

Case 3:

The costs of mixers and splitter are taken into consideration in case 3 to study the potential impacts of cost of splitters and mixers on the complexity of network structure. An improvement network configuration is obtained due to the trade-offs between operating and capital cost. The complexity of this network (Figure 12.7b) is greatly reduced and only 4 splitters, 3 mixers, and 12 pipelines are required, though the number of variables have been increased to capture the configuration in SEN-based SS superstructures. It should be noted that the cost of mixers and splitters are not included in the objective value, even though the objective value is larger than that of case 3. The equivalent and revised objective value of case 2 should be 952.38, which expels the cost of mixers and splitters. To simplify the network structure, Cheng and Chang (2007) limited the numbers of pipelines attached to the mixing and splitting nodes in the actual pipeline network. It is worth noting that these inequality constraints may lead to the preclusion of a class of optimal network structures, where the best cost-optimal scheme actually lies.

Case 4:

In case 4, the STS superstructure by Li et al. (2010) is adopted to represent batch schedule network as well as the WN. In this superstructure, junctions are designed for mixing/splitting purposes and can significantly improve splitting and mixing opportunities. Moreover, the cost of pipelines and junctions are considered at the same time. The initial number of junctions is chosen heuristically as the total number of units, that is, 4 junctions are embedded in the superstructure for this case. Obviously, placing sufficient junctions could provide enough opportunities to match all streams with different concentration grades. The optimal schedule scheme is the same as the case 1 (see Figure 12.6); whereas the network configuration (Figure 12.7c) is quite different as the multistage mixing and splitting stream from jun1 to jun2 can be clearly identified. Compared to case 3, the installation cost decreases by 50% and the objective value increases by 15%. Only 2 junctions and 11 pipelines are required to complete all the WN tasks since all the junctions and pipelines can be shared by different units during the whole time horizon in a flexible batch schedule mechanism. In other words, the STS superstructure could fully explore the water reuse opportunities between water-using units in each time event to trade-off various costs and achieve a higher integrated WAN. It can be concluded from the results that the STS superstructure is clearly superior to the former M–U–S and SEN-based framework. Furthermore, as evident from the network structure, the buffer tank is not embedded in the optimal network, which indicates that buffer tanks are not necessary in some batch WAN if optimal production sequences and operating policies of water flows are adopted.

Case 5:

In the previous cases, Δt is assumed to be 1 and all constant terms of processing time ($\alpha_{i,j}$) are set to the average processing time, while the variable terms ($\beta_{i,j}$) are set to be 0. Case 5 assumes a variation of 33% around the mean value of the average processing time $\tilde{t}_{i,j}$ and $\alpha_{i,j}$ takes the value of $(2/3)\,\tilde{t}_{i,j}$ which corresponds to the minimum processing time $t_{i,j}^{\min}$. Δt assumes to be 0.33 h in this case. Compared to cases 1 through 4, case 5 combines the discrete-time representation with continuous-time representation and it is much more flexible as operating time dependent on the amount processed. Unlike the continuous-time formulation, it can provide unified time duration to identify the overall number of time intervals. In this case, the objective function is the maximum of net profit, which equals to the total income in each batch period minus the cost of equipment, raw material, freshwater, and water treatment. The network structure and the optimal batch production schedule are illustrated in Figures 12.7d and 12.8, respectively. Within the time horizon of 8 h, only 3 pipelines are optimized to be increased, whereas the total income and objective value are increased 15% and 61% (see Table 12.6), respectively. This case indicates that the STS framework proposed above is more effective than previous approaches, as all basic elements (states, tasks, equipment, and time) are illustrated in one single and comprehensive representation.

In summary, the example provides five cases in which the batch schedules and water configurations are considered simultaneously. The detailed comparison of these five cases is provided in Table 12.6. It can be seen from the table that the objective value changes with the tuning of the schedule policy, network configurations, cost formulation, structure of configurations, and so on. Finally, the introduction of junctions and multistage mixing largely reduces the network cost and increases the compactness of the network. Table 12.7 provides the computational statistics from case 2 to 5. Besides, the number of variables is related to the time interval, which is partitioned in advance in different cases even though the configurations are the same. Moreover, the number of variables and the computation time are also determined by the time representation.

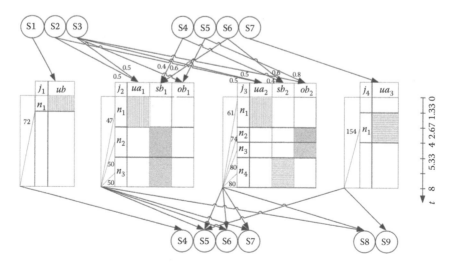

FIGURE 12.8   Optimal batch schedule scheme of case 5.

TABLE 12.6
Comparison of Optimization Results of Cases 1 through 5

	Case 1	Case 2	Case 3	Case 4	Case 5
Objective value (cost unit)	987.815	1056.38 (952.38)	978.39	1123.76	1810.27
Total income (cost unit)	2733.4	2733.4	2733.4	2773.44	3163.11
Purchasing cost (cost unit)	1366.7	1366.7	1366.7	1366.7	1054.3
Freshwater cost (cost unit)	105.1	83.35	86.16	123.3	61.33
Treatment cost (cost unit)	228.705	156.96	185.28	201.68	108.04
Installation cost of buffer tanks (cost unit)	45.08	70.01	51.87	0	71.17
Installation cost of junctions/splitters and mixers (cost unit)	–	104	65	58	58
Cost of network without splitters and mixers (cost unit)	–	310.32	323.31	323.31	298.34
Cost of network with splitters and mixers (cost unit)	–	414.32	388.31	381.31	356.34
Number of buffer tanks	2	1	1	0	1
Number of pipelines	16	16	12	11	14
Number of junctions/splitters and mixers	7	10	7	2	2
Yield of product S8 (kg)	46.67	46.67	46.67	46.67	72
Yield of product S9 (kg)	90	90	90	90	138.86
Purchasing amount of S1 (kg)	46.67	46.67	46.67	46.67	72
Purchasing amount of S2 (kg)	35	35	35	35	54
Purchasing amount of S3 (kg)	55	55	55	55	84.86
Volume of buffer tank $b1$ (m^3)	1	53.33	30.97	0	43.37
Volume of buffer tank $b2$ (m^3)	14.2	0	0	0	0
Total discharging volume (m^3)	121.77	100	102.83	140	87.28

TABLE 12.7
Comparison of Constraints and Variables

	Example			
	Case 2	Case 3	Case 4	Case 5
No. of constraints	1855–2581	1996–2722	2318	14,132
No. of variables	1601–2215	1685–2295	1820	10,005
Continuous variables	1225–1639	1274–1684	1235	6824
Binary variables	376–576	411–611	585	3181

12.6 CONCLUSIONS

An overview of the recent developments in the batch WN design is presented. Owing to the complexity in both modeling and consequent solution procedure, most of the work in batch WN has focused on single- and multiple-contaminant systems with a predefined schedule. An integrated methodology based on STS superstructures has been presented in this study to formulate an MINLP model for one-step optimization of batch schedules, water-reuse subsystems, and wastewater-treatment subsystems. A flexible schedule model combining the merits of discrete- and continuous-time formulations is also introduced to represent the precedence order of all operations. The advantage of this methodology is that not only all possible alternative network topologies can be captured, but also the sequential order of all operations as well can also be properly addressed. In addition, the complexity of the network can always be reduced by introducing the additional costs of junctions and pipelines. As demonstrated by the results of the benchmark example, the interaction between the process schedule and WN will possibly allow greater water recovery and capital cost reduction. The comprehensive comparison study of previous integrated design approaches highlights the limitations of current models and solution strategies, and indicates that the research in this area still commands attention and more works are expected in the near future.

ACKNOWLEDGMENTS

The authors gratefully acknowledge financial support from the National Natural Science Foundation of China, under Grant Nos. 20876020 and 21276039.

NOMENCLATURE

SETS

B	Set of buffer tanks
E	Set of all units, $E = U \cup J$
EQ	Set of all units and junctions, $EQ = SA \cup TR \cup B \cup OA \cup J \cup JUN$
I	Set of all operations in batch production, $I = SB \cup UA \cup UB \cup OB$
I_j^n	Set of operation i performed in $j(\,j \in J_i)$ at event point n
I_j	Set of operations which can be performed in unit j
I_s	Set of operations which either produce or consume state s
J	Set of all involved units in batch production
J_i	Set of units where operation i can be performed
JUN	Set of junctions
K	Set of all pollutants in water-allocation network
N	Set of all event points within the time horizon
OA	Set of water sinks
OB	Set of operations in batch production which only consume water
S	Set of all involved material states
S_m	Set of all raw materials
S_p	Set of all final products
SA	Set of freshwater sources
SB	Set of operations which generate wastewater in batch production
Tr	Set of wastewater-treatment units
T	Set of all intervals of equal durations, $T = \{t \mid t = 1,2,3,\dots,T(H/\Delta t)\}$
U	Set of all units not involved in batch production, $U = SA \cup TR \cup B \cup OA$
UA	Set of operations in batch production with nonidentical charging and discharging time intervals

UB — Set of operations in batch production with identical charging and discharging time intervals

PARAMETERS

$\chi_{i_j^n}, \varphi_{i_j^n}$	positive integers, the number of intervals of the starting and ending points of i_j^n
ΔT	the length of each variable interval
$v_{b,0}, c_{b,k,0}^{\text{out}}$	the initial amount of water and the initial concentration of pollutant k in b
$\rho_{s,i}^c, \rho_{s,i}^p, \rho_{s,i}^p$	proportion of state s consumed, produced from operation i
$a_{i,j}, \beta_{i,j}\, b_{i,j}$	the constant and variable term of processing time of operation i at unit j
$b_{i_j^n}$	the number of variable term needed for operation i_j^n to be performed
$BC_{i,j}$	the total amount processed at each variable term of operation i at unit j
$BC_{i,j}^{\min}, BC_{i,j}^{\max}$	the minimum and maximum capacities of unit j used for performing operation i
$C_{sa,k}^{\text{out}}$	the concentration of pollutant k in water generated from sa
$C_{ub,k}^{\text{loss}}$	fixed concentration of the pollutant k in water loss from ub
$C_{i,k}^{\text{in,max}}, C_{i,k}^{\text{in,min}}$	the upper and lower bounds of the concentration of pollutant k in water consumed by operation i
$C_{u,k}^{\text{in,max}}, C_{u,k}^{\text{out,max}}$	the upper bounds of the inlet and outlet concentration of pollutant k of u
$C_{i,k}^{\text{out,max}}, C_{i,k}^{\text{out,min}}$	the upper and lower bounds of the concentration of pollutant k in water generated from operation i
dv_{ub}^{loss}	the fixed water loss rate and relevant concentrations in ub
$D_{s,n}^{\min}$	the minimum market demand for state s at event point n
$DV_{ua}^{\text{loss}}, C_{ua,k}^{\text{loss}}$	the fixed amount of water loss and relevant concentrations in ua
$f_{tr,t}^{\text{loss}}, c_{tr,k,t}^{\text{loss}}$	the fixed water loss rate and relevant concentrations in tr
$F_e^{\text{in,max}}, F_e^{\text{out,max}}$	the upper bounds of the inlet and outlet flow rate of e
$F_i^{\text{in,max}}, F_i^{\text{in,min}}$	the upper and lower bounds of water-consuming rates of the operation i
$F_i^{\text{out,max}}, F_i^{\text{out,min}}$	the upper and lower bounds of water generating rates of the operation i
$F_{jun}^{\text{in,min}}, F_{jun}^{\text{in,max}}$	the upper and lower bounds of the inlet flow rate of jun
$F_{obj_j^n,t}^{\text{in}}$	the nominal water-consuming rate of ob_j^n
$F_{sa}^{\text{out,max}}$	the maximum allowable supply rate of sa
$F_{sb_j^n}^{\text{out}}$	the nominal water generating rate of sb_j^n
$Fs^{\max}, Fs^{\min}$	the upper and lower bounds of the flow rates in the system
H	the fixed time horizon
$N_{jun}^{\min}, N_{jun}^{\max}$	the upper and lower bounds of the number of streams connected to jun
$pr_{b1}, pr_{b2}, pr_{b3}$	the cost coefficients of buffer tanks
$pr_{jun}, pr_{\text{pipe}}$	the cost coefficients of junctions and pipelines
pr_{sa}, pr_{tr}	the cost coefficients of freshwater and wastewater treatment
pr_{s_m}, pr_{s_p}	the cost coefficients of raw materials and products, respectively

$R_{s,n}^{\max}$	the maximum supply of state s at the event point n
$R_{tr,k}$	the removal ratio of pollutant k in tr
ST_s^{in}	the initial amount of the state s
$ST_s^{\max}$	the maximum storage capacity for state s
$\tilde{t}_{i,j}$	the average process time of operation i at unit j
$t_{i,j}^{\max}$	the maximum process time of operation i at unit j
$V_b^{\max}, V_b^{\min}$	the upper and lower bounds of the designing volumes of b

CONTINUOUS VARIABLES

$\mu_{ub_j^n,k}$	the nominal instantaneous mass load of pollutant k in ub_j^n
$B_{i_j^n}$	amount of the material undertaking in operation i_j^n
$c_{eq,k,t}^{\mathrm{in}}, c_{eq,k,t}^{\mathrm{out}}$	the concentration of pollutant k in waters consumed and generated by eq in interval t
$c_{i_j^n,k,t}^{\mathrm{in}}, c_{i_j^n,k,t}^{\mathrm{out}}$	the concentration of pollutant k in the input and output streams of i_j^n in interval t
$c_{j,k,t}^{\mathrm{in}}, c_{j,k,t}^{\mathrm{out}}$	the inlet and outlet concentration of pollutant k in interval t for j
$c_{jun,t}^{\mathrm{in}}, c_{jun,t}^{\mathrm{out}}$	the inlet and outlet concentration of pollutant k in interval t for jun
$c_{oa,k,t}^{\mathrm{in}}$	the concentration of pollutant k in waters discharged to oa in interval t
$c_{ob_j^n,k,t}^{\mathrm{in}}$	the concentration of the pollutant k in waters consumed by $ob_j^n v_{b,t}$ the water volume of b in interval t
$c_{sa,k,t}^{\mathrm{out}}$	the concentration of pollutant k in waters generated from sa
$f_{eq,t}^{\mathrm{in}}, f_{eq,t}^{\mathrm{out}}$	the inlet and outlet flow rate of eq in interval t
$f_{i_j^n,t}^{\mathrm{in}}, f_{i_j^n,t}^{\mathrm{out}}$	the water generating and consuming rates of i_j^n in interval t
$f_{j,t}^{\mathrm{in}}, f_{j,t}^{\mathrm{out}}$	the inlet and outlet flow rates of j in interval t
$f_{jun,t}^{\mathrm{in}}, f_{jun,t}^{\mathrm{out}}$	the inlet and outlet flow rates of jun in interval t
$f_{oa,t}^{\mathrm{in}}$	the inlet rate of oa in interval t
$f_{sa,t}^{\mathrm{out}}$	the outlet flow rate of sa in interval t
$fs_{eq,eq',t}$	the flow rate in stream from eq to eq' in interval t
$F_{i_j^n}^{\mathrm{in}}, F_{i_j^n}^{\mathrm{out}}$	the nominal inlet and outlet flow rate of i_j^n
$m_{ub_j^n,k,t}$	the instantaneous mass load of pollutant k in ub_j^n in interval t
$M_{ua_j^n,k}$	the accumulated mass load of pollutant k in ua_j^n
$r_{s,n}, d_{s,n}$	amount of the state s purchased and sold at event point n
ST_s^{in}	initial amount of the state s
$ST_{s,n}$	amount of the state s at event point n
$T_{i_j^n}^M$	time boundaries of charging and discharging for i_j^n
$T_{i_j^n}^s, T_{i_j^n}^f$	operation i_j^n starting and finishing time

BINARY VARIABLES

$n(b), n(tr), n(jun)\ n(tr)$	binary variable used to denote if b, tr, or jun is selected in the optimal network configuration
$ne(eq, eq')$	binary variable used to denote if pipeline between eq and eq' is selected in optimal network configuration

$nfs(eq, eq', t)$	binary variable used to denote if actual stream exists between eq and eq' in interval t
$w(jun, t)$	binary variable used to denote if jun is occupied in interval t
$wv(i,n)$	binary variable used to signify if operation i is performed at event point n
$w^{\text{in}}(i_j^n,t), w^{\text{out}}(i_j^n,t)$	binary variables used to denote if i_j^n is consuming or generating water in interval t
$w(i_j^n,t)$	binary variable used to denote if i_j^n is being performed in interval t
$yv(j,n)$	binary variable used to signify if unit j is utilized at event point n

REFERENCES

Adekola, O., and T. Majozi. 2011. Wastewater minimization in multipurpose batch plants with a regeneration unit: Multiple contaminants. *Comput. Chem. Eng.* 35:2824–2836.

Ahmetović, E., and I.E. Grossmann. 2011. Global superstructure optimization for the design of integrated process water networks. *AIChE J.* 57:434–457.

Almato, M., A. Espuna, and L. Puigjaner. 1999. Optimisation of water use in batch process industries. *Comput. Chem. Eng.* 23:1427–1437.

Alva-Argaez, A., A. Kokossis, and R. Smith. 2007. The design of waterusing systems in petroleum refining using a water-pinch decomposition. *Chem. Eng. J.* 128:33–46.

Bagajewicz, M.J. 2000. A review of recent design procedures for water networks in refineries and process plants. *Comput. Chem. Eng.* 24:2093–2113.

Bagajewicz, M.J., M. Rivas, and M.J. Savelski. 2000. A robust method to obtain optimal and sub-optimal design and retrofit solutions of water utilization systems with multiple contaminants in process plants. *Comput. Chem. Eng.* 24:1461–1466.

Bagajewicz, M.J., and M. Savelski. 2001. On the use of linear models for the design of water utilization systems in process plants with a single contaminant. *Chem. Eng. Res. Des.* 79:600–610.

Bandyopadhyay, S., M.D. Ghanekar, and H.K Pillai. 2006. Process water management. *Ind. Eng. Chem. Res.* 45:5287–5297.

Castro, P., A.P.F.D. Barbosa-Povoa, and H. Matos. 2001. An improved RTN continuous-time formulation for the short-term scheduling of multipurpose batch plants. *Ind. Eng. Chem. Res.* 40:2059–2068.

Chaturvedi, N.D., and S. Bandyopadhyay. 2014. Optimization of multiple freshwater resources in a flexible schedule batch water network. *Ind. Eng. Chem. Res.* 53:5996–6005.

Chen, C.L., and C.Y. Chang. 2009. A resource-task network approach for optimal short-term/periodic scheduling and heat integration in multipurpose batch plants. *Appl. Therm. Eng.* 29:1195–1208.

Chen, C.L., C.Y. Chang, and J.Y. Lee. 2008b. Continuous-time formulation for the synthesis of water-using networks in batch plants. *Ind. Eng. Chem. Res.* 47:7818–7832.

Chen, C.L., C.Y. Chang, and J.Y. Lee. 2011. Resource-task network approach to simultaneous scheduling and water minimization of batch plants. *Ind. Eng. Chem. Res.* 50:3660–3674.

Chen, C.L., and J.Y. Lee. 2008a. A graphical technique for the design of water-using networks in batch processes. *Chem. Eng. Sci.* 63:3740–3754.

Chen, C.L., J.Y. Lee, D.K.S. Ng, and D.C.Y. Foo. 2010. A unified model of property integration for batch and continuous processes. *AIChE J.* 56:1845–1858.

Chen, C.L., J.Y. Lee, J.W. Tang, and Y.J. Ciou. 2009. Synthesis of waterusing network with central reusable storage in batch processes. *Comput. Chem. Eng.* 33:267–276.

Cheng, K.F., and C.T. Chang. 2007. Integrated water network designs for batch processes. *Ind. Eng. Chem. Res.* 46:1241–1253.

Chew, I.M.L., R.R. Tan, D.C.Y. Foo, and A.S.F. Chiu. 2009. Game theory approach to the analysis of inter-plant water integration in an eco-industrial park. *J. Clean. Prod.* 17:1611–1619.

Dhole, V.R., N. Ramchandani, R.A. Tainsh, and M. Wasilewski. 1996. Make your process water pay for itself. *Chem. Eng.* 103:100–103.

Dogaru E.L., and V Lavric. 2011. Dynamic water network topology optimization of batch processes. *Ind. Eng. Chem. Res.* 50:3636–3652.

Dong, H.G., C.Y. Lin, and C.T. Chang. 2008. Simultaneous optimization approach for integrated water-allocation and heat-exchange networks. *Chem. Eng. Sci.* 63:3664–3678.

Faria, D.C., and M.J. Bagajewicz. 2010. On the appropriate modeling of process plant water systems. *AIChE J.* 56:668–689.

Floudas, C.A., I.G. Akrotirianakis, S. Caratzoulas, C.A. Meyer, and J. Kallrath. 2005. Global optimization in the 21st century: Advances and challenges. *Comput. Chem. Eng.* 29:1185–1202.

Foo, D.C.Y. 2009. State-of-the-art review of pinch analysis techniques for water network synthesis. *Ind. Eng. Chem. Res.* 48:5125–5159.

Foo, D.C.Y., Z.A. Manan, and Y. L. Tan. 2005. Synthesis of maximum water recovery network for batch process systems. *J. Clean. Prod.* 13:1381–1394.

Gouws, J.F., and T. Majozi. 2008. Impact of multiple storage in wastewater minimization for multicomponent batch plants: Toward zero effluent. *Ind. Eng. Chem. Res.* 47:369–379.

Gouws, J.F., and T. Majozi. 2009. Usage of inherent storage for minimisation of wastewater in multipurpose batch plants. *Chem. Eng. Sci.* 64:3545–3554.

Gouws, J.F., T. Majozi, D.C.Y. Foo, C.L. Chen, and J.Y. Lee. 2010. Water minimization techniques for batch processes. *Ind. Eng. Chem. Res.* 49:8877–8893.

Gunaratnam, M., A. Alva-Argaez, A. Kokossis, J.K. Kim, and R. Smith. 2005. Automated design of total water systems. *Ind. Eng. Chem. Res.* 44:588–599.

Hallale, N. 2002. A new graphical targeting method for water minimisation. *Adv. Environ. Res.* 6:377–390.

Hernandez-Suarez, R., J. Castellanos-Fernandez, and J.M. Zamora. 2004. Superstructure decomposition and parametric optimization approach for the synthesis of distributed wastewater treatment networks. *Ind. Eng. Chem. Res.* 43:2175–2191.

Ierapetritou, M.G., and C.A. Floudas. 1998. Effective continuous-time formulation for short-term scheduling. 1. Multipurpose batch processes. *Ind. Eng. Chem. Res.* 37:4341–4359.

Jezowski, J. 2010. Review of water network design methods with literature annotations. *Ind. Eng. Chem. Res.* 49:4475–4516.

Karimi, I.A., and C.M. McDonald. 1997. Planning and scheduling of parallel semicontinuous processes. 2. Short-term scheduling. *Ind. Eng. Chem. Res.* 36:2701–2714.

Kim, J.K., and R. Smith. 2004. Automated design of discontinuous water systems. *Process Saf. Environ. Prot.* 82:238–248.

Kondili, E., C.C. Pantelides, and R.W.H. Sargent. 1993. A general algorithm for short-term scheduling of batch operations—I. MILP formulation. *Comput. Chem. Eng.* 17:11–227.

Kuo, W.C.J., and R. Smith. 1997. Effluent treatment system design. *Chem. Eng. Sci.* 52:4273–4290.

Kuo, W.C.J., and R. Smith. 1998. Designing for the interactions between water-use and effluent treatment. *Chem. Eng. Res. Des.* 76:287–301.

Lavric, V., P. Iancu, and V. Pleşu. 2005. Genetic algorithm optimisation of water consumption and wastewater network topology. *J. Clean. Prod.* 13:1405–1415.

Lee, J.Y., C.L. Chen, and C.Y. Lin. 2013. A mathematical model for water network synthesis involving mixed batch and continuous units. *Ind. Eng. Chem. Res.* 52:7047–7055.

Li, B.H., and C.T. Chang. 2006. A mathematical programming model for discontinuous water-reuse system design. *Ind. Eng. Chem. Res.* 45:5027–5036.

Li, B.H., and C.T. Chang. 2007. A simple and efficient initialization strategy for optimizing water-using network designs. *Ind. Eng. Chem. Res.* 46:8781–8786.

Li, B.H., and C.T. Chang. 2011. A model-based search strategy for exhaustive identification of alternative water network designs. *Ind. Eng. Chem. Res.* 50:3653–3659.

Li, L.J., R.J. Zhou, and H.G. Dong. 2010. State-time-space superstructure based MINLP formulation for batch water-allocation network design. *Ind. Eng. Chem. Res.* 49:236–251.

Liu, Y.J., X.G. Yuan, and Y.Q. Luo. 2007. Synthesis of water utilisation system using concentration interval analysis method (II). Discontinuous process. *Chin. J. Chem. Eng.* 15:369–375.

Majozi, T. 2005a. An effective technique for wastewater minimization in batch processes. *J. Clean. Prod.* 13:1374–1380.

Majozi, T. 2005b. Wastewater minimisation using central reusable water storage in batch plants. *Comput. Chem. Eng.* 29:1631–1646.

Majozi, T., C.J. Brouckaert, and C.A. Buckley. 2006. A graphical technique for wastewater minimization in batch processes. *J. Environ. Manage.* 78:317–329.

Majozi, T., and J.F. Gouws. 2009. A mathematical optimization approach for wastewater minimization in multiple contaminant batch plants. *Comput. Chem. Eng.* 33:1826–1840.

Majozi, T., and X.X. Zhu. 2001. A novel continuous-time MILP formulation for multipurpose batch plants. 1. Short-term scheduling. *Ind. Eng. Chem. Res.* 40:5935–5949.

Mann, J.G., and Y.A. Liu, 1999. *Industrial Water Reuse and Wastewater Minimization.* NewYork: McGraw Hill.

Manan, Z.A., Y.L. Tan, and D.C.Y. Foo. 2004. Targeting the minimum water flowrate using water cascade analysis technique. *AIChE J.* 50:3169–3183.

Maravelias, C.T., and I.E. Grossmann. 2003. New general continuous-time state-task network formulation for short-term scheduling of multipurpose batch plants. *Ind. Eng. Chem. Res.* 42:3056–3074.

Ng, D.K.S., D.C.Y. Foo, A. Rabie, and M.M. El-Halwagi. 2008. Simultaneous synthesis of property-based water reuse/recycle and interception networks for batch processes. *AIChE J.* 54:2624–2632.

Ng, D.K.S., D.C.Y. Foo, and R.R. Tan. 2007a. Targeting for total water network. 1. Waste stream identification. *Ind. Eng. Chem. Res.* 46:9107–9113.

Ng, D.K.S., D.C.Y. Foo, and R.R. Tan. 2007b. Targeting for total water network. 2. Waste treatment targeting and interactions with water system elements. *Ind. Eng. Chem. Res.* 46:9114–9125.

Pinto, J.M., and I.E. Grossmann. 1994. Optimal cyclic scheduling of multistage continuous multiproduct plants. *Comput. Chem. Eng.* 18:797–816.

Prakash, R., and U.V. Shenoy. 2005. Targeting and design of water networks for fixed flowrate and fixed contaminant load operations. *Chem. Eng. Sci.* 60:255–268.

Savulescu, L.E., J.K. Kim, and R. Smith. 2005a. Studies on simultaneous energy and water minimization—Part I: Systems with no water re-use. *Chem. Eng. Sci.* 60:3279–3290.

Savulescu, L.E., J.K. Kim, and R. Smith. 2005b. Studies on simultaneous energy and water minimization—Part II: Systems with maximum re-use of water. *Chem. Eng. Sci.* 60:3291–3308.

Shaik, M., and C.A. Floudas. 2008. Short-term scheduling of batch and continuous processes. In: L.G. Papageorgiou and M.C. Georgiadis (Eds.), *Supply Chain Optimization. Part II.* WILEY-VCH: Weinheim, pp. 173–217.

Shoaib, A.M., S.M. Aly, M.E. Awad, D.C.Y. Foo, and M.M. El-Halwagi. 2008. A hierarchical approach for the synthesis of batch water network. *Comput. Chem. Eng.* 32:530–539.

Sorin, M., and S. Bédard. 1999. The global pinch point in water reuse networks. *Process Saf. Environ. Prot.* 77(5):305–308.

Takama, N., T. Kuriyama, K. Shiroko, and T. Umeda. 1980. Optimal water allocation in a petroleum refinery. *Comput. Chem. Eng.* 4:251–258.

Teles, J., P.M. Castro, and A.Q. Novais. 2008. LP-based solution strategies for the optimal design of industrial water networks with multiple contaminants. *Chem. Eng. Sci.* 63:376–394.

Tokos, H., and Z.N. Pintaric. 2009. Synthesis of batch water network for a brewery plant. *J. Clean. Prod.* 17:1465–1479.

Tsai, M.J., and C.T. Chang. 2001. Water usage and treatment network design using genetic algorithms. *Ind. Eng. Chem. Res.* 40:4874–4888.

Wang, Y.P., and R. Smith. 1994a. Wastewater minimization. *Chem. Eng. Sci.* 49:981–1006.

Wang, Y.P., and R. Smith. 1994b. Design of distributed effluent treatment systems. *Chem. Eng. Sci.* 49:3127–3145.

Wang, Y.P., and R. Smith. 1995. Time pinch analysis. *Trans. Inst. Chem. Eng.* 73(A):905–912.

Zamora, J.M., and I.E. Grossmann. 1998. Continuous global optimization of structured process systems models. *Comput. Chem. Eng.* 22:1749–1770.

Zhou, R.J., L.J. Li, W. Xiao, and H.G. Dong. 2009. Simultaneous optimization of batch process schedules and water-allocation network. *Comput. Chem. Eng.* 33:1153–1168.

13 Synthesis of Water Networks with Mixed Batch and Continuous Process Units

Jui-Yuan Lee and Cheng-Liang Chen

CONTENTS

13.1 INTRODUCTION

Water is an essential resource widely used in the process industries as a raw material, a heat-transfer medium, or a mass separating agent. Common uses in the latter case include steam stripping, liquid–liquid extraction, and various washing operations. Rapid industrial growth has increased fresh water consumption as well as wastewater production, and consequently caused serious water pollution in the world. Concurrently, the scarcity of industrial water (partly due to climate change) and increasingly stringent discharge regulations have led to rising costs of fresh water and effluent treatment. These environmental and economic issues stimulated the recent development of systematic methodologies for efficient and responsible use of water in industry. Particularly, design of water recovery systems through *process integration* techniques, also known as *water network synthesis*, has been commonly accepted as an effective means in this regard, with *reuse, recycling,* and

319

regeneration as possible options for reducing fresh water and wastewater (Wang and Smith 1994).

In the past decades, numerous research papers on water network synthesis for both continuous and batch processes have been published (Foo 2009; Gouws et al. 2010; Jeżowski 2010). The techniques developed can be broadly classified as *pinch analysis* (Chen and Lee 2008; El-Halwagi et al. 2003; Foo et al. 2005; Hallale 2002; Majozi et al. 2006; Manan et al. 2004; Prakash and Shenoy 2005; Wang and Smith 1994, 1995) and *mathematical optimization* (Ahmetović and Grossmann 2011; Almató et al. 1999; Chen et al. 2008, 2009; Cheng and Chang 2007; Gunaratnam et al. 2005; Karuppiah and Grossmann 2006; Kim and Smith 2004; Li and Chang 2006; Majozi 2005a,b, 2006; Shoaib et al. 2008; Takama et al. 1980). Pinch techniques, generally a two-step approach involving targeting and network design, offer good insights for problem decomposition and system debottlenecking. Although not providing such insights, mathematical techniques are useful in dealing with complex cases, such as multiple contaminants, cost considerations, topological constraints, and limited piping connections.

While most water network studies are intended for either completely continuous or completely batch processes, there are also processes with mixed continuous and batch operations (e.g., breweries, sugar mills, and tire production plants). This calls for the development of effective methods to handle such processes. In the following sections, a formal problem statement is given first. The integration of process units of different operation modes is discussed next, and the mathematical model is then presented. Two examples are used to illustrate the proposed approach.

13.2   PROBLEM STATEMENT

The problem addressed in this chapter can be formally stated as follows. Given is a set of batch and continuous water-using units $i \in I$. The former may operate in *truly batch* ($i \in I^b$) or *semicontinuous* mode ($i \in I^{sc}$). Furthermore, all batch operations are assumed to be carried out cyclically with a fixed schedule. These water-using units require water to remove a set of contaminants $c \in C$ with fixed mass loads from the process materials. Available for service are different qualities of fresh water from a set of water sources $w \in W$. The used water from units will be sent to a set of wastewater disposal systems $d \in D$ for final discharge, or can be partially reutilized for reduction of fresh water usage and wastewater generation. To facilitate water allocation and recovery, a set of storage tanks $s \in S$ for intermediate water storage may be used. The objective is to synthesize an optimal water network that achieves the minimum fresh water consumption, while satisfying all process constraints. A mathematical model is therefore to be developed in this study for determining the scope for water recovery, the corresponding water network configuration, and the water storage policy.

13.3   INTEGRATION OF CONTINUOUS AND BATCH UNITS

Figure 13.1 illustrates three common operation modes for water-using units. While continuous units operate uninterruptedly for long durations (e.g., 8000 h/y) spanning

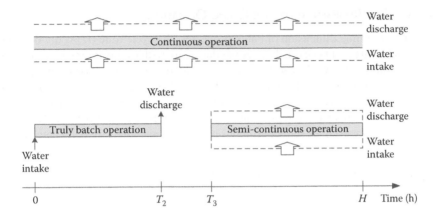

FIGURE 13.1   Types of water-using operations.

many operation cycles, batch units are scheduled to operate within certain periods of time (as short as a few hours or days) and often in a cyclic manner. In addition, the operating period of a truly batch or semicontinuous unit is always shorter than the batch cycle time. In terms of water usage, a truly batch unit takes in water at the start and discharges wastewater at the end of its operation. A typical example of such operations is the batch reaction for agrochemicals production in which water is used as the reaction solvent and for product washing. By contrast, semicontinuous and continuous units have steady water intake and discharge during the course of the operation. Typical examples include various extraction and washing processes in the chemical industry.

For water network synthesis involving process units of different operation modes, the main challenge would be the integration of these units. It is proposed in this study to treat a continuous process as a special case of a semicontinuous process lasting for the whole batch cycle time, as shown in Figure 13.1, assuming that slight changes in water supply are acceptable even for continuous water-using operations. Since water intake and discharge for truly batch units take place at time points, and those for semicontinuous units in time intervals, there will be no direct water transfers between truly batch and semi-continuous units. Water integration between these two types of units can only be carried out indirectly through water storage tanks.

13.4   MATHEMATICAL FORMULATION

Having treated all continuous processes as semicontinuous ones, the problem becomes to synthesize a water network of truly batch and semicontinuous units. A mathematical model is then developed for the remaining problem based on a superstructure including all feasible network connections between water-using units (Figures 13.3a and 13.3b) and storage tanks (Figure 13.4). The formulation consists mainly of mass balance equations. The notation used is given in the nomenclature.

13.4.1 Time Representation for Cyclic Operation

To address the time dimension of batch processes, the cycle time (H) is divided into $|T|$ time intervals (not necessarily of equal duration), with events taking place only at interval boundaries. As shown in Figure 13.2, the interval boundaries (time points) in each cycle are numbered from $t = 1$ to $t = |T| + 1$, with the latter coinciding with the start of the next cycle, that is, $t = 1$. With this time representation, both time intervals and time points can be represented by the same index and set $t \in T$. Note that time intervals are used for semicontinuous units and time points for truly batch units.

13.4.2 Mass Balance for Truly Batch Units

Figure 13.3a shows a schematic diagram of a truly batch unit $i \in I^b$. The inlet water may come from other truly batch units i' ($\in I^b$), storage tanks s or fresh water sources w, and the outlet water may be sent to other truly batch units i', storage tanks s, or wastewater disposal systems d. Equations 13.1 and 13.2 describe the inlet and outlet water flow balances for unit $i \in I^b$ at time point t, respectively. Assuming no water losses or gains during operation, the overall water flow balance for unit $i \in I^b$ is given by Equation 13.3. Note that T_i^{se} is the set of time point pairs corresponding to the start and end times for the operation of unit $i \in I^b$.

$$q_{it}^{in} = \sum_{i' \in I^b} q_{i'it} + \sum_{s \in S} q_{sit} + \sum_{w \in W} q_{wit} \quad \forall i \in I^b, t \in T \tag{13.1}$$

$$q_{it}^{out} = \sum_{i' \in I^b} q_{ii't} + \sum_{s \in S} q_{ist} + \sum_{d \in D} q_{idt} \quad \forall i \in I^b, t \in T \tag{13.2}$$

$$q_{it}^{in} = q_{it'}^{out} \quad \forall i \in I^b, (t,t') \in T_i^{se} \tag{13.3}$$

The lower and upper bounds for the inlet and outlet water flows of truly batch units are given in Equations 13.4 and 13.5:

$$Q_i^L Y_{it}^s \leq q_{it}^{in} \leq Q_i^U Y_{it}^s \quad \forall i \in I^b, t \in T \tag{13.4}$$

$$Q_i^L Y_{it}^e \leq q_{it}^{out} \leq Q_i^U Y_{it}^e \quad \forall i \in I^b, t \in T \tag{13.5}$$

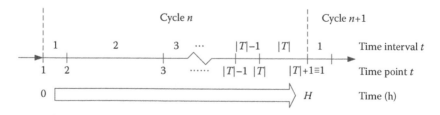

FIGURE 13.2 Continuous-time representation over a cycle.

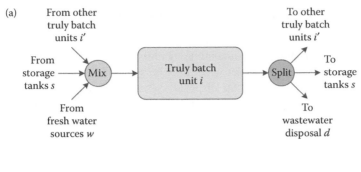

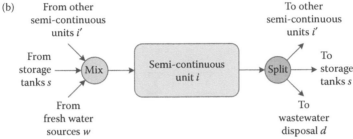

FIGURE 13.3   Schematics of (a) truly batch and (b) semicontinuous water-using units.

where Y_{it}^s and Y_{it}^e are binary parameters indicating the start and end times of truly batch operations, respectively. These two constraints ensure a reasonable amount of water entering and leaving the unit, and that no water enters or leaves when it is not the start ($Y_{it}^s = 0$) or end of the operation ($Y_{it}^e = 0$).

In addition to the water flow balances, contaminant balances are also considered for truly batch units. Equation 13.6 describes the inlet contaminant balance for unit $i \in I^b$ at time point t, and Equation 13.7 depicts the overall contaminant balance. The maximum inlet and outlet concentrations for each truly batch unit are specified by Equations 13.8 and 13.9. Note that these two constraints force the inlet and outlet concentrations to be zero when there is no water input ($Y_{it}^s = 0$) or output ($Y_{it}^e = 0$).

$$q_{it}^{in} c_{ict}^{in} = \sum_{i' \in I^b} q_{i'it} c_{i'ct}^{out} + \sum_{s \in S} q_{sit} c_{sct}^{out} + \sum_{w \in W} q_{wit} C_{wc} \quad \forall c \in C, i \in I^b, t \in T \tag{13.6}$$

$$q_{it}^{in} c_{ict}^{in} + M_{ic} = q_{it'}^{out} c_{ict'}^{out} \quad \forall c \in C, i \in I^b, (t,t') \in T_i^{se} \tag{13.7}$$

$$c_{ict}^{in} \le C_{ic}^{in,max} Y_{it}^s \quad \forall c \in C, i \in I^b, t \in T \tag{13.8}$$

$$c_{ict}^{out} \le C_{ic}^{out,max} Y_{it}^e \quad \forall c \in C, i \in I^b, t \in T \tag{13.9}$$

13.4.3   MASS BALANCE FOR SEMICONTINUOUS UNITS

Figure 13.3b shows a schematic diagram of a semicontinuous unit $i \in I^{sc}$. The inlet water may come from other semicontinuous units i' ($\in I^{sc}$), storage tanks s, or fresh water sources w, while the outlet water may be sent to other semicontinuous units i', storage tanks s, or wastewater disposal systems d. Equations 13.10 and 13.11 describe the inlet and outlet water flow rate balances for unit $i \in I^{sc}$ in time interval t, respectively. Note that Y_{it}^{op} is a binary parameter indicating the operating periods of semicontinuous units. With the same assumption made for truly batch units, the overall water flow rate balance for unit $i \in I^{sc}$ is given by Equation 13.12.

$$f_i^{in} Y_{it}^{op} = \sum_{i' \in I^{sc}} f_{i'it} + \sum_{s \in S} f_{sit} + \sum_{w \in W} f_{wit} \quad \forall i \in I^{sc}, t \in T \tag{13.10}$$

$$f_i^{out} Y_{it}^{op} = \sum_{i' \in I^{sc}} f_{ii't} + \sum_{s \in S} f_{ist} + \sum_{d \in D} f_{idt} \quad \forall i \in I^{sc}, t \in T \tag{13.11}$$

$$f_i^{in} = f_i^{out} \quad \forall i \in I^{sc} \tag{13.12}$$

Contaminant balances are also considered for semicontinuous units. Equation 13.13 describes the inlet contaminant balance for unit $i \in I^{sc}$ in time interval t, and Equation 13.14 depicts the overall contaminant balance, which is only performed in the operating periods. The maximum inlet and outlet concentrations for each semicontinuous unit are specified by Equations 13.15 and 13.16. These two constraints, similar to Equations 13.8 and 13.9, force the inlet and outlet concentrations to be zero when the unit is not operating ($Y_{it}^{op} = 0$).

$$f_i^{in} \bar{c}_{ict}^{in} = \sum_{i' \in I^{sc}} f_{i'it} \bar{c}_{i'ct}^{out} + \sum_{s \in S} f_{sit} \bar{c}_{sct}^{out} + \sum_{w \in W} f_{wit} C_{wc} \quad \forall c \in C, i \in I^{sc}, t \in T$$

$$\tag{13.13}$$

$$f_i^{in} \bar{c}_{ict}^{in} + \bar{M}_{ic} = f_i^{out} \bar{c}_{ict}^{out} \quad \forall c \in C, i \in I^{sc}, t \in T_i^{op} \tag{13.14}$$

$$\bar{c}_{ict}^{in} \leq C_{ic}^{in,max} Y_{it}^{op} \quad \forall c \in C, i \in I^{sc}, t \in T \tag{13.15}$$

$$\bar{c}_{ict}^{out} \leq C_{ic}^{out,max} Y_{it}^{op} \quad \forall c \in C, i \in I^{sc}, t \in T \tag{13.16}$$

13.4.4   MASS BALANCE FOR STORAGE TANKS

Figure 13.4 shows a schematic diagram of a storage tank s. Its inlet water may come from truly batch units $i \in I^b$ at time points or semicontinuous units $i \in I^{sc}$ in time

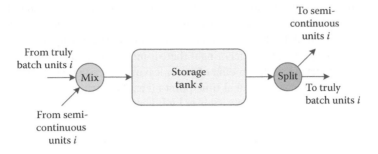

FIGURE 13.4   Schematic of a storage tank.

intervals; the outlet water of tank s may be sent to both types of units $i \in I^b \cup I^{sc}$. Equations 13.17 and 13.18 describe the water flow balances between tank s and truly batch units at time point t. Note that if there are inlet and outlet water flows for a tank at the same point, the inlet flow is assumed to occur before the outlet flow. The water flow rate balances between tank s and semicontinuous units in time interval t are given by Equations 13.19 and 13.20.

$$q_{st}^{in} = \sum_{i \in I^b} q_{ist} \quad \forall s \in S, t \in T \tag{13.17}$$

$$q_{st}^{out} = \sum_{i \in I^b} q_{sit} \quad \forall s \in S, t \in T \tag{13.18}$$

$$f_{st}^{in} = \sum_{i \in I^{sc}} f_{ist} \quad \forall s \in S, t \in T \tag{13.19}$$

$$f_{st}^{out} = \sum_{i \in I^{sc}} f_{sit} \quad \forall s \in S, t \in T \tag{13.20}$$

Equation 13.21 describes the overall water balance for tank s—the amount of water stored in tank s at a time point (t) is equal to that at the previous time point ($t-1$) adjusted by the inlet and outlet water flows during time interval ($t-1$) (i.e., the interval between time points [$t-1$] and t) and those at time point t. Note that Equation 13.21 is applicable at time points $t > 1$. The overall water balance for tank s at $t = 1$ is given by Equation 13.22, where $|T|$ denotes the last time interval of an operation cycle.

$$q_{st} = q_{s,t-1} + \left(f_{s,t-1}^{in} - f_{s,t-1}^{out} \right) \Delta_{t-1} + q_{st}^{in} - q_{st}^{out} \quad \forall s \in S, t \in T, t > 1 \tag{13.21}$$

$$q_{st} = q_{s|T|} + \left(f_{s|T|}^{in} - f_{s|T|}^{out} \right) \Delta_{|T|} + q_{st}^{in} - q_{st}^{out} \quad \forall s \in S, t \in T, t = 1 \tag{13.22}$$

To provide sufficient time for water to be well mixed in a tank, such that the outlet concentration of the tank can be constant in any time interval, Equations 13.23 and 13.24 are introduced to forbid the inlet and outlet water flows from occurring in the same time interval, with binary variable y_{st}^{in} denoting if there is inlet water to tank s in time interval t. Equation 13.25 states that the amount of water leaving tank s in time interval t cannot exceed the amount of water stored at time point t (i.e., the start of time interval t). The capacity constraint for storage tanks is given by Equations 13.26 and 13.27.

$$F_s^L y_{st}^{in} \leq f_{st}^{in} \leq F_s^U y_{st}^{in} \quad \forall s \in S, t \in T \tag{13.23}$$

$$f_{st}^{out} \leq F_s^U \left(1 - y_{st}^{in}\right) \quad \forall s \in S, t \in T \tag{13.24}$$

$$f_{st}^{out} \Delta_t \leq q_{st} \quad \forall s \in S, t \in T \tag{13.25}$$

$$q_{s,t-1} + \left(f_{s,t-1}^{in} - f_{s,t-1}^{out}\right)\Delta_{t-1} + q_{st}^{in} \leq Q_s^{cap} \quad \forall s \in S, t \in T, t > 1 \tag{13.26}$$

$$q_{s|T|} + \left(f_{s|T|}^{in} - f_{s|T|}^{out}\right)\Delta_{|T|} + q_{st}^{in} \leq Q_s^{cap} \quad \forall s \in S, t \in T, t = 1 \tag{13.27}$$

Equations 13.28 and 13.29 describe the inlet contaminant balances for tank s for mass flows from truly batch units at time point t and from semicontinuous units in time interval t, respectively. The overall contaminant balance for tank s is given by Equations 13.30 and 13.31. Since the inlet and outlet water flows are not allowed to exist at the same time, the outlet concentration of tank s in time interval t will be the same as the concentration in tank s at time point t, as given in Equation 13.32.

$$q_{st}^{in} c_{sct}^{in} = \sum_{i \in I^b} q_{ist} c_{ict}^{out} \quad \forall c \in C, s \in S, t \in T \tag{13.28}$$

$$f_{st}^{in} \overline{c}_{sct}^{in} = \sum_{i \in I^{sc}} f_{ist} \overline{c}_{ict}^{out} \quad \forall c \in C, s \in S, t \in T \tag{13.29}$$

$$q_{st} c_{sct}^{out} = q_{s,t-1} c_{sc,t-1}^{out} + \left(f_{s,t-1}^{in} \overline{c}_{sc,t-1}^{in} - f_{s,t-1}^{out} \overline{c}_{sc,t-1}^{out}\right)\Delta_{t-1} + q_{st}^{in} c_{sct}^{in} - q_{st}^{out} c_{sct}^{out}$$
$$\forall c \in C, s \in S, t \in T, t > 1 \tag{13.30}$$

$$q_{st} c_{sct}^{out} = q_{s|T|} c_{sc|T|}^{out} + \left(f_{s|T|}^{in} \overline{c}_{sc|T|}^{in} - f_{s|T|}^{out} \overline{c}_{sc|T|}^{out}\right)\Delta_{|T|} + q_{st}^{in} c_{sct}^{in} - q_{st}^{out} c_{sct}^{out}$$
$$\forall c \in C, s \in S, t \in T, t = 1 \tag{13.31}$$

$$\overline{c}_{sct}^{out} = c_{sct}^{out} \quad \forall c \in C, s \in S, t \in T \tag{13.32}$$

13.4.5 OBJECTIVE FUNCTIONS

For maximum water recovery, the primary objective is to minimize the fresh water consumption (FWC) of the water-using operations:

$$\min_{x_1 \in \Omega_1} FWC = \sum_{w \in W} \sum_{i \in I^b} \sum_{t \in T} q_{wit} + \sum_{w \in W} \sum_{i \in I^{sc}} \sum_{t \in T} f_{wit} \Delta_t \tag{13.33}$$

$$\mathbf{x}_1 \equiv \left\{ \begin{array}{l} c_{ict}^{in}, c_{ict}^{out}, c_{sct}^{in}, c_{sct}^{out}, \overline{c}_{ict}^{in}, \overline{c}_{ict}^{out}, \overline{c}_{sct}^{in}, \overline{c}_{sct}^{out}, \\ f_i^{in}, f_i^{out}, f_{idt}, f_{ii't}, f_{ist}, f_{sit}, f_{st}^{in}, f_{st}^{out}, f_{wit}, \\ q_{idt}, q_{ii't}, q_{ist}, q_{it}^{in}, q_{it}^{out}, q_{sit}, q_{st}, q_{st}^{in}, q_{st}^{out}, q_{wit}, y_{st}^{in} \\ \forall c \in C, d \in D, i, i' \in I, s \in S, t \in T, w \in W \end{array} \right\} \tag{13.34}$$

$$\Omega_1 = \left\{ \mathbf{x}_1 \middle| \text{Equations 13.1--13.32} \right\} \tag{13.35}$$

where $\mathbf{x}_1$ is a vector of the variables, and Ω_1 a feasible solution space defined by the constraints. With the presence of bilinear terms in the contaminant balance equations and the use of binary variables for storage tanks, the model is a mixed integer nonlinear programme (MINLP).

After minimizing the use of fresh water, the second objective is to minimize the water storage capacity (WSC) for tank size determination (Equation 13.36). In this case, the capacity of each tank (Q_s^{cap} in Equations 13.26 and 13.27) is treated as a variable and rewritten as q_s^{cap}. Additionally, the earlier determined minimum fresh water consumption (FWC^{min}) is added as a constraint. The resulting model is also an MINLP.

$$\min_{x_2 \in \Omega_2} WSC = \sum_{s \in S} q_s^{cap} \tag{13.36}$$

$$\mathbf{x}_2 \equiv \mathbf{x}_1 \cup \left\{ q_s^{cap}, \forall s \in S \right\} \tag{13.37}$$

$$\Omega_2 = \left\{ \mathbf{x}_2 \middle| \begin{array}{l} \text{Equations 13.1--13.32;} \\ \sum_{w \in W} \sum_{i \in I^b} \sum_{t \in T} q_{wit} + \sum_{w \in W} \sum_{i \in I^{sc}} \sum_{t \in T} f_{wit} \Delta_t = FWC^{min} \end{array} \right\} \tag{13.38}$$

Similar sequential approaches can be found in the literature. Majozi (2006) presented a two-stage algorithm for fresh water and reusable water storage minimization in batch plants. In the first stage, the fresh water requirement is minimized; in the second stage, the amount of water stored over the time horizon is minimized subject to the earlier set fresh water target. Shoaib et al. (2008) proposed a hierarchical approach for the synthesis of batch water networks. This approach involves three stages in which the fresh water flow, the number of tanks, and the number of connections are minimized sequentially. Chen et al. (2008) considered the minimization of water throughput or piping connections to refine the network configuration after minimizing the fresh water consumption and the WSC.

13.5   ILLUSTRATIVE EXAMPLES

Two modified literature examples are solved to demonstrate the application of the proposed model. In both examples, the model is implemented in the GAMS environment (Rosenthal 2013) on a Core 2, 2.00 GHz processor with BARON (Tawarmalani and Sahinidis 2005) as the MINLP solver.

EXAMPLE 13.1: INTEGRATION OF CONTINUOUS AND TRULY BATCH UNITS

The first example is adapted from Majozi (2005a,b) by adding a continuous unit (F) to the facility of five truly batch units (A–E). Table 13.1 shows the operating data for these water-using units, with Figure 13.5 showing the corresponding Gantt chart. It can be seen that the cycle time for repeated batch operation is 7.5 h. According to the scheduled start and end times for units A–E, the cycle time is divided into six time intervals with six time points, and the values of binary parameters Y_{it}^s, Y_{it}^e, and Y_{it}^{op} are determined as in Table 13.2. Note that unit F is treated as a semicontinuous unit operating in all the time intervals. In this example, a single, uncontaminated fresh water source ($C_{wc} = 0$) is available.

Before any integration for water recovery, in which case all the water-using units use fresh water, the minimum fresh water requirement is calculated as 2635.49 tons per cycle. With two storage tanks available to facilitate water reuse/recycle, the model involves 486 constraints, 615 continuous variables, and 12 binary variables. The solution is obtained in 4 CPUs with a minimum fresh water consumption of 1150 tons per cycle. This corresponds to a 56.36% reduction in the fresh water use compared to the situation without water integration. Figure 13.6 shows the optimal water network configuration. In this arrangement, only units A and F use fresh water; both tanks are employed for water reuse, and there is no direct water transfer between units. Tank 1 (ST1) stores effluents from units A, C, and E for reuse in units C, E, and F, while tank 2 (ST2) allows effluents from units

TABLE 13.1
Operating Data for Example 13.1

Unit	Limiting Flow (ton)	Limiting Concentration (ppm) $C_{ic}^{in,max}$	$C_{ic}^{out,max}$	Mass Load	Time (h) Start	End
A	–	0	100	100 kg	0	3
B	–	250	510	72.8 kg	0	4
C	[300,400]	100	100	0 kg	4	5.5
D	–	250	510	72.8 kg	2	6
E	[300,400]	100	100	0 kg	6	7.5
F	–	100	250	25 kg/h	0	7.5

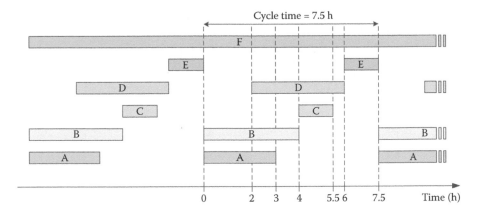

FIGURE 13.5 Gantt chart for repeated batch operation for Example 13.1.

A and F to be reused in units B-D. Figure 13.7 shows the water storage profiles of the tanks. The capacities required are 720 tons for ST1 and 280 ton for ST2.

EXAMPLE 13.2: INTEGRATION OF CONTINUOUS AND SEMICONTINUOUS UNITS

The second example is adapted from Kim and Smith (2004) by adding a continuous unit (U5) to the water system with four semicontinuous units (U1–U4). Table 13.3 shows the limiting conditions and timing data for the water-using operations, with Figure 13.8 showing the corresponding Gantt chart. The cycle time for repeated batch operation is 5 h. According to the scheduled operating periods of units U1–U4, the cycle time is divided into four time intervals with four time

TABLE 13.2

Operation Parameters for Example 13.1

		Time Interval/Point					
Parameter	Unit	$t = 1$	$t = 2$	$t = 3$	$t = 4$	$t = 5$	$t = 6$
Y_{it}^s	A	1	0	0	0	0	0
	B	1	0	0	0	0	0
	C	0	0	0	1	0	0
	D	0	1	0	0	0	0
	E	0	0	0	0	0	1
Y_{it}^e	A	0	0	1	0	0	0
	B	0	0	0	1	0	0
	C	0	0	0	0	1	0
	D	0	0	0	0	0	1
	E	1	0	0	0	0	0
Y_{it}^{op}	F	1	1	1	1	1	1

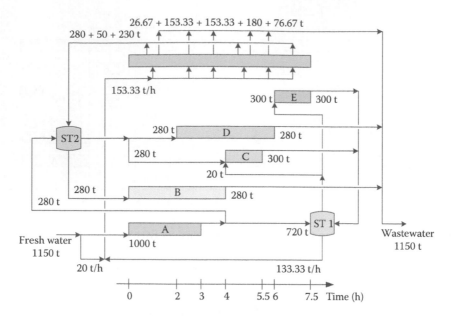

FIGURE 13.6 Optimal water network configuration for Example 13.1.

points, and the values of binary parameter Y_{it}^{op} are determined as in Table 13.4. In this example, a single pure fresh water source is available.

Before water integration, the minimum fresh water requirement is calculated as 480 tons per cycle. With the use of one storage tank, the model involves 160 constraints, 255 continuous variables, and 4 binary variables. The solution is obtained in 1 CPU s with a minimum fresh water consumption of 356.25 tons per cycle.

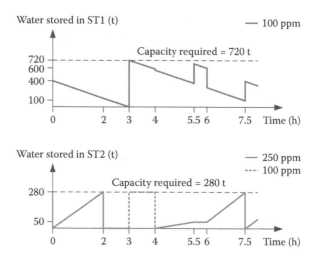

FIGURE 13.7 Water storage profiles for Example 13.1.

TABLE 13.3
Limiting Water Data for Example 13.2

Unit	Limiting Concentration (ppm) $C_{ic}^{in,max}$	$C_{ic}^{out,max}$	Mass Load (kg/h)	Time (h) Start	End
U1	0	100	2	0	1
U2	50	100	5	1	3.5
U3	50	800	30	3	5
U4	400	800	4	1	3
U5	200	400	20	0	5

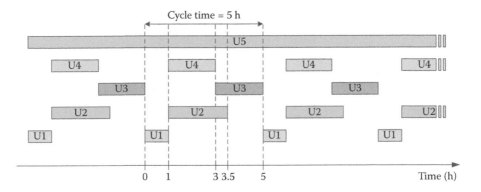

FIGURE 13.8   Gantt chart for repeated batch operation for Example 13.2.

This corresponds to a 25.78% reduction in the fresh water use compared to the case without water recovery. Figure 13.9 shows the optimal water network configuration. Note that most water reuse is carried out directly between units; only part of the effluent from U2 is stored for reuse in U5. Figure 13.10 shows the water storage profile of the tank. The capacity required is 47.5 tons.

TABLE 13.4
Operation Parameters for Example 13.2

Parameter	Unit	Time Interval/Point t = 1	t = 2	t = 3	t = 4
Y_{it}^{op}	U1	1	0	0	0
	U2	0	1	1	0
	U3	0	0	1	1
	U4	0	1	0	0
	U5	1	1	1	1

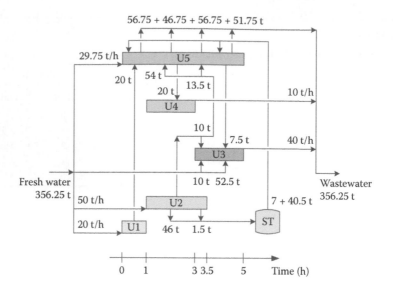

FIGURE 13.9   Optimal water network configuration for Example 13.2.

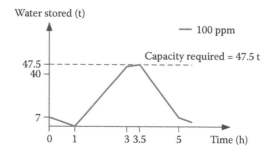

FIGURE 13.10   Water storage profile for Example 13.2.

13.6   SUMMARY

A mathematical model has been developed in this chapter for the synthesis of water networks with continuous and batch process units. The model assumes cyclic operation of batch units with a fixed schedule, while a continuous process is treated as a special case of a semicontinuous process. Two modified literature examples were solved to illustrate the proposed approach. The results show that significant reductions in both fresh water consumption and wastewater generation can be achieved through appropriate water integration. In addition, the model is generic in nature and can be readily applied to interplant water integration and multicontaminant water systems (Lee et al. 2013, 2014). Future work will require the incorporation of regeneration processes (Adekola and Majozi 2011) and a robust scheduling framework into the model. It should be noted that having a flexible schedule can reduce the need for storage. Instead of taking water reduction as the primary concern, the minimum fresh water constraint may sometimes be relaxed to simplify the water network (i.e., trading off fresh water consumption against network complexity). There are also cases with trade-offs between water recovery potential and the requirement for storage and/or regeneration systems to be analyzed (Nun et al. 2011). Hence, a more comprehensive formulation would be needed to take into account the conflicting objectives in water network synthesis.

NOMENCLATURE

INDICES AND SETS

$c \in C$	contaminants
$d \in D$	wastewater disposal systems
$i \in I$	water-using units
$i \in I^b \subset I$	truly batch units
$i \in I^{sc} \subset I$	semicontinuous units
$s \in S$	storage tanks
$t \in T$	time intervals/points
$(t,t') \in T_i^{se}$	start and end time points for the operation of truly batch unit i
$t \in T_i^{op}$	operating periods of semicontinuous unit i
$w \in W$	fresh water sources

PARAMETERS

Δ_t	length of time interval t
$C_{ic}^{in,max}$	maximum inlet concentration of contaminant c for unit i
$C_{ic}^{out,max}$	maximum outlet concentration of contaminant c for unit i
C_{wc}	concentration of contaminant c in fresh water source w
F_s^L	lower bound for the inlet water flow rate to tank s
F_s^U	upper bound for the inlet/outlet water flow rate of tank s
M_{ic}	mass load of contaminant c in unit $i \in I^b$
$\bar{M}_{ic}$	mass load of contaminant c in unit $i \in I^{sc}$
Q_i^L	lower bound for the inlet/outlet water flow of unit $i \in I^b$
Q_i^U	upper bound for the inlet/outlet water flow of unit $i \in I^b$
Q_s^{cap}	storage capacity of tank s
Y_{it}^e	binary parameter denoting if unit $i \in I^b$ ceases to operate at time point t
Y_{it}^{op}	binary parameter denoting if unit $i \in I^{sc}$ operates in time interval t
Y_{it}^s	binary parameter denoting if unit $i \in I^b$ begins to operate at time point t

VARIABLES

c_{ict}^{in}	inlet concentration of contaminant c to unit $i \in I^b$ at time point t
c_{ict}^{out}	outlet concentration of contaminant c from unit $i \in I^b$ at time point t
$\bar{c}_{ict}^{in}$	inlet concentration of contaminant c to unit $i \in I^{sc}$ in time interval t
$\bar{c}_{ict}^{out}$	outlet concentration of contaminant c from unit $i \in I^{sc}$ in time interval t
c_{sct}^{in}	inlet concentration of contaminant c to tank s at time point t

c_{sct}^{out} outlet concentration of contaminant c from tank s at time point t

$\bar{c}_{sct}^{in}$ inlet concentration of contaminant c to tank s in time interval t

$\bar{c}_{sct}^{out}$ outlet concentration of contaminant c from tank s in time interval t

f_i^{in} inlet water flow rate to unit $i \in I^{sc}$ in time interval t

f_i^{out} outlet water flow rate from unit $i \in I^{sc}$ in time interval t

f_{idt} water flow rate from unit $i \in I^{sc}$ to disposal system d in time interval t

$f_{ii't}$ water flow rate from unit $i \in I^{sc}$ to unit $i' \in I^{sc}$ in time interval t

f_{ist} water flow rate from unit $i \in I^{sc}$ to tank s in time interval t

f_{sit} water flow rate from tank s to unit $i \in I^{sc}$ in time interval t

f_{st}^{in} inlet water flow rate to tank s in time interval t

f_{st}^{out} outlet water flow rate from tank s in time interval t

f_{wit} water flow rate from fresh water source w to unit $i \in I^{sc}$ in time interval t

q_{it}^{in} inlet water flow to unit $i \in I^{b}$ at time point t

q_{it}^{out} outlet water flow from unit $i \in I^{b}$ at time point t

q_{idt} water flow from unit $i \in I^{b}$ to disposal system d at time point t

$q_{ii't}$ water flow from unit $i \in I^{b}$ to unit $i' \in I^{b}$ at time point t

q_{ist} water flow from unit $i \in I^{b}$ to tank s at time point t

q_{sit} water flow from tank s to unit $i \in I^{b}$ at time point t

q_{st} amount of water stored in tank s at time point t

q_{st}^{in} inlet water flow to tank s at time point t

q_{st}^{out} outlet water flow from tank s at time point t

q_{wit} water flow from fresh water source w to unit $i \in I^{b}$ at time point t

y_{st}^{in} binary variable indicating if there is inlet water flow to tank s in time interval t

REFERENCES

Adekola, O. and T. Majozi. 2011. Wastewater minimization in multipurpose batch plants with a regeneration unit: Multiple contaminants. *Comput. Chem. Eng.* 35: 2824–36.

Ahmetović, E. and I.E. Grossmann. 2011. Global superstructure optimization for the design of integrated process water networks. *AIChE J.* 57: 434–57.

Almató, M., A. Espñna, and L. Puigjaner. 1999. Optimisation of water use in batch process industries. *Comput. Chem. Eng.* 23: 1427–37.

Chen, C.-L., C.-Y. Chang, and J.-Y. Lee. 2008. Continuous-time formulation for the synthesis of water-using networks in batch plants. *Ind. Eng. Chem. Res.* 47: 7818–32.

Chen, C.-L. and J.-Y. Lee. 2008. A graphical technique for the design of water-using networks in batch processes. *Chem. Eng. Sci.* 63: 3740–54.

Chen, C.-L., J.-Y. Lee, J.-W. Tang, and Y.-J. Ciou. 2009. Synthesis of water-using network with central reusable storage in batch processes. *Comput. Chem. Eng.* 33: 267–76.

Cheng, K.-F. and C.-T. Chang. 2007. Integrated water network designs for batch processes. *Ind. Eng. Chem. Res.* 46: 1241–53.

El-Halwagi, M.M., F. Gabriel, and D. Harell. 2003. Rigorous graphical targeting for resource conservation via material recycle/reuse networks. *Ind. Eng. Chem. Res.* 42: 4319–28.

Foo, D.C.Y. 2009. State-of-the-art review of pinch analysis techniques for water network synthesis. *Ind. Eng. Chem. Res.* 48: 5125–59.

Foo, D.C.Y., Z.A. Manan and Y.L. Tan. 2005. Synthesis of maximum water recovery network for batch process systems. *J. Clean. Prod.* 13: 1381–94.

Gouws, J.F., T. Majozi, D.C.Y. Foo, C.-L. Chen, and J.-Y. Lee. 2010. Water minimization techniques for batch processes. *Ind. Eng. Chem. Res.* 49: 8877–93.

Gunaratnam, M., A. Alva-Argáez, A. Kokossis, J.-K. Kim, and R. Smith. 2005. Automated design of total water systems. *Ind. Eng. Chem. Res.* 44: 588–99.

Hallale, N. 2002. A new graphical targeting method for water minimisation. *Adv. Environ. Res.* 6: 377–90.

Jeżowski, J. 2010. Review of water network design methods with literature annotations. *Ind. Eng. Chem. Res.* 49: 4475–516.

Karuppiah, R. and I.E. Grossmann. 2006. Global optimization for the synthesis of integrated water systems in chemical processes. *Comput. Chem. Eng.* 30: 650–73.

Kim, J.-K. and R. Smith. 2004. Automated design of discontinuous water systems. *Proc. Safe. Environ. Prot.* 82: 238–48.

Lee, J.-Y., C.-L. Chen, and C.-Y. Lin. 2013. A mathematical model for water network synthesis involving mixed batch and continuous units. *Ind. Eng. Chem. Res.* 52: 7047–55.

Lee, J.-Y., C.-L. Chen, C.-Y. Lin, and D.C.Y. Foo. 2014. A two-stage approach for the synthesis of inter-plant water networks involving continuous and batch units. *Chem. Eng. Res. Des.* 92: 941–53.

Li, B.-H. and C.-T. Chang. 2006. A mathematical programming model for discontinuous water-reuse system design. *Ind. Eng. Chem. Res.* 45: 5027–36.

Majozi, T. 2005a. An effective technique for wastewater minimization in batch processes. *J. Clean. Prod.* 13: 1374–80.

Majozi, T. 2005b. Wastewater minimisation using central reusable water storage in batch plants. *Comput. Chem. Eng.* 29: 1631–46.

Majozi, T. 2006. Storage design for maximum wastewater reuse in multipurpose batch plants. *Ind. Eng. Chem. Res.* 45: 5936–43.

Majozi, T., C.J. Brouckaert, and C.A. Buckley. 2006. A graphical technique for wastewater minimisation in batch processes. *J. Environ. Manage.* 78: 317–29.

Manan, Z.A., Y.L. Tan, and D.C.Y. Foo. 2004. Targeting the minimum water flow rate using water cascade analysis technique. *AIChE J.* 50: 3169–83.

Nun, S., D.C.Y. Foo, R.R. Tan, L.F. Razon, and R. Egashira. 2011. Fuzzy automated targeting for trade-off analysis in batch water networks. *Asia-Pac. J. Chem. Eng.* 6: 537–51.

Prakash, R. and U.V. Shenoy. 2005. Targeting and design of water networks for fixed flow rate and fixed contaminant load operations. *Chem. Eng. Sci.* 60: 255–68.

Rosenthal, R.E. 2013. *GAMS—A User's Guide.* Washington, DC: GAMS Development Corporation.

Shoaib, A.M., S.M. Aly, M.E. Awad, D.C.Y. Foo, and M.M. El-Halwagi. 2008. A hierarchical approach for the synthesis of batch water network. *Comput. Chem. Eng.* 32: 530–9.

Tawarmalani, M. and N.V. Sahinidis. 2005. A polyhedral branch-and-cut approach to global optimization. *Math. Program.* 103: 225–49.

Takama, N., T. Kuriyama, K. Shiroko, and T. Umeda. 1980. Optimal water allocation in a petroleum refinery. *Comput. Chem. Eng.* 4: 251–8.

Wang, Y.P. and R. Smith. 1994. Wastewater minimisation. *Chem. Eng. Sci.* 49: 981–1006.

Wang, Y.-P. and R. Smith. 1995. Time pinch analysis. *Chem. Eng. Res. Des.* 73: 905–12.

14 Water Conservation in Fixed Scheduled Batch Processes

Nitin Dutt Chaturvedi and
Santanu Bandyopadhyay

CONTENTS

14.1 INTRODUCTION

This chapter provides the detailed algebraic methodologies to set the targets for a fixed scheduled batch process involving fixed flow operations with only a single contaminant. These methodologies are applicable to both semicontinuous and truly batch processes. The concept of prioritized cost, originally introduced by Shenoy and Bandyopadhyay (2007) for continuous process, is extended for targeting multiple resources in batch processes. Initially, targeting methodology for special cases when there is only one external resource is explained. Subsequently, a general methodology for targeting multiple resources in batch processes is given. These methodologies are proven mathematically and hence, guarantee optimality. The applicability of the methodology is demonstrated with various illustrative examples.

14.2 PROBLEM DEFINITION

The general problem of targeting multiple resources for a fixed-scheduled, fixed-flow batch process (Figure 14.1) may be given as follows:

- A set of internal demands N_d is given. Each demand accepts a flow F_{dj} in a given fixed interval of time with a quality that has to be less than a pre-determined limit q_{dj}. It may be noted that quality follows the inverse scale (Bandyopadhyay, 2006).
- A set of N_s internal sources is given. Each source generates a flow F_{si} with a given quality q_{si} for a fixed interval of time. The flow from an internal source can be reused/recycled to any other internal demand that appears during or at some later time.
- A set of N_r external resources is also given. Each resource is available during the entire time horizon of the batch process with a quality q_{rk} and a cost per unit flow C_{rk}. The availability of each resource may be limited to a specified maximum of $F_{rk,max}$.
- The objective is to develop an algorithm to determine the minimum operating cost.

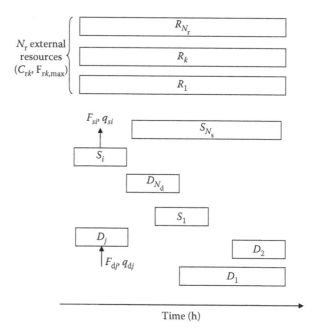

FIGURE 14.1 A schematic of batch process showing demand, sources, resources, and so on. (From Chaturvedi, N.D., Bandyopadhyay, S., 2013. *Chem. Eng. Sci.* 104, 1081–1089. With permission.)

It may be noted that while defining the sources and demands, the schedule of the processing scheme is assumed to be known. It may be further noted that for the special case when only a single external resource is available, the minimum operating cost and the minimum external resource are equivalent.

The entire time horizon of the batch process is subdivided into several time intervals (say I_1, I_2, I_3, ...) such that all sources and/or demands must start or end at these time intervals. In other words, neither any source nor any demand is allowed to end or to start inbetween these time intervals. For a batch process, the overall resource requirement and, equivalently, the overall waste generation is the sum total of the resource requirements and the waste generations in individual intervals.

$$R_k = \sum_{i=1}^{n} R_{I_{ik}} \qquad (14.1)$$

$$W = \sum_{i=1}^{n} W_{I_i} \qquad (14.2)$$

14.3   SINGLE-BATCH PROCESS AND CYCLIC BATCH PROCESS

In a single-batch process, only one cycle of production is to be carried out. It is obvious that sources available in later time interval could not supply to the demands required in some prior time intervals. Therefore, it can be concluded that for a single-batch process the sources available in interval I_i can be used in an interval I_l if and only if $i \leq l$.

On the other hand, in the cyclic batch process, processes are normally operated in repeated batches. It is worth noting that the targeting procedure for a cyclic batch process is equivalent to that of a continuous process (Foo et al., 2005). As, in the case of a cyclic batch, flow can be transferred from a general interval I_i to I_l in both directions for $i \leq k$ as well as for $i > k$ (in the next batch). On the other hand, in the case of single-batch process, transfer can be done if and only if $i \leq l$. So in a cyclic batch process a source can supply to all intervals, to its own interval directly, and other intervals indirectly. In the case of a single-batch process, a source can do an indirect supply to its preceding intervals only. The following theorem can be concluded for resource targeting in single-batch process.

Theorem 14.1

For targeting a cyclic batch process, all the time intervals may be collapsed as a single interval and the overall minimum resource requirement can be determined for that single interval. ∎

14.4   WATER TARGETING FOR A SINGLE-BATCH PROCESS

As discussed in the previous section (Section 14.3), the targeting problem for a single-batch process is additionally limited by time constraints. Initially, the problem is analyzed for the case when a single external resource is available; furthermore, the problem is analyzed for cases of multiple external resources.

14.4.1   WATER TARGETING FOR SINGLE-BATCH PROCESS-SINGLE RESOURCE

In this section, the methodology for calculating the minimum water requirement when there is only one external resource is explained. Initially, batch process containing two intervals is analyzed. In this case, sources available in interval I_1 can supply the demands in either interval I_1 or I_2, while the sources available in interval I_2 can only supply demands in interval I_2. Subsequent to the analysis of the batch process containing two intervals, a batch process containing three intervals is analyzed. These results are generalized for any single-batch process and the targeting algorithm is proposed.

Pillai and Bandyopadhyay (2007) proved that for any RAN, minimization of the external resource requirement (R) is equivalent to the minimization of the total waste generation (W). Therefore, it is sufficient to minimize the overall waste generation in order to minimize the overall resource requirement. To calculate the change in waste flow, the following expression for the minimum waste generation, as derived by Pillai and Bandyopadhyay (2007), may be utilized:

$$W = \sum_{\substack{j=1 \\ q_{sjl_i} > q_{Pl_i}}}^{N_{sl_i}} F_{sjl_i} - \sum_{\substack{k=1 \\ q_{djl_i} > q_{Pl_i}}}^{N_{dl_i}} F_{dkl_i} + \sum_{\substack{j=1 \\ q_{sjl_i} \le q_{Pl_i}}}^{N_{sl_i}} F_{sjl_i} \frac{\left(q_{sjl_i} - q_{rs}\right)}{\left(q_{Pl_i} - q_{rs}\right)} - \sum_{\substack{k=1 \\ q_{djl_i} \le q_{Pl_i}}}^{N_{dl_i}} F_{dkT_i} \frac{\left(q_{dkl_i} - q_{rs}\right)}{\left(q_{Pl_i} - q_{rs}\right)}$$

(14.3)

It may be noted that the change in waste generation depends on the pinch quality of individual intervals. It should also be noted that Equation 14.3 does not assume that the resource quality has to be the least. This equation for calculating waste can be applied to any problem where quality load inequality holds for every internal demand. Using Equation 14.3, it can be derived that transfer of flow from interval I_1 to interval I_2 in a two-interval batch process, which can lead to a reduction in overall waste generation as well as in overall resource requirement, if and only if the pinch quality index of interval I_1 is lower than that of interval I_2. This result can be concluded as the following lemma.

Lemma 14.1

Transfer of flow from interval I_1 to interval I_2 in a two-interval batch process can lead to a reduction in overall waste generation as well as in overall resource requirements, if and only if the pinch quality index of interval I_1 is lower than that of interval I_2. ∎

For a batch process containing only three time intervals, there could be two options for transferring the waste; first, sequentially and second, directly. In the sequential transfer, waste of interval I_1 is transferred to interval I_2 and then the combined waste at the end of interval I_2 is transferred to interval I_3. On the other hand, in direct transfer, the waste of interval I_1 and I_2 are directly transferred to interval I_3. Let δ_{I1} amount of flow from pinch quality source (q_{PI1}) of interval I_1 and δ_{I2} amount of flow from pinch quality source (q_{PI2}) of interval I_2 are transferred to interval I_3 using both of these options. The change in waste flow for direct transfer (ΔW_D) can be calculated directly using Equation 14.3. However, in the case of sequential transfer, the flow from interval I_1 (i.e., δ_{I1} at a quality of q_{PI1}) to interval I_2 may get redistributed. The distributed amount at various qualities can be obtained using Equation 14.3. A combined flow now can be transferred to interval I_3 and the change in waste flow (ΔW_S) can be determined. The differences of the waste flows, calculated for both these options, are summarized as follows for various possible cases.

$$\Delta W_S - \Delta W_D = \begin{cases} 0 & \text{Case 1}: q_{PI3} \geq q_{PI2} \geq q_{PI1} \\ -\delta_{I1}q_{PI1}\left(\dfrac{q_{PI2} - q_{PI3}}{(q_{PI2} - q_{rs})(q_{PI3} - q_{rs})}\right) & \text{Case 2}: q_{PI2} \geq q_{PI3} \geq q_{PI1} \\ -\delta_{I1}\left(1 - \dfrac{q_{PI2} - q_{PI1}}{q_{PI2} - q_{rs}}\right) & \text{Case 3}: q_{PI2} \geq q_{PI1} \geq q_{PI3} \\ 0 & \text{Case 4}: q_{PI1} \geq q_{PI2} \geq q_{PI3} \end{cases}$$

$$(14.4)$$

There are two more cases to be considered: Case 5 ($q_{PI1} \geq q_{PI3} \geq q_{PI2}$) and Case 6 ($q_{PI3} \geq q_{PI1} \geq q_{PI2}$). In these cases, there is a possibility of a change in the waste profile without any change in overall waste generation (illustrated in Chaturvedi and Bandyopadhyay, 2012); these cases are considered separately.

As $q_{PI1} \geq q_{PI2}$ for these cases, Lemma 14.1 implies that there is no change in overall waste generation, when waste of interval I_1 is transferred to interval I_2, but this transfer may lead to a change in the waste profile. It is worth noting that if there is no change in the waste profile then ΔW_S and ΔW_D are same.

It may be observed that when waste redistributes completely below or completely above the pinch quality of interval I_3 (q_{PI3}) then for both options, that is, for sequential and direct transfer, the change in waste flow is identical. However, when waste is redistributed across the pinch quality of interval I_3 (q_{PI3}) then the sequential transfer leads to more or the same reduction in waste as compared to a direct transfer. Therefore, the following statement can be concluded.

Lemma 14.2

For a batch process containing only three time intervals, sequential transfer of the waste profile always leads to minimum waste generation and, equivalently, the minimum resource requirement.

The principle of induction can be used to extend the formulation to any number of intervals and prove the following theorem. For detailed proof, Chaturvedi and Bandyopadhyay (2012) can be referred. ∎

Theorem 14.2

In a batch resource allocation network, targeting via the sequential transfer of the waste profile always leads to the overall minimum resource requirement. ∎

14.4.1.1   Targeting Algorithm-Single Resource

Based on Theorem 14.2, an algorithm is proposed to target the minimum resource requirement for a single-batch process with a given schedule.

> *Step* 1: The entire time horizon of a batch process is subdivided into several time intervals (say I_1, I_2, I_3, ..., I_n), such that all sources and/or demands must start or end at the end points of these time intervals. In other words, neither any source nor any demand is allowed to end or start inbetween these time intervals.
>
> *Step* 2: Integrate the sources and demands within the first interval. Calculate the minimum resource requirement R_{I_1} and the waste profile $W_1(C_{1i})$ of interval T_1.
>
> *Step* 3: Add the waste profile $W_1(C_{1i})$ as sources to interval I_2. And then calculate minimum resource requirement R_{I_2} and determine the waste profile $W_2(C_{2i})$.
>
> *Step* 4: Continue the same procedure until the last interval I_n.
>
> *Step* 5: The total resource requirement is calculated as a sum of the resource requirements of all the intervals and the overall waste generation is the waste at the end of last interval I_n.

It may be noted that this chapter source composite method, proposed by Bandyopadhyay (2006), is used to calculate the minimum resource requirement and waste profile.

14.4.2   WATER TARGETING FOR SINGLE-BATCH PROCESS-MULTIPLE RESOURCES

After establishing the methodology for single resource targeting in Section 14.4.1, we will now understand the methodology for targeting operating costs when multiple resources are available. It is important to note that when multiple resources are present, minimizing the resource requirement may not be equivalent to minimizing operating costs (Shenoy and Bandyopadhyay, 2007). Consider a batch process having only two intervals I_1 and I_2. Let q_{PI1} and q_{PI2} be the individual pinch qualities of these two intervals when solved using the sequential transfer of the waste profile (as described in previous section) considering the purest resource only. There could be two possible cases according to pinch qualities of two intervals.

$$\text{Case 1: } q_{PI1} \geq q_{PI2}$$
$$\text{Case 2: } q_{PI1} \leq q_{PI2}$$

As per Lemma 14.1, the waste transfer from a higher pinch quality interval to lower pinch quality interval does not affect the resource requirement; however, it may change the waste profile. Hence, for case 1, the overall operating cost remains undisturbed when wastes of interval I_1 are transferred to interval I_2 (as the pinch quality of interval I_1 is higher than the pinch quality of interval I_2). Therefore, the minimum operating cost for the entire batch process is the sum of the operating cost of individual intervals. The minimum operating cost of an interval can be targeted using the methodology proposed by Shenoy and Bandyopadhyay (2007).

For case 2; let R_1 and R_2 be the two resources (without loss of generality, we can assume that $q_{r1} \leq q_{r2}$). In this case, the pinch quality of the first interval is lower than that of the second interval, so the waste transfer from the first interval to the second interval will affect the overall resource requirement. The introduction of the second resource will change the waste profile of the first interval and this will affect the overall resource requirements and hence, the overall operating cost. Let 'δ' amount of resource R_2 get added in the first interval. The change in waste (ΔW_{I1}) and the change in cost (ΔC_{I1}) due to this perturbation in the first interval are determined to be:

$$\Delta W_{I1} = \delta \frac{q_{r2} - q_{r1}}{q_{PI1} - q_{r1}} \tag{14.4}$$

$$\Delta C_{I1} = C_{r2}\delta + C_{r1}\delta \left(1 - \frac{q_{r2} - q_{r1}}{q_{PI1} - q_{r1}} \right) \tag{14.5}$$

Equations 14.4 and 14.5 can be derived based on the algebraic expression developed in Pillai and Bandyopadhyay (2007). Similarly, change in the cost of the second interval (ΔC_{I2}) due to the extra waste transfer is expressed as

$$\Delta C_{I2} = -C_{r1}\delta \left(\frac{q_{PI1} - q_{r1}}{q_{PI2} - q_{r1}} \right) \left(\frac{q_{r2} - q_{r1}}{q_{PI1} - q_{r1}} \right) \tag{14.6}$$

The addition of resource R_2 in the first interval is beneficial if only if the overall change in operating cost (ΔC) is negative. Using Equations 14.5 and 14.6, this condition may be expressed as

$$\frac{C_{r2}}{q_{PI2} - q_{r2}} \leq \frac{C_{r1}}{q_{PI2} - q_{r1}} \tag{14.7}$$

Exactly the same expression can be obtained for cost optimal introduction of R_2 in the second interval.

The quantity $C_{ri}/(q_P - q_{ri})$ in Equation 14.7 is called the prioritized cost of a resource (Shenoy and Bandyopadhyay, 2007). Prioritized cost of a resource is proportional to its actual cost and inversely proportional to the difference between the pinch quality and the quality of the resource. It is interesting to note that Equation 14.7 is independent of q_{PI1}. This implies that cost prioritization of resources for both intervals I_1 and I_2 is governed according to the pinch quality of interval I_2. This observation can be summarized as follows:

Lemma 14.3

Minimum operating cost of a batch process containing two intervals can be determined via the cost prioritization of resources according to the pinch quality of the last interval, whenever the pinch quality of the last interval is greater than that of the first interval, otherwise cost prioritization of resources should be done independently.   ■

For a batch process containing three time intervals, there could be six possible cases according to pinch qualities of individual time intervals. Exploring all the cases in a similar way, the following lemma can be established. Detailed derivations are not included in this chapter, however further reading of the paper by Chaturvedi and Bandyopadhyay (2013) is suggested.

Lemma 14.4

For a batch process containing three time intervals, the overall minimum operating cost can be determined via the cost prioritization of resources according to the pinch quality of the last interval whenever the pinch quality of the last interval is the highest.   ■

Combining Lemmas 14.3 and 14.4, the following generalized result can be proven using the principle of mathematical induction. (For detailed proof refer to Chaturvedi and Bandyopadhyay, 2013.)

Theorem 14.3

The cost prioritization of resources for an interval is to be determined based on the highest pinch quality of all the subsequent intervals, including itself.   ■

14.4.2.1 Targeting Algorithm-Multiple Resources

Based on the above theorem and lemmas, the following algorithm is proposed to target the minimum operating cost for a batch process within a given schedule. Figure 14.2 shows the flow chart of the algorithm.

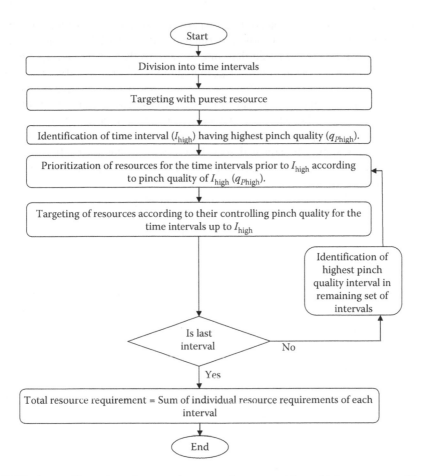

FIGURE 14.2 Flow chart for targeting operating cost with multiple resources. (From Chaturvedi, N.D., Bandyopadhyay, S., 2013. *Chem. Eng. Sci.* 104, 1081–1089. With permission.)

1. The entire time horizon of a batch process is subdivided into several time intervals (say $I_1, I_2, I_3, ..., I_m$), such that all sources and/or demands must start or end at the end points of these time intervals. In other words, neither any source nor any demand is allowed to end or start in between these time intervals.

2. Calculate the pinch quality (q_{PIi}) of each interval using the purest resource following the sequential transfer of the waste profile (Theorem 14.2).

3. Identify the time interval with the highest pinch quality (say I_{high}). Determine the prioritized cost of resources for the intervals prior to I_{high} according to the highest pinch quality (q_{Phigh}). After arranging resources in terms of increasing quality, the prioritized cost for each resource identifies the order in which resources have to be targeted. For minimum operating costs, the acceptable resources will form a sequence such that their qualities are in increasing order while their prioritized costs are in decreasing order.

4. Introduce the resources according to a prioritized sequence using the methodology proposed by Shenoy and Bandyopadhyay (2007) for this set of intervals. It should be noted that, due to the introduction of an intermediate resource, the pinch point of I_{high} may jump to a lower quality. In such a case, the prioritized cost for the new resource has to be calculated based on the new pinch quality. If the prioritized cost is still less than that of other purer resources, the algorithm may be continued with the new pinch point. However, if the prioritized cost increases due to the pinch jump, then the waste flow should be adjusted such that the waste composite line passes through both the original pinch point and the new pinch point (Shenoy and Bandyopadhyay, 2007).
5. Consider the next set of intervals and repeat Steps 3 and 4 until the last interval. The total resource requirement is the sum individual resource requirement of each interval (Equation 14.1).

It has already been explained that the targeting procedure for a cyclic batch process is equivalent to that of a continuous process. Therefore, targeting multiple resources for a cyclic batch process can be carried out applying the procedure proposed by Shenoy and Bandyopadhyay (2007) after collapsing all the time intervals into a single time interval.

14.5   ILLUSTRATIVE EXAMPLES

The algorithms may be applied to minimize the operating cost in different batch RANs. Applicability of the proposed algorithms is demonstrated through the following illustrative examples.

EXAMPLE 14.1: SINGLE AND MULTIPLE RESOURCES EXAMPLE

The limiting water data for this example (Chaturvedi and Bandyopadhyay, 2012) are given in Table 14.1. Figure 14.3 shows the corresponding Gantt chart for the production planning. The five sources and five demands of this example can be categorized in four time intervals (see Table 14.2). Prior to the exploration of water reuse opportunities, the total fresh water requirement and wastewater generation are calculated to be 287.5 t and 262.5 t, respectively.

Single Resource Targeting

For where single resource (pure fresh water) is available, the algorithm proposed in Section 14.4.1 is applied. For the first interval, 10 t of wastewater is generated at 37.5 ppm by applying the procedure of the source composite curve (Bandyopadhyay, 2006). From the mass balance of the first interval, the fresh water requirement (R_{f1}) is targeted to be 2.5 t. Taking the waste from the first interval as source for the second interval, the fresh water requirement for this interval is targeted to be zero and a total of 12.5 t of wastewater is generated (2.5 t at 60 ppm and 10 t at 30 ppm). Following the proposed procedure of the sequential transfer of the waste profile, fresh water requirements for the next two intervals

TABLE 14.1
Limiting Water Data for Example 14.1

Source	Concentration (ppm)	Duration (h)	Flow rate (t/h)	Flow (t)
S1	37.5	0.0–1.0	20	20
S2	60	1.0–4.0	12.5	37.5
S3	30	1.0–2.0	90	90
S4	40	2.0–3.0	15	15
S5	20	2.0–3.0	100	100
Demand	**Concentration (ppm)**	**Duration (h)**	**Flow Rate (t/h)**	**Flow (t)**
D1	30	0.0–1.0	12.5	12.5
D2	33.75	1.0–2.0	100	100
D3	10	3.0–4.0	35	35
D4	22.5	2.0–3.0	100	100
D5	0	2.0–4.0	20	40

Source: From Chaturvedi, N.D., Bandyopadhyay, S., 2012. *Ind. Eng. Chem. Res.* 51, 8015–8024.

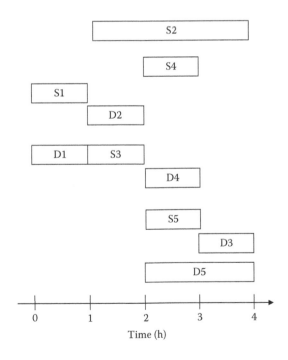

FIGURE 14.3 Gantt chart for production planning for Example 14.1. (From Chaturvedi, N.D., Bandyopadhyay, S., 2012. *Ind. Eng. Chem. Res.* 51, 8015–8024. With permission.)

TABLE 14.2

Interval Flow for Example 14.1

I_1(0.0–1.0) h		I_2(1.0–2.0) h		I_3(2.0–3.0) h		I_4(3.0–4.0) h	
Conc. (ppm)	Flow (t)	Conc. (ppm)	Flow (t)	Conc. (ppm)	Flow (t)	Conc. (ppm)	Flow (t)
37.5	20	60	12.5	60	12.5	60	12.5
30	–12.5	33.75	–100	40	15	10	–35
		30	90	22.5	–100	0	–20
				20	100		
				0	–20		

Source: From Chaturvedi, N.D., Bandyopadhyay, S., 2012. *Ind. Eng. Chem. Res.* 51, 8015–8024. With permission.

are calculated (detailed calculations are not shown for brevity). The waste profile at the end of each interval and the fresh water requirement of each interval are tabulated in Table 14.3. The overall waste generation is the waste at the end of the final interval, that is, 35 t. The overall resource requirement is calculated as a sum of the resource requirement of all the intervals, that is, 60 t. One of the possible water allocation networks to achieve this target is shown in Figure 14.4. However, an alternative water network can be obtained using the nearest neighbor algorithm (NNA) proposed by Prakash and Shenoy (2005) in each interval, I, including the waste from the previous interval.

In this example, the waste transfer does not affect the individual minimum resource requirement of interval I_2 and I_3, but the waste gets redistributed in both intervals. Table 14.4 shows the individual interval resource requirement compared for both sequential and direct approaches. Where, in the direct transfer, waste profile of interval I_1, I_2, and I_3 is directly transferred to interval T_4. The waste profile

TABLE 14.3

Interval Wise Fresh Water Requirement and Waste Profile for Example 14.1

	Time Interval							
	I_1		I_2		I_3		I_4	
Waste-Water	Conc. (ppm)	Flow (t)	Conc. (ppm)	Flow (t)	Conc. (ppm)	Flow (t)	Conc. (ppm)	Flow (t)
	37.5	10	60	2.5	60	15	60	27.5
			30	10	40	7.5	40	7.5
					20	17.5		
Fresh-water (t)	2.5		0		20		37.5	

Source: From Chaturvedi, N.D., Bandyopadhyay, S., 2012. *Ind. Eng. Chem. Res.* 51, 8015–8024. With permission.

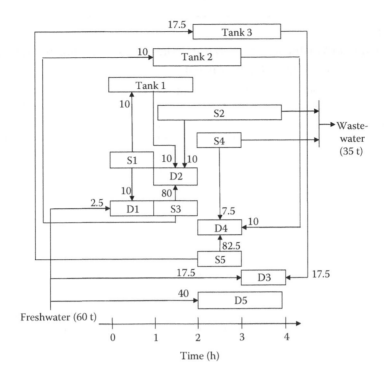

FIGURE 14.4 Water allocation network for Example 14.1 (single-batch process). (From Chaturvedi, N.D., Bandyopadhyay, S., 2012. *Ind. Eng. Chem. Res.* 51, 8015–8024. With permission).

TABLE 14.4

Individual Interval Resource Requirement Compared for Sequential and Direct Approaches for Example 14.1

Interval	Resource Requirement (t)	
	Direct	Sequential
I_1	2.5	2.5
I_2	0	0
I_3	20	20
I_4	42.5	37.5
Total	65	60

Source: From Chaturvedi, N.D., Bandyopadhyay, S., 2012. *Ind. Eng. Chem. Res.* 51, 8015–8024. With permission.

TABLE 14.5

Calculation of Wastewater Generation Using Source Composite Method for Example 14.1 (Cyclic Batch Process)

Conc. (ppm)	Net Interval Flow (t)	Cum. Flow (t)	Mass Load (kg)	Cum. Mass Load (kg)	Wastewater Flow (t)
60	37.5	37.5	0	0	
40	15	52.5	750	750	30
37.5	20	72.5	131.25	881.25	28.5
33.75	−100	−27.5	271.87	1153.12	23.61
30	77.5	50	−103.12	1050	30.00
22.5	−100	−50	375	1425	23.33
20	100	50	−125	1300	32.5
10	−35	15	500	1800	15
0	−40	−25	150	1950	

Source: From Chaturvedi, N.D., 2013. Water and energy targeting for batch process, PhD thesis.

transfer from interval I_1 to I_2 and then to I_3 does not change the individual resource requirement of interval I_2 and I_3, but the waste of interval I_1 gets redistributed in each of interval. If the interval I_2 and I_3 are bypassed and waste redistribution is not allowed, a penalty of 5 t in resource requirements and equivalency in waste transfer are observed.

For a cyclic batch process, the net interval flow is calculated. Using the source composite method (see Table 14.5), the minimum wastewater generation is targeted to be 32.5 t and from the overall mass balance of the batch process the minimum fresh water requirement is calculated to be 57.5 t. Figure 14.5 shows a possible water allocation to achieve these targets. To obtain an alternative allocation network, the NNA may be applied.

TWO RESOURCES

Now, two external resources are assumed to be available for this example (Table 14.6). The operating cost is $4800, when targeting is carried out using only the purest resource (R_1), applying the sequential transfer of the waste profile. It may be observed that the pinch quality (at 60 ppm) of the last interval is highest, and hence, prioritized costs of resources for all intervals are governed by the pinch quality of the last interval I_4. Accordingly, prioritized costs R_1 and R_2 are calculated to be 1.34 and 0.4 $/t ppm, respectively.

The prioritized cost of R_2 is less than the prioritized cost of R_1, therefore, R_2 is introduced from I_1 to I_4 following the sequential transfer of the waste profile to substitute R_1. It should be noted that, due to the addition of resource R_2, the pinch quality of the fourth interval jumps to a lower pinch quality of 20 ppm. However, a new pinch point prioritized cost of R_2 (2 $/t ppm) is still less than that of other R_1 (4 $/t ppm); the algorithm may be continued with the new pinch point. The overall cost for the resource supply is calculated to be $3968 and resource requirements of R_1 and R_2 are calculated to be 40 t and 38.4 t, respectively. The reduction of 17.3% is observed from the base case. Figure 14.6 shows one of the possible water allocation networks to achieve this target.

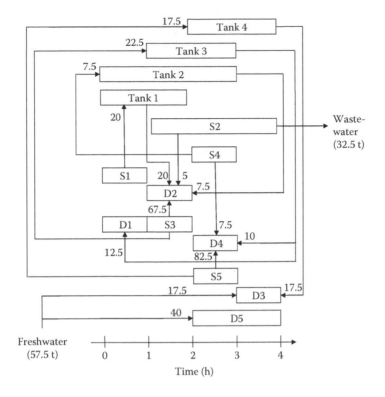

FIGURE 14.5 Water allocation network for Example 14.1 (cyclic batch). (From Chaturvedi, N.D., Bandyopadhyay, S., 2012. *Ind. Eng. Chem. Res.* 51, 8015–8024. With permission.)

The cyclic batch resource requirements for R_1 and R_2 are calculated to be 40 t and 35 t, respectively. The cost of the resource supply for this case is $3900.

EXAMPLE 14.2: THREE RESOURCES EXAMPLE

The limiting data for this example (Chaturvedi and Bandyopadhyay, 2013) are given in Table 14.7 and the resource specifications are listed in Table 14.8. The four sources and four demands of this example can be categorized in four time intervals (see Table 14.9).

TABLE 14.6

Resource Data for Example 14.1

	Cost ($/t)	Quality (Contaminant Conc; ppm)
R_1	80	0
R_2	20	10

Source: From Chaturvedi, N.D., Bandyopadhyay, S., 2013. *Chem. Eng. Sci.* 104, 1081–1089.

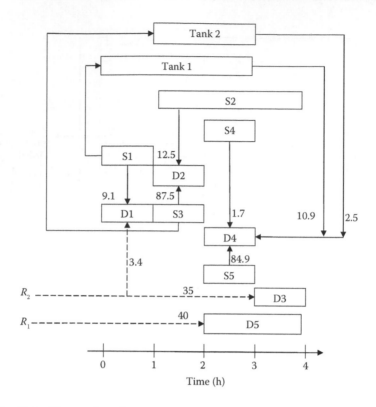

FIGURE 14.6 Water allocation network for Example 14.1 (two resources). (From Chaturvedi, N.D., Bandyopadhyay, S., 2013. *Chem. Eng. Sci.* 104, 1081–1089. With permission.)

TABLE 14.7
Limiting Water Data for Example 14.2

Source	Duration (h)	Concentration (ppm)	Flow (t)
S1	0.0–1.0	12	7
S2	1.0–2.0	15	9
S3	2.0–3.0	25	20
S4	3.0–4.0	10	10
Demand	**Duration (h)**	**Concentration (ppm)**	**Flow (t)**
D1	0.0–1.0	5	2
D2	0.0–2.0	11	10
D3	1.0–2.0	12	5
D4	2.0–3.0	15	10
D5	3.0–4.0	5	5

Source: From Chaturvedi, N.D., Bandyopadhyay, S., 2013. *Chem. Eng. Sci.* 104, 1081–1089.

TABLE 14.8
Resource Data for Example 14.2

	Quality (Contaminant Conc; ppm)	Cost ($/t)
FW1	0	45
FW2	10	25
FW3	20	15

Source: From Chaturvedi, N.D., Bandyopadhyay, S., 2013. *Chem. Eng. Sci.* 104, 1081–1089. With permission.

TABLE 14.9
Various Time Intervals for Example 14.2

I_1(0–1.0 h)		I_2(1.0–2.0 h)		I_3(2.0–3.0 h)		I_4(3.0–4.0 h)	
Conc. (ppm)	Flow (t)	Conc. (ppm)	Flow (t)	Conc. (ppm)	Flow (t)	Conc. (ppm)	Flow (t)
12	7	15	9	25	20	10	10
11	−5	12	−5	15	−10	5	−5
5	−2	11	−5				

Source: From Chaturvedi, N.D., Bandyopadhyay, S., 2013. *Chem. Eng. Sci.* 104, 1081–1089. With permission.

Initially, targeting is carried out using the purest resource (FW1) and applying the sequential transfer of the waste profile. The calculated resource requirement, operating cost, and pinch quality for each interval are shown in Table 14.10. The operating cost is calculated to be $408.6.

The pinch quality interval I_3 is the highest; the prioritized cost of resources for intervals I_1, I_2, and I_3 should be governed according to the pinch quality of I_3 (Theorem 14.3). The prioritized cost for I_4 is to be calculated independently. The prioritized costs of the resources FW1, FW2, and FW3 are calculated (prioritized

TABLE 14.10
Interval Wise Cost Requirement for Example 14.2 Using Purest Resource

	I_1	I_2	I_3	I_4	Total
Pinch quality (ppm)	12	15	25	10	
Resource requirement (t), FW1	1.58	2	3	2.5	9.08
Total operating cost ($)					408.6

Source: From Chaturvedi, N.D., Bandyopadhyay, S., 2013. *Chem. Eng. Sci.* 104, 1081–1089. With permission.

TABLE 14.11

Resource Requirement Calculation for Time Interval I_1 for Example 14.2

Contaminant conc. (ppm)	Net Flow (t)	Cum. Flow (t)	Quality Load Mass Load (kg)	Cum. Mass Load (kg)	Waste Flow for Purest Resource FW1 Only (t)	Waste Flow for First Resource FW1 (t)	Waste Flow for Second Resource FW2 (t)
12	7	7	0	0	1.58		4.5
11	−5	2	7	7	1.09		2
10	0	2	2	9	1	1	
5	−2	0	10	19	0		
0	0	0	0	19			

Source: From Chaturvedi, N.D., Bandyopadhyay, S., 2013. *Chem. Eng. Sci.* 104, 1081–1089. With permission.

cost = $C_{ri}/(q_p − q_{ri})$) to be 1.8, 1.67, and 3 $/t ppm, respectively, for the first three intervals. As the prioritized cost of FW3 is higher than that of FW2, FW3 cannot substitute FW2. Therefore, for the first three intervals, only FW2 and FW1 should be used. For the last interval, FW1 is the only resource that can be used.

For time interval I_1, the resource requirements of FW1 and FW2 are calculated (Table 14.11) using the source composite method (Shenoy and Bandyopadhyay, 2007) to be 1 and 3.5 t respectively and a 4.5 t waste is generated at 12 ppm. It may be noted that, in this chapter, the source composite method is used. However, the proposed methodology is generic in nature and any established methodology can be used for multiple resources targeting for a particular interval. Similarly, with transferring the waste profile sequentially, resource requirements of intervals I_2 and I_3 are calculated (not shown due to brevity). As there are no other resources that can be utilized in I_4, the minimum FW1 requirement remains the same as reported in Table 14.10, that is, 2.5 t. Results are summarized in Table 14.12. The minimum operating cost is $389, a 20.5% reduction from using only the purest resource. One of the possible water allocation networks to achieve this target is shown in Figure 14.7.

For cyclic batch resource requirements can be calculated by collapsing all the intervals into a single interval. The resource requirements for FW1 and FW2 are

TABLE 14.12

Interval Wise Cost Requirement for Example 14.2

	I_1	I_2	I_3	I_4	Total
Resource requirement, FW1(t)	1	0	0	2.5	
Resource requirement, FW2(t)	3.5	4.3	1.47	0	
Cost ($)	132.5	107.5	36.75	113	*389*

Source: From Chaturvedi, N.D., Bandyopadhyay, S., 2013. *Chem. Eng. Sci.* 104, 1081–1089. With permission.

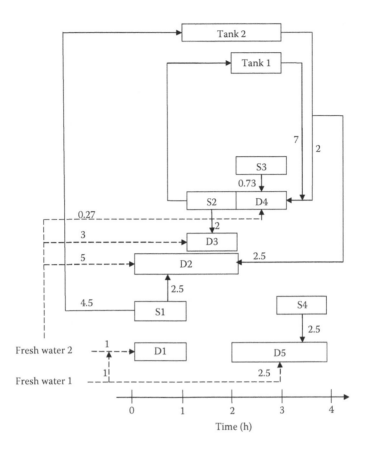

FIGURE 14.7 Water allocation network for Example 14.2. (From Chaturvedi, N.D., Bandyopadhyay, S., 2013. *Chem. Eng. Sci.* 104, 1081–1089.)

calculated to be 3.5 t and 1.76 t, respectively. The minimum operating cost is calculated to be $201.67.

EXAMPLE 14.3: TWO-PRODUCT BATCH PLANT

The methodologies given in this chapter are applied to calculate fresh water requirement and operating cost (single resource and multiple resources) for the two-product batch plant (Kondili et al., 1993).

Two products, Product 1 and Product 2, are to be produced from three raw materials Feed A, Feed B, and Feed C (corresponding to the flow sheet represented in Figure 14.8). The production recipe is as follows:

Feed A is heated inside in a heater (HR). A mixture of 50% feed B and 50% feed C, on a mass basis, is reacted (reaction 1). The product of this reaction is intermediate BC. A mixture of 40% hot A and 60% intermediate BC is reacted (reaction 2) to form product 1 (40%) and intermediate AB (60%). A mixture of 20% feed C and 80% intermediate AB is reacted (reaction 3). The reaction produces impure E. Impure E is purified to produce product 2 (90%) and intermediate AB (10%). A heater (HR) is used to heat feed A. Two reactors, RR1 and RR2, are

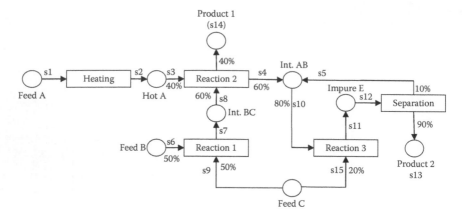

FIGURE 14.8   Flow sheet corresponding to two product batch plant. (From Chaturvedi, N.D., Bandyopadhyay, S., 2014a. *J. Clean. Prod.* 77, 105–115. With permission.)

available to perform three different chemical reactions, reaction 1, reaction 2, and reaction 3 and separator exists to purify Impure E. Table 14.13 shows information regarding the production process. The processing time of each task varies with the batch size. One of the possible schedules (Majozi and Zhu, 2001) for 8 h time of horizon to maximize production is shown in Figure 14.9. Total production (product 1 and product 2) is 151.3 units comprising 70.9 units of product 1 and 80.4 units of product 2. Feed A is fed to the heater, and feed B and feed C are fed to RR1 and RR2 at starting point of horizon. The feeds are processed as per recipe, product 1 is produced via reaction 2 and product 2 is produced after separation.

Water is required for washing RR1 and RR2 at the end of any reaction. Table 14.14 shows information regarding the washing requirement. Washing time is

---

TABLE 14.13

Information Regarding Production Process for Two Product Batch Plant

Task	Unit	Max Batch Size (kg)	Processing Time (h) $\alpha + \beta*$Batch Size	
			α (h)	$\beta \times 10^2$ (h/kg)
Heating (H)	HR	100	0.667	0.6667
Reaction 1 (R1)	RR1	50	1.334	2.6640
	RR2	80	1.334	1.6650
Reaction 2 (R2)	RR1	50	1.334	2.6640
	RR2	80	1.334	1.6650
Reaction 3 (R3)	RR1	50	0.667	1.3320
	RR2	80	0.667	0.8325
Separation (S)	SR	200	1.334	0.6667

Source: From Kondili, E., Pantelides, C.C., Sargent, R.W.H., 1993. *Comput. Chem. Eng.* 17, 211–227.

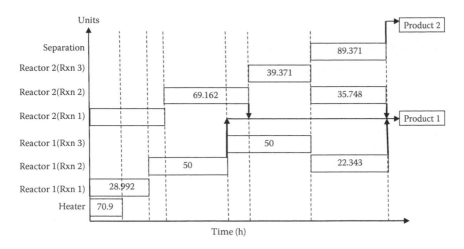

FIGURE 14.9 Production schedule for case study. (From Majozi, T., Zhu, X.X., 2001. *Ind. Eng. Chem. Res.* 40, 5935–5949. With permission.)

included in the total operation time of the process unit. The start and end times of washing are extracted from the schedule shown in Figure 14.9 (Table 14.15). The minimum fresh water requirement can be calculated using the proposed algorithm. The total fresh water requirement is calculated to be 340 t. For the cyclic batch, the fresh water requirement is calculated to be 265 t.

TWO RESOURCES

It should be noted that when a single resource is available, the minimum fresh water requirement and the fresh water requirement to achieve minimum operating cost are same. However, when more than one resource is available then it may not be same. Two resources are available and their specifications are listed

TABLE 14.14
Data Pertaining Water Requirement for Illustrative Example 14.3

Task	Unit	Washing Time (h)	Contaminant Conc. (ppm)		Flow (t)
Heating (H)	HR	0	Demand	Source	
Reaction 1 (R1)	RR1	0.2	250	600	80
	RR2	0.2	250	600	80
Reaction 2 (R2)	RR1	0.2	500	800	100
	RR2	0.2	500	800	100
Reaction 3 (R3)	RR1	0.2	400	850	120
	RR2	0.2	400	850	120
Separation (S)	SR	0			

Source: From Chaturvedi, N.D., Bandyopadhyay, S., 2014a. *J. Clean. Prod.* 77, 105–115. With permission.

TABLE 14.15
Limiting Water Data for Illustrative Example 14.3

Task	Unit	Start Time (h)	End Time (h)	Contaminant Conc. (ppm) Inlet	Outlet	Flow (t)
Rxn 1	RR1	1.89	2.09	250	600	80
Rxn 1	RR2	2.41	2.61	250	600	80
Rxn 2	RR1	4.55	4.75	500	800	100
Rxn 2	RR2	4.88	5.08	500	800	100
Rxn 3	RR1	5.877	6.077	400	850	120
Rxn 3	RR2	5.877	6.077	400	850	120
Rxn 2	RR1	7.8	8	500	800	100
Rxn 2	RR2	7.8	8	500	800	100

Source: From Chaturvedi, N.D., 2013. Water and energy targeting for batch process, PhD thesis.

in Table 14.16. Applying the methodology given in Section 14.4.2, the total cost requirement for the water supply is calculated to be $2054. The operating cost when only the purest resource is used is $3060. Hence, a reduction of 32% is observed. Detailed results are shown in Table 14.17. Figure 14.10 shows one of the possible water allocation networks.

TABLE 14.16
Fresh Water Resources Specifications for Example 14.3

	Cost ($/t)	Quality (Contaminant Conc; ppm)
FW1	9	0
FW2	4	350

Source: From Chaturvedi, N.D., Bandyopadhyay, S., 2014b. *Ind. Eng. Chem. Res.* 53, 5996–6005.

TABLE 14.17
Results for Example 14.3: Water Minimization

	Resource	FW1 (t)	FW2 (t)	Total Operating Cost ($)
Without integration	Single resource	800	–	7200
Fixed schedule	Single resource	340	–	3060
	Multiple resources	45.71	523.16	2504.12

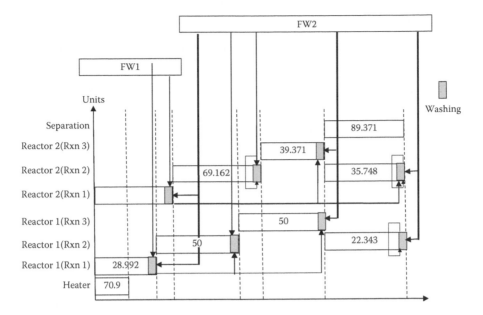

FIGURE 14.10   Possible water allocation network for given schedule Example 14.3. (From Chaturvedi, N.D., Bandyopadhyay, S., 2014b. *Ind. Eng. Chem. Res.* 53, 5996–6005. With permission.)

14.6   CONCLUSIONS

Methodologies to minimize the operating cost for fixed flow rate and fixed schedule batch process with single quality are explained in this chapter for the special case, when only a single resource is available. It is established that the sequential transfer of the waste profile always leads to an overall minimum resource requirement and any interval jump during the transfer of waste may increase the overall resource requirement. In the case of multiple resources, it has been proven that minimizing operating cost prioritization of resources for an interval is determined based on the highest pinch quality of all the subsequent intervals, including itself.

This chapter proposes the methodologies for resource targeting for fixed schedule batch processes. However, further minimization in operating costs can be done via exploring flexibilities in the schedule. Chaturvedi and Bandyopadhyay (2014a) proposed a multiple objective formulation to simultaneously target the minimization of the fresh water requirement and the maximization of production in a batch process while exploring flexibilities in the schedule for when a single resource is available. Chaturvedi and Bandyopadhyay (2014b) proposed a mathematical formulation to minimize the operating cost of the water allocation network in a batch process when multiple freshwater resources are available. The methodology incorporates the variable scheduling of a batch process for a given production. These works are suggested for further reading.

REFERENCES

Bandyopadhyay, S., 2006. Source composite curve for waste reduction. *Chem. Eng. J.* 125, 99–110.

Chaturvedi, N.D., 2013. Water and energy targeting for batch process, PhD thesis.

Chaturvedi, N.D., Bandyopadhyay, S., 2012. A rigorous targeting to minimize resource requirement in batch processes. *Ind. Eng. Chem. Res.* 51, 8015–8024.

Chaturvedi, N.D., Bandyopadhyay, S., 2013. Targeting for multiple resources in batch processes. *Chem. Eng. Sci.* 104, 1081–1089.

Chaturvedi, N.D., Bandyopadhyay, S., 2014a. Simultaneously targeting for the minimum water requirement and the maximum production in a batch process. *J. Clean. Prod.* 77, 105–115.

Chaturvedi, N.D., Bandyopadhyay, S., 2014b. Optimization of multiple freshwater resources in a flexible-schedule batch water network. *Ind. Eng. Chem. Res.* 53, 5996–6005.

Foo, D.C.Y., Manan, Z.A., Tan, Y.L., 2005. Synthesis of maximum water recovery network for batch process systems. *J. Clean. Prod.* 13, 1381–1394.

Kondili, E., Pantelides, C.C., Sargent, R.W.H., 1993. A general algorithm for short-term scheduling of batch operations—I. MILP formulation. *Comput. Chem. Eng.* 17, 211–227.

Majozi, T., Zhu, X.X., 2001. A novel continuous-time MILP formulation for multipurpose batch plants. 1. Short-term scheduling. *Ind. Eng. Chem. Res.* 40, 5935–5949.

Pillai, H.K., Bandyopadhyay, S., 2007. A rigorous targeting algorithm for resource allocation networks. *Chem. Eng. Sci.* 62, 6212–6221.

Prakash, R., Shenoy, U.V., 2005. Targeting and design of water networks for fixed flowrate and fixed contaminant load operations. *Chem. Eng. Sci.* 60, 255–268.

Shenoy, U. V, Bandyopadhyay, S., 2007. Targeting for multiple resources. *Ind. Eng. Chem. Res.* 46, 3698–3708.

15 Simultaneous Optimization of Energy and Water Use in Multipurpose Batch Plants

Esmael Reshid Seid, Jui-Yuan Lee, and Thokozani Majozi

CONTENTS

15.1 INTRODUCTION

This chapter presents a mathematical technique for simultaneous optimization of energy and water in multipurpose batch facilities. The technique is based on time average model (TAM) and time slice model (TSM), where the time slice is a variable determined by optimization to keep the flexibility of the schedule, as compared to previous models based on fixed schedule and fixed time slice for heat integration. The scheduling mathematical formulation that forms the basis of the entire integrated framework is presented in detail in Chapter 2 of this book. The benefits of addressing this complex problem within a comprehensive framework are demonstrated through comparisons with published work.

15.2   PROBLEM STATEMENT

The problem addressed in this chapter can be stated as follows:

i. Given the production recipe (state task network [STN] or state sequence network [SSN] representation)
ii. The capacity of units and the type of tasks each unit can perform
iii. The maximum storage capacity for each material
iv. The task-processing times
v. Hot duties for tasks require heating and cold duties for tasks that require cooling
vi. Operating temperatures of heat sources and heat sinks
vii. Minimum allowable temperature differences
viii. The material heat capacities
ix. The units' washing time
x. The mass load of each contaminant
xi. The concentration limits of each contaminant
xii. The costs of raw materials, products, and utilities
xiii. The scheduling time horizon (for profit maximization problem)
xiv. Production demand (for makespan minimization problem)

Determine:

i. The optimum production schedule, that is, allocation of tasks to units, timing of all tasks, and batch sizes
ii. Optimum energy requirement and associated heat exchange configuration
iii. Optimum water requirement and associated water-reuse network

15.3   MATHEMATICAL FORMULATION

The scheduling model by Seid and Majozi (2012) was adopted as a scheduling platform since it has proven results in better CPU time and optimal objective value compared to other scheduling models. Uneven discretization of the time horizon, the so called continuous time, was used.

15.3.1   HEAT INTEGRATION MODEL

The mathematical model is based on the superstructure in Figure 15.1. Each task may operate using either direct or stand-alone mode by using only external utilities. If direct integration is not sufficient to satisfy the required duty, external utilities may make up for any deficit.

Constraints (15.1) and (15.2) are simultaneously active and ensure that one hot unit will be integrated with one cold unit when direct heat integration takes place in

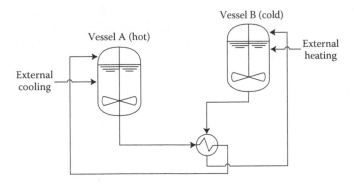

FIGURE 15.1   Superstructure for the energy integration.

order to simplify operation of the process. It is worth noting that mathematically it is also possible for one unit to integrate with more than one unit at a given time point when the summation notation is not used. However, this is very difficult to implement. Moreover, if two units are to be heat integrated at a given time point, they must both be active at that time point. For better understanding, the difference between time point p and extended time point pp is explained using Figure 15.2. If a unit j that is active at time point p is integrated with more than one unit in different temperature and time intervals, an extended time point pp must be defined. Unit $j1$ active at

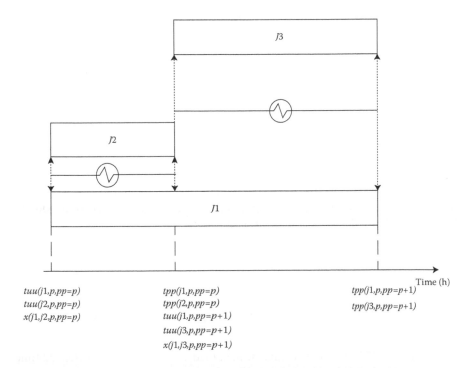

FIGURE 15.2   Differentiating time point p and extended time point pp.

time point p can be integrated with units $j2$ and $j3$ in different time and temperature intervals. At the beginning, unit $j1$ is integrated with unit $j2$ at time point p and the extended time point pp is the same as time point p. Later $j1$ is integrated with unit $j3$ in another time interval where extended time point pp equals to $p + 1$, pp is equal to or greater than time point p and less than or equal to $n + p$, where n is a parameter which is greater than or equal to zero. If n equals 2, then a unit that is active at time point p can be integrated in three different intervals. The model should be solved starting from n equals zero and adding one at a time until no better objective value is achieved.

$$\sum_{s_{inj_c}} x\left(s_{inj_c}, s_{inj_h}, p, pp\right) \leq y\left(s_{inj_h}, p\right), \quad \forall\, p, pp \in P,\ s_{inj_h} \in S_{inj_h},\ s_{inj_c} \in S_{inj_c} \tag{15.1}$$

$$\sum_{s_{inj_h}} x\left(s_{inj_c}, s_{inj_h}, p, pp\right) \leq y\left(s_{inj_c}, p\right), \quad \forall\, p, pp \in P,\ s_{inj_h} \in S_{inj_h},\ s_{inj_c} \in S_{inj_c} \tag{15.2}$$

Constraint (15.3) describes the amount of cooling load required by the hot unit from its initial temperature to its target temperature. In a situation where the temperature in the reactor unit is fixed during exothermic reaction, the heat load becomes the product of the amount of mass that undergoes reaction and the heat of reaction.

$$cl\left(s_{inj_h}, p\right) = mu\left(s_{inj_h}, p\right)cp\left(s_{inj_h}\right)\left(T^{in}_{s_{inj_h}} - T^{out}_{s_{inj_h}}\right), \quad \forall\, p \in P,\ s_{inj_h} \in S_{inj_h} \tag{15.3}$$

Constraint (15.4) describes the amount of heating load required by the cold unit from its initial temperature to its target temperature. In a situation where the temperature in the reactor unit is fixed during endothermic reaction, the heat load becomes the product of the amount of mass that undergoes reaction and the heat of reaction.

$$hl\left(s_{inj_c}, p\right) = mu\left(s_{inj_c}, p\right)cp\left(s_{inj_c}\right)\left(T^{out}_{s_{inj_c}} - T^{in}_{s_{inj_c}}\right), \quad \forall\, p \in P,\ s_{inj_c} \in S_{inj_c} \tag{15.4}$$

Constraints (15.5) and (15.6) describe the average heat flow for the hot and cold unit, respectively, during the processing time, which is the same as TAM, to address the energy balance during heat integration properly.

$$cl\left(s_{inj_h}, p\right) = avcl\left(s_{inj_h}, p\right)\left(tp\left(s_{inj_h}, p\right) - tu\left(s_{inj_h}, p\right)\right), \quad \forall\, p \in P,\ s_{inj_h} \in S_{inJ_h} \tag{15.5}$$

$$hl\left(s_{inj_c}, p\right) = avhl\left(s_{inj_c}, p\right)\left(tp\left(s_{inj_c}, p\right) - tu\left(s_{inj_c}, p\right)\right), \quad \forall\, p \in P,\ s_{inj_c} \in S_{inj_c} \tag{15.6}$$

Constraints (15.7) and (15.8) define the heat load at time point p and extended time point pp for the hot and cold unit.

$$hlp\left(s_{inj_c}, p, pp\right) = avhl\left(s_{inj_c}, p\right)\left(tpp\left(s_{inj_c}, p, pp\right) - tuu\left(s_{inj_c}, p, pp\right)\right),$$
$$\forall\, p, pp \in P,\ s_{inj_c} \in S_{inj_c} \tag{15.7}$$

$$clp\left(s_{inj_h}, p, pp\right) = avcl\left(s_{inj_h}, p\right)\left(tpp\left(s_{inj_h}, p, pp\right) - tuu\left(s_{inj_h}, p, pp\right)\right),$$
$$\forall\, p, pp \in P,\ s_{inj_h} \in S_{inj_h} \tag{15.8}$$

Constraints (15.9) and (15.10) are used to calculate the temperature of the hot and cold unit at the intervals.

$$clp\left(s_{inj_h}, p, pp\right) = mu\left(s_{inj_h}, p\right)cp\left(s_{inj_h}\right)\left(T^{in}\left(s_{inj_h}, p, pp\right) - T^{out}\left(s_{inj_h}, p, pp\right)\right),$$
$$\forall\, p, pp \in P,\ s_{inj_h} \in S_{inj_h} \tag{15.9}$$

$$hlp\left(s_{inj_c}, p, pp\right) = mu\left(s_{inj_c}, p\right)cp\left(s_{inj_c}\right)\left(T^{out}\left(s_{inj_c}, p, pp\right) - T^{in}\left(s_{inj_c}, p, pp\right)\right),$$
$$\forall\, p, pp \in P,\ s_{inj_c} \in S_{inj_c} \tag{15.10}$$

Constraint (15.11) states that the amount of heat exchanged by the hot unit with the cold units should be less than the cooling load required by the hot unit during the interval.

$$\sum_{s_{inj_c}} qe\left(s_{inj_c}, s_{inj_h}, p, pp\right) \le clp\left(s_{inj_h}, p, pp\right), \quad \forall\, p, pp \in P, s_{inj_h} \in S_{inj_h},\ s_{inj_c} \in S_{inj_c}$$
$$\tag{15.11}$$

Constraint (15.12) states that the amount of heat exchanged by the cold unit with the hot units should be less than the heat load required by the cold unit during the interval

$$\sum_{s_{inj_h}} qe\left(s_{inj_c}, s_{inj_h}, p, pp\right) \le hlp\left(s_{inj_c}, p, pp\right), \quad \forall\, p, pp \in P, s_{inj_h} \in S_{inj_h},\ s_{inj_c} \in S_{inj_c}$$
$$\tag{15.12}$$

Constraints (15.13) and (15.14) state that the temperature of the unit at the start of an interval should be equal to the temperature at the end of the previous interval.

$$T^{in}\left(s_{inj_h}, p, pp\right) = T^{out}\left(s_{inj_h}, p, pp - 1\right), \quad \forall\, p, pp \in P,\ s_{inj_h} \in S_{inj_h} \tag{15.13}$$

$$T^{in}\left(s_{inj_c}, p, pp\right) = T^{out}\left(s_{inj_c}, p, pp - 1\right), \quad \forall\, p, pp \in P,\ s_{inj_c} \in S_{inj_c} \tag{15.14}$$

Constraints (15.15) and (15.16) state that the temperature at the start of the first interval, which is time point p, which is also pp, should be equal to the initial temperature of the task.

$$T^{in}\left(s_{inj_h}, p, pp\right) = T^{in}_{s_{inj_h}}, \quad \forall \, p, pp \in P, \, p = pp, \, s_{inj_h} \in S_{inj_h} \qquad (15.15)$$

$$T^{in}\left(s_{inj_c}, p, pp\right) = T^{in}_{s_{inj_c}}, \quad \forall \, p, pp \in P, \, p = pp, \, s_{inj_c} \in S_{inj_c} \qquad (15.16)$$

Constraints (15.17) and (15.18) ensure that the minimum thermal driving forces are obeyed when there is direct heat integration between a hot and a cold unit.

$$T^{in}\left(s_{inj_h}, p, pp\right) - T^{out}\left(s_{inj_c}, p, pp\right) \geq \Delta T - \Delta T^U\left(1 - x\left(s_{inj_c}, s_{inj_h}, p, pp\right)\right),$$
$$\forall \, p, pp \in P, \, s_{inj_h} \in S_{inj_h}, \, s_{inj_c} \in S_{inj_c} \qquad (15.17)$$

$$T^{out}\left(s_{inj_h}, p, pp\right) - T^{in}\left(s_{inj_c}, p, pp\right) \geq \Delta T - \Delta T^U\left(1 - x\left(s_{inj_c}, s_{inj_h}, p, pp\right)\right),$$
$$\forall \, p, pp \in P, \, s_{inj_h} \in S_{inj_h}, \, s_{inj_c} \in S_{inj_c} \qquad (15.18)$$

Constraints (15.19) through (15.22) ensure that the times at which units are active are synchronized when direct heat integration takes place.

$$tuu\left(s_{inj_h}, p, pp\right) \geq tuu\left(s_{inj_c}, p, pp\right) - M\left(1 - x\left(s_{inj_c}, s_{inj_h}, p, pp\right)\right),$$
$$\forall \, p, pp \in P, \, s_{inj_h} \in S_{inj_h}, \, s_{inj_c} \in S_{inj_c} \qquad (15.19)$$

$$tuu\left(s_{inj_h}, p, pp\right) \leq tuu\left(s_{inj_c}, p, pp\right) + M\left(1 - x\left(s_{inj_c}, s_{inj_h}, p, pp\right)\right),$$
$$\forall \, p, pp \in P, \, s_{inj_h} \in S_{inj_h}, \, s_{inj_c} \in S_{inj_c} \qquad (15.20)$$

$$tpp\left(s_{inj_h}, p, pp\right) \geq tpp\left(s_{inj_c}, p, pp\right) - M\left(1 - x\left(s_{inj_c}, s_{inj_h}, p, pp\right)\right),$$
$$\forall \, p, pp \in P, \, s_{inj_h} \in S_{inj_h}, \, s_{inj_c} \in S_{inj_c} \qquad (15.21)$$

$$tpp\left(s_{inj_h}, p, pp\right) \leq tpp\left(s_{inj_c}, p, pp\right) + M\left(1 - x\left(s_{inj_c}, s_{inj_h}, p, pp\right)\right),$$
$$\forall \, p, pp \in P, \, s_{inj_h} \in S_{inj_h}, \, s_{inj_c} \in S_{inj_c} \qquad (15.22)$$

Constraints (15.23) and (15.24) stipulate that the starting time of the heating load required for the cold unit and cooling load required for the hot unit at the first interval should be equal to the starting time of the hot and cold unit.

$$tuu\left(s_{inj_h}, p, pp\right) = tu\left(s_{inj_h}, p\right), \quad \forall \, p, pp \in P, \, p = pp, \, s_{inj_h} \in S_{inj_h} \qquad (15.23)$$

$$tuu\left(s_{\text{inj}_c}, p, pp\right) = tu\left(s_{\text{inj}_c}, p\right), \quad \forall \ p, pp \in P, \ p = pp, \ s_{\text{inj}_c} \in S_{\text{inj}_c} \quad (15.24)$$

Constraints (15.25) and (15.26) state that the starting time of heating and cooling in an interval should be equal to the finishing time at the previous interval.

$$tuu\left(s_{\text{inj}_h}, p, pp\right) = tpp\left(s_{\text{inj}_h}, p, pp - 1\right), \quad \forall \ p, pp \in P, \ s_{\text{inj}_h} \in S_{\text{inj}_h} \quad (15.25)$$

$$tuu\left(s_{\text{inj}_c}, p, pp\right) = tpp\left(s_{\text{inj}_c}, p, pp - 1\right), \quad \forall \ p, pp \in P, \ s_{\text{inj}_c} \in S_{\text{inj}_c} \quad (15.26)$$

Constraint (15.27) ensures that if heat integration occurs, the heat load should have a value that is less than the maximum amount of heat exchangeable. When the binary variable associated to heat integration takes a value zero, no heat integration occurs and the associated heat load is zero.

$$qe\left(s_{\text{inj}_c}, s_{\text{inj}_h}, p, pp\right) \leq Q^U x\left(s_{\text{inj}_c}, s_{\text{inj}_h}, p, pp\right), \quad \forall \ p, pp \in P, \ s_{\text{inj}_h} \in S_{\text{inj}_h}, \ s_{\text{inj}_c} \in S_{\text{inj}_c}$$

$$(15.27)$$

Constraints (15.28) and (15.29) state that if the binary variable associated with heat integration is active, then the binary variable associated with heating and cooling must be active at that interval.

$$x\left(s_{\text{inj}_c}, s_{\text{inj}_h}, p, pp\right) \leq y_{\text{int}}\left(s_{\text{inj}_h}, p, pp\right), \quad \forall \ p, pp \in P, \ s_{\text{inj}_h} \in S_{\text{inj}_h}, \ s_{\text{inj}_c} \in S_{\text{inj}_c} \quad (15.28)$$

$$x\left(s_{\text{inj}_c}, s_{\text{inj}_h}, p, pp\right) \leq y_{\text{int}}\left(s_{\text{inj}_c}, p, pp\right), \quad \forall \ p, pp \in P, \ s_{\text{inj}_h} \in S_{\text{inj}_h}, \ s_{\text{inj}_c} \in S_{\text{inj}_c} \quad (15.29)$$

Constraints (15.30) and (15.31) state that the heating and cooling loads take on a value for a certain duration when the binary variables associated with heating and cooling are active.

$$tpp\left(s_{\text{inj}_h}, p, pp\right) - tuu\left(s_{\text{inj}_h}, p, pp\right) \leq Hy_{\text{int}}\left(s_{\text{inj}_h}, p, pp\right), \quad \forall \ p, pp \in P, \ s_{\text{inj}_h} \in S_{\text{inj}_h}$$

$$(15.30)$$

$$tpp\left(s_{\text{inj}_c}, p, pp\right) - tuu\left(s_{\text{inj}_c}, p, pp\right) \leq Hy_{\text{int}}\left(s_{\text{inj}_c}, p, pp\right), \quad \forall \ p, pp \in P, \ s_{\text{inj}_c} \in S_{\text{inj}_c}$$

$$(15.31)$$

Constraints (15.32) and (15.33) state that temperatures change in the heating and cooling unit when the binary variables associated with heating and cooling are active.

$$T^{\text{in}}\left(s_{\text{inj}_h}, p, pp\right) - T^{\text{out}}\left(s_{\text{inj}_h}, p, pp\right) \le \Delta T^U\left(s_{\text{inj}_h}\right) y_{\text{int}}\left(s_{\text{inj}_h}, p, pp\right),$$
$$\forall \; p, pp \in P, \; s_{\text{inj}_h} \in S_{\text{inj}_h} \tag{15.32}$$

$$T^{\text{out}}\left(s_{\text{inj}_c}, p, pp\right) - T^{\text{in}}\left(s_{\text{inj}_c}, p, pp\right) \le \Delta T^U\left(s_{\text{inj}_c}\right) y_{\text{int}}\left(s_{\text{inj}_c}, p, pp\right),$$
$$\forall \; p, pp \in P, \; s_{\text{inj}_c} \in S_{\text{inj}_c} \tag{15.33}$$

Constraint (15.34) states that the cooling of a hot unit will be satisfied by direct heat integration and external cooling utility if required.

$$cl\left(s_{\text{inj}_h}, p\right) = cw\left(s_{\text{inj}_h}, p\right) + \sum_{s_{\text{inj}_c}} qe\left(s_{\text{inj}_c}, s_{\text{inj}_h}, p, pp\right),$$
$$\forall \; p, pp \in P, \; s_{\text{inj}_h} \in S_{\text{inj}_h}, \; s_{\text{inj}_c} \in S_{\text{inj}_c} \tag{15.34}$$

Constraint (15.35) states that the heating of a cold unit will be satisfied by direct heat integration and external heating utility if required.

$$hl\left(s_{\text{inj}_c}, p\right) = st\left(s_{\text{inj}_h}, p\right) + \sum_{s_{\text{inj}_h}} qe\left(s_{\text{inj}_c}, s_{\text{inj}_h}, p, pp\right),$$
$$\forall \; p, pp \in P, \; s_{\text{inj}_h} \in S_{\text{inj}_h}, \; s_{\text{inj}_c} \in S_{\text{inj}_c} \tag{15.35}$$

15.3.2 Wastewater Minimization Model

The superstructure on which the wastewater minimization model is based is depicted in Figure 15.3. Only the water-using operations which are part of a complete batch process are depicted. Unit j represents a water-using operation in which the water

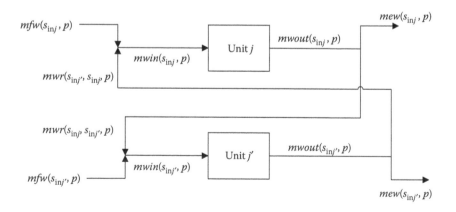

FIGURE 15.3 Superstructure for water usage.

used can consist of freshwater, reuse water, or a combination of reuse water and freshwater. Water from unit j can be reused elsewhere or sent to effluent treatment.

Constraint (15.36) defines the amount of water entering the unit as the sum of freshwater and reuse water from other units.

$$mwin(s_{inj}, p) = mfw(s_{inj}, p) + \sum_{s_{inj'}} mrw(s_{inj'}, s_{inj}, p), \quad \forall\ p \in P,\ s_{inj}, s_{inj'} \in S_{inj}$$

(15.36)

Constraint (15.37) states that the amount of water leaving the unit is equal to the sum of reuse water sent to other units and water sent to effluent treatment.

$$mwout(s_{inj}, p) = \sum_{s_{inj'}} mrw(s_{inj}, s_{inj'}, p) + mew(s_{inj}, p), \quad \forall\ p \in P,\ s_{inj}, s_{inj'} \in S_{inj}$$

(15.37)

Constraint (15.38) is the water balance around the unit and states that the amount of water entering the unit equals the amount of water leaving the unit.

$$mwin(s_{inj}, p) = mwout(s_{inj}, p), \quad \forall\ p \in P,\ s_{inj} \in S_{inj} \qquad (15.38)$$

Constraint (15.39) defines the inlet contaminant load as the mass of contaminant entering with reuse water.

$$cin(s_{inj}, c, p)mwin(s_{inj}, p) = \sum_{s_{inj'}} cout(s_{inj'}, c, p)mrw(s_{inj'}, s_{inj}, p),$$

$$\forall\ p \in P,\ s_{inj}, s_{inj'} \in S_{inj},\ c \in C \qquad (15.39)$$

Constraint (15.40) states that the amount of contaminant leaving the unit equals the sum of the contaminant entering into the unit and the contaminant removed from the process.

$$mwout(s_{inj}, p)cout(s_{inj}, c, p) = SMC(s_{inj})mu(s_{inj}, p) + cin(s_{inj}, c, p)mwin(s_{inj}, p),$$

$$\forall\ p \in P,\ s_{inj} \in S_{inj},\ c \in C \qquad (15.40)$$

Constraint (15.41) ensures that the amount of reused water from unit j to other units does not exceed the maximum allowable water in the receiving units. It also indicates whether water from unit j is reused or not.

$$mrw(s_{inj}, s_{inj'}, p) \leq W_{in}^{U}(s_{inj'})y_{re}(s_{inj}, s_{inj'}, p), \quad \forall\ p \in P,\ s_{inj}, s_{inj'} \in S_{inj} \qquad (15.41)$$

Constraint (15.42) ensures that the reuse of water from unit j in other units can occur only if the units are active.

$$y_{re}(s_{inj}, s_{inj'}, p) \leq y(s_{inj'}, p), \quad \forall p \in P, \; s_{inj}, s_{inj'} \in S_{inj} \tag{15.42}$$

Constraint (15.43) gives the upper bound on the water entering into unit j. It also ensures that water enters into the unit only if it is active.

$$mwin(s_{inj}, p) \leq W_{in}^{U}(s_{inj})y(s_{inj}, p), \quad \forall p \in P, \; s_{inj} \in S_{inj} \tag{15.43}$$

In constraints (15.44) and (15.45), wastewater can only be directly reused if the finishing time of the unit producing wastewater and the starting time of the unit receiving wastewater coincide.

$$tuw(s_{inj}, p) \geq tpw(s_{inj'}, p) - M * y_{re}(s_{inj}, s_{inj'}, p), \quad \forall p \in P, \; s_{inj}, s_{inj'} \in S_{inj} \tag{15.44}$$

$$tuw(s_{inj}, p) \leq tpw(s_{inj'}, p) + M * y_{re}(s_{inj}, s_{inj'}, p), \quad \forall p \in P, \; s_{inj}, s_{inj'} \in S_{inj} \tag{15.45}$$

Constraint (15.46) defines the finishing time of the washing operation as the starting time of the washing operation added to the duration of washing.

$$tpw(s_{inj}, p) \geq tuw(s_{inj}, p) + \tau w(s_{inj})y(s_{inj}, p), \quad \forall p \in P, \; s_{inj} \in S_{inj} \tag{15.46}$$

Constraint (15.47) ensures that the staring time of a task in a unit is greater than the finishing time of the washing operations.

$$tu(s_{inj}, p) \geq tpw\left(s'_{inj}, p - 1\right), \quad \forall p \in P, \; s_{inj}, \; s'_{inj} \in S_{inj}, S_{inj}^{*} \tag{15.47}$$

Constraint (15.48) stipulates that the starting time of the washing operation in a unit occurs after the completion of the task in the unit.

$$tuw(s_{inj}, p) \geq tp(s_{inj}, p), \quad \forall p \in P, \; s_{inj} \in S_{inj} \tag{15.48}$$

Constraints (15.49) and (15.50) ensure that the inlet and outlet concentrations do not exceed the maximum allowable concentration.

$$cin(s_{inj}, c, p) \leq cin^{U}(s_{inj}, c), \quad \forall p \in P, \; s_{inj} \in S_{inj}, \; c \in C \tag{15.49}$$

$$cout(s_{inj}, c, p) \leq cout^{U}(s_{inj}, c), \quad \forall p \in P, \; s_{inj} \in S_{inj}, \; c \in C \tag{15.50}$$

Constraint (15.51) is the objective function in terms of profit maximization, with profit defined as the difference between revenue from product, cost of utility, raw material cost, freshwater cost, and effluent treatment cost.

$$
\max \left(
\begin{array}{l}
\displaystyle\sum_{s^p} price(s^p)d(s^p) - \sum_{p}\sum_{s_{inj_h}} costcw * cw\left(s_{inj_h}, p\right) - \sum_{p}\sum_{s_{inj_c}} costst * st\left(s_{inj_c}, p\right) \\[2ex]
\displaystyle - \sum_{p}\sum_{s_{inj}} costfw * mfw\left(s_{inj}, p\right) - \sum_{p}\sum_{s_{inj}} costew * mew(s_{inj}, p)
\end{array}
\right)
$$

$$
\forall\; p \in P,\; s_{inj_h} \in S_{inj_h},\; s_{inj_c} \in S_{inj_c},\; s_{inj} \in S_{inj} \tag{15.51}
$$

Constraint (15.52) defines minimization of energy and wastewater if the product demand is known.

$$
\min \left(
\begin{array}{l}
\displaystyle\sum_{p}\sum_{s_{inj_h}} costcw * cw\left(s_{inj_h}, p\right) + \sum_{p}\sum_{s_{inj_c}} costst * st\left(s_{inj_c}, p\right) \\[2ex]
\displaystyle + \sum_{p}\sum_{s_{inj}} costfw * mfw(s_{inj}, p) + \sum_{p}\sum_{s_{inj}} costew * mew(s_{inj}, p)
\end{array}
\right),
$$

$$
\forall\; p, pp \in P,\; s_{inj_h} \in S_{inj_h},\; s_{inj_c} \in S_{inj_c},\; s_{inj} \in S_{inj} \tag{15.52}
$$

15.4 CASE STUDIES

Case studies from published literature were selected to demonstrate the application of the proposed model. The results from the proposed models were obtained using CPLEX 9 as MILP solver and CONOPT 3 as NLP solver in DICOPT interface of GAMS 22.0 and were solved using a 2.4 GHz, 4 GB of RAM, Acer TravelMate 5740 G computer.

15.4.1 CASE STUDY I

This case study has been investigated extensively in published literature (Halim and Srinivasan, 2011). It is a simple batch plant requiring only one raw material to yield a product as depicted in the STN representation in Figure 15.4. The plant is comprised of five units and two intermediate storage units. The conversion of the raw material

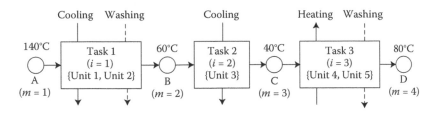

FIGURE 15.4 STN representation of a simple batch plant producing one product.

TABLE 15.1

Scheduling Data for Case Study I

Task (i)	Unit (j)	Max Batch Size (kg)	Total Operation Time (h)	Washing Time (h)	Material State (m)	Initial Inventory (kg)	Max Storage (kg)	Revenue or Cost ($/kg or $/MJ)
Task 1	Unit 1	100	1.5	0.25	A	1000	1000	0
	Unit 2	150	2	0.3	B	0	200	0
Task 2	Unit 3	200	1.5	0	C	0	250	0
Task 3	Unit 4	100	1	0.25	D	0	1000	5
	Unit 5	150	1.5	0.3	Wash water			0.1
					Wastewater			0.05
					Cooling water			0.02
					Steam			1

Note: Total operation time includes processing time and washing time.

into product is achieved through three sequential processes. The first task can be performed in two units ($j1$ and $j2$), the second task can be performed only in unit $j3$ and the third task can be performed in units $j4$ and $j5$. Tasks 1 and 2 require cooling during their operation, while task 3 requires heating. The cooling and heating demands are satisfied by external utilities and heat integration. The operational philosophy requires that the units are cleaned before the next batch is processed. Both freshwater and reuse water can be used as cleaning agents. Table 15.1 gives the capacities of the units, durations of processing and washing tasks, initial availability of states, storage capacities, and selling prices and costs for the states. Table 15.2 gives data pertaining to initial and target temperatures for the tasks, specific heat capacities for the states, and maximum inlet and outlet contaminant concentrations which are unit dependent and the specific contaminant loads.

TABLE 15.2

Energy and Cleaning Requirements for Case Study I

Task (i)	T^{in} (°C)	T^{out} (°C)	Unit (j)	C_p (kJ/kg°C)	Max Inlet Concentration (ppm)	Max Outlet Concentration (ppm)	Contaminant Loading (g Contaminant/ kg Batch)
Task 1	140	60	Unit 1	4	500	1000	0.2
			Unit 2	4	50	100	0.2
Task 2	60	40	Unit 3	3.5			0.2
Task 3	40	80	Unit 4	3	150	300	0.2
			Unit 5	3	300	2000	0.2
Cooling water	20	30					
Steam	170	160					

15.4.1.1 Results and Discussion

The computational results for case study I using the proposed model for the different scenarios and results obtained from literature are presented in Table 15.3. For the scenario without energy and water integration, the total cost of utilities was $293.5. Applying only water integration, the total cost obtained was $259.5, which is an 11.6% reduction, compared to the stand-alone operation without energy and water integration. For the scenario with energy integration only, a total cost of $208 was obtained, which is a 29.1% reduction compared to the stand-alone operation. The fifth column shows the results obtained with combined energy and water integration solved simultaneously giving a total cost of $191.8, which is a 34.6% savings compared to the stand-alone operation. These results show that in order to achieve the best economic performance, the scheduling problem has to be solved simultaneously considering both water and energy integration.

The performance of the proposed model was compared with the sequential optimization technique by Halim and Srinivasan (2011) and resulted in an overall cost of $239.5, which is an 18.4% savings, much less than the 34.6% savings obtained by the proposed model. This work also gives better results compared to the resent

TABLE 15.3
Computational Results for Case Study I

	Proposed Formulation without Water and Energy Integration	Proposed Formulation with Water Integration	Proposed Formulation with Energy Integration	Proposed Formulation with Water and Energy Integration	Halim and Srinivasan (2011) with Water and Energy Integration	Adekola et al. (2013) with Water and Energy Integration
Profit ($)	4706.5	4740.5	4791.5	4808.2	4764.1	4777.3
Steam (MJ)	120	120	36.63	39	43.9	66
Cooling water (MJ)	390	390	281.2	309	313.9	336
Total freshwater (kg)	1105	878.2	1105	977.7	1238.4	1013.3
Revenue from product ($)	5000	5000	5000	5000	5000	5000
Cost of steam ($)	120	120	36.63	39	43.9	66
Cost of cooling water ($)	7.8	7.8	5.623	6.2	6.3	6.72
Cost of freshwater ($)	110.5	87.8	110.5	97.7	123.8	101.3
Cost of wastewater ($)	55.25	43.9	55.2	48.9	61.9	50.7
Total cost ($)	293.5	259.5	208	191.8	235.9	224.72
CPU time (s)	2.3	5000	5000	5000	Not reported	28,797

simultaneous optimization technique of Adekola et al. (2013) with a cost savings of 23.4% compared to 34.6% obtained by the proposed model. The suboptimal results of the method by Adekola et al. (2013) are attributed to two basic drawbacks. The first drawback is due to restricting the flexibility of the schedule by forcing the heat-integrated units to start at the same time. The second drawback is that the model is not based on TAM and the possibility of heat integration between pairs of tasks as well as possible ΔT violations was investigated for each pair of hot and cold tasks before optimization using the initial and target temperatures of the heat-integrated tasks. This limits the chance a unit is able to be integrated in multiple intervals with different intermediate temperatures, with other units. Using the proposed model, we keep the schedule flexible by allowing the heat-integrated units to start anywhere between the start and finish times of the heat-integrated tasks. This benefit can be demonstrated in Figure 15.5, for example unit 2 during processing a task from 3.2 to 4.9 h is integrated

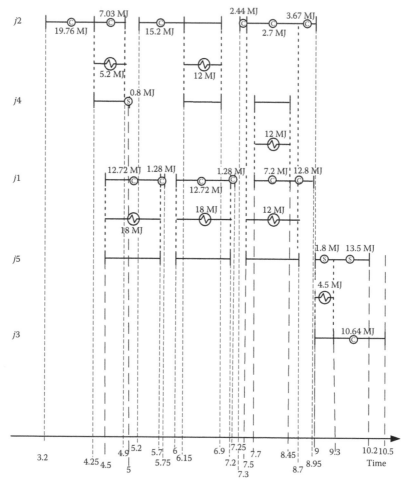

FIGURE15. 5   Possible energy integration within the time horizon of 12 h for case study I.

to exchange heat with unit 4 during processing a task from 4.25 to 5 h. These two units are integrated from 4.25 to 4.9 h to exchange heat, which is not possible by the method of Adekola et al. (2013). Consequently, this work reduces the steam requirement by 40.9% compared to the technique by Adekola et al. (2013). The efficiency of the proposed model can be attributed to solving the scheduling problem while incorporating water and energy into the same framework, and also using the recent efficient scheduling technique by Seid and Majozi (2012). Figure 15.5 details the possible amount of energy integration between the cold and hot units and the time intervals during which energy integration occurred.

The energy requirement of units $j2$ and $j4$ during the interval 3.2–5 h is emphasized to elaborate on the application of the proposed model. The cooling load of unit $j2$ between 3.2 and 4.9 h was 32 MJ. This is partly satisfied through energy integration with unit $j4$ in the same time interval, resulting in an external cooling requirement of 26.8 MJ rather than 32 MJ if it operated in stand-alone mode. At the beginning of the operation of unit $j2$ from 3.2 to 4.25 h, the cooling requirement was 19.76 MJ. This value was obtained using the TAM by multiplying the duration (4.25–3.2 h) and the energy demand per hour (32 MJ/1.7 h [total duration of the task] = 18.823 MJ) where the cooling requirement is fully satisfied by external cooling. For the rest of its operation between 4.25 and 4.9 h, the cooling requirement was 12.24 MJ, satisfied partly with energy integration (5.2 MJ) and the difference by external cooling. The heating requirement of unit $j4$ when it is operated during the interval 4.25–5 h was 6 MJ. From 4.25 to 4.9 h, the steam requirement was 5.2 MJ obtained from the TAM. This heating requirement was fully satisfied during the interval, by integrating with the hot unit $j2$. The rest of the heating, 0.8 MJ, required during its operation between 4.9 and 5 h was satisfied by external steam.

Figure 15.6 shows the amount of contaminant removed, freshwater usage, amount of reused water, and wastewater produced from washing the necessary units. The washing operation of unit $j2$ between 4.9 and 5.2 h required 200 kg of freshwater to remove a contaminant load of 20 g, producing water with a contaminant concentration of 100 ppm. Part of this water produced from unit $j2$, 50 kg, was used for cleaning unit $j4$ to remove a contaminant load of 10 g. This was possible because the outlet concentration from unit $j2$ (100 ppm) was lower than the maximum inlet contaminant concentration (150 ppm) for unit $j4$. From Figure 15.6 the total amount of reused water was 358.23 kg, thereby reducing the water usage from 1105 kg (without water integration) to 977.7 kg (with water integration). This resulted in a savings of 11.5% freshwater usage and wastewater produced.

The amount of material produced, the starting and finishing times of the processes, and washing tasks are shown in Figure 15.7 in the form of a Gantt chart.

15.4.2 CASE STUDY II

A case study taken from Halim and Srinivasan (2011) is used in this section to further demonstrate the performance of the model. This case study is, in essence, the adaptation of a problem from Kondili et al. (1993) to include energy and water integration. The batch plant produces two different products sharing the same processing units, where Figure 15.8 shows the plant flow sheet. The unit operations consist

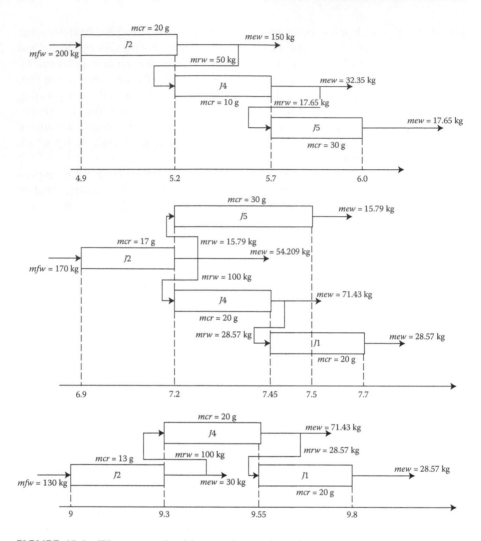

FIGURE 15.6 Water network with water integration within the time horizon of 12 h for case study I. *Note: mfw* = freshwater, *mcr* = contaminant removed, *mrw* = recycled water, *mew* = water sent to effluent.

of preheating, three different reactions, and separation. The plant accommodates many common features of multipurpose batch plants such as units performing multiple tasks, multiple units suitable for a task, and dedicated units for specific tasks. The STN and SSN representations of the flow sheet are shown in Figure 15.9. Tables 15.4 and 15.5 give the required data to solve the scheduling problem. The production recipe is as follows:

1. Raw material, Feed A, is heated from 50°C to 70°C to form HotA used in reaction 2.

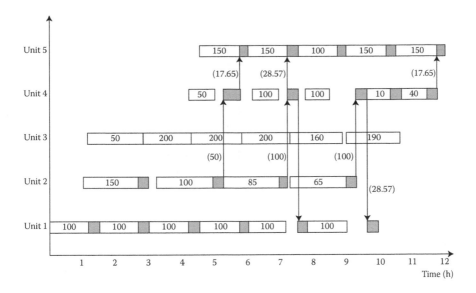

FIGURE 15.7 Gantt chart for the time horizon of 12 h incorporating energy and water integration for case study I.

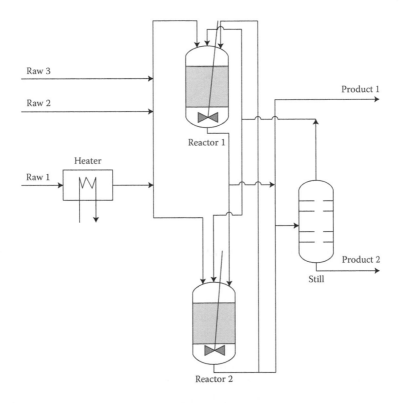

FIGURE 15.8 Flow sheet for case study II.

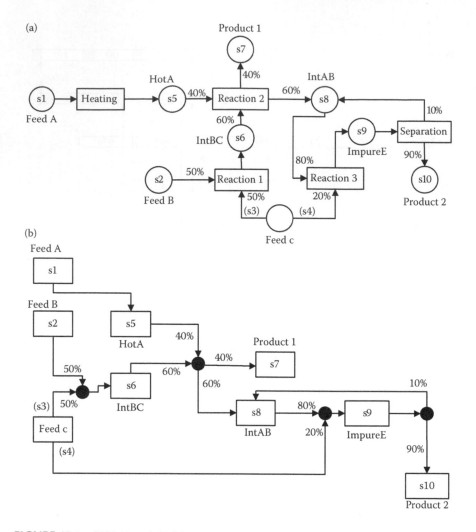

FIGURE 15.9   STN (a) and (b) SSN representation for case study II.

2. Reactant materials, 50% Feed B and 50% Feed C are used in reaction 1 to produce IntBC. During the reaction, the material has to be cooled from 100°C to 70°C.

3. Sixty percent of the intermediate material, IntBC, and 40% of HotA are used in reaction 2 to produce product 1 and IntAB. The process needs to be heated from 70°C to 100°C during its operation.

4. Twenty percent of the reactant, Feed C, and 80% of intermediate, IntAB, from reaction 2 are used in reaction 3 to produce ImpureE. The reaction needs its temperature to be raised from 100°C to 130°C during its operation.

5. The separation process produces 90% product 2 and 10% IntAB from Impure E. Cooling water is used to lower its temperature from 130°C to 100°C.

TABLE 15.4
Scheduling Data for Case Study II

Task (i)	Unit (j)	Max Batch Size (kg)	$\alpha(s_{inj})$	$\beta(s_{inj})$	Washing Time	Material State (s)	Initial Inventory	Max Storage (kg)	Revenue or Cost ($/kg or $/MJ)
Heating (H)	HR	100	0.667	0.007	0	Feed A	1000	1000	0
Reaction-1 (R1)	RR1	50	1.334	0.027	0.25	Feed B	1000	1000	0
	RR2	80	1.334	0.017	0.3	Feed C	1000	1000	0
Reaction-2 (R2)	RR1	50	1.334	0.027	0.25	HotA	0	100	0
	RR2	80	1.334	0.017	0.3	IntAB	0	200	0
Reaction-3 (R3)	RR1	50	0.667	0.013	0.25	IntBC	0	150	0
	RR2	80	0.667	0.008	0.3	ImpureE	0	200	0
Separation (S)	SR	200	1.334	0.007	0	Prod1	0	1000	20
						Prod2	0	1000	20
						Wash water			0.1
						Wastewater			0.05
						Cooling water			0.02
						Steam			1

The processing time of a task i in unit j is assumed to be linearly dependent on its batch size B, that is, $\alpha_i + \beta_i B$ where α_i is a constant term of the processing time of task i and β_i is a coefficient of variable processing time of task i. The batch-dependent processing time makes this case study more complex. Table 15.4 gives the relevant data on coefficients of processing times, the capacity of the processing units, duration of washing, initial inventory of raw materials, storage capacity, and relevant costs. Four contaminants are considered in the case study. The maximum inlet and outlet concentrations are given in Table 15.5. The production demand is given as 200 kg for both Prod1 and Prod2. The objective here is to optimize with respect to makespan, energy, and water consumption.

15.4.2.1 Results and Discussion

The computational statistics for this case study using the proposed model and results obtained from literature are presented in Table 15.6. For makespan minimization, an objective value of 19.5 h was obtained using the proposed model, which is better than 19.96 h obtained by Halim and Srinivasan (2011) and 19.93 h obtained by Adekola et al. (2013). Using the makespan obtained, the case study was solved using the different scenarios for water minimization, energy minimization, and the simultaneous minimization of energy and water by setting customer requirements for Product 1

TABLE 15.5

Data Required for Energy and Water Integration for Case Study II

Task (i)	$T^{in}_{S_{uj}}$ (°C)	$T^{out}_{S_{uj}}$ (°C)	Unit (j)	C_p (kJ/kg°C)	Max Inlet Concentration (ppm)				Max Outlet Concentration (ppm)				Contaminants (ar,br,cp, and dw) Loading (g Contaminant/kg Batch)
					ar	Br	cp	dw	ar	br	cp	dw	
Heating (H)	50	70	HR	2.5									
Reaction-1	100	70	RR1	3.5	300	500	800	400	700	800	1200	900	0.2
			RR2	3.5	300	500	800	400	700	800	1200	900	0.2
Reaction-2	70	100	RR1	3.2	700	600	300	400	1200	1000	600	800	0.2
			RR2	3.2	700	600	300	400	1200	1000	600	800	0.2
Reaction-3	100	130	RR1	2.6	500	200	400	300	800	500	700	900	0.2
			RR2	2.6	500	200	400	300	800	500	700	900	0.2
Separation	130	100	SR	2.8									
Cooling water	20	30											
Steam	170	160											

TABLE 15.6

Computational Results for Case Study II

	Proposed Formulation without Water and Energy Integration	Proposed Formulation with Water Integration	Proposed Formulation with Energy Integration	Propsed Formulation with Water and Energy Integration	Halim and Srinivasan (2011) with Water and Energy Integration	Adekola et al. (2013) with Water and Energy Integration
Makespan (h)	19.5	19.5	19.5	19.5	19.96	19.93
Objective ($)	127.5	112	96.4	94.3	103.3	96.4594
Steam (MJ)	75.3	75.3	44.9	43.3	61.4	44.88
Cooling water (MJ)	50.2	50.2	19.7	18.1	35.4	19.72
Total freshwater (kg)	357.94	238.1	341.3	337.7	275.1	341.2
Revenue from product ($)	8000	8000	8000	8000	8000	8000
Cost of steam ($)	75.3	75.3	44.9	43.3	61.4	44.88
Cost of cooling water ($)	1	1	0.4	0.36	0.7	0.3994
Cost of freshwater ($)	35.8	23.8	34.1	33.8	27.5	34.12
Cost of wastewater ($)	17.9	11.9	17.1	16.9	13.8	17.06
Number of time points/slots	11	11	11	11	N/A	17
CPU time (s)	5000	5000	5000	6074	Not reported	24,532

and Product 2. The total energy and freshwater required for the stand-alone operation were 125.5 MJ and 357.94 kg, respectively.

For the scenario of water integration only allowing the use of reuse water, the
total cost was $112, resulting in 12.2% savings when compared to the stand-alone
operation, which had a total cost of $127.52. By using only energy integration, the
total energy requirement was reduced from 125.5 J in the stand-alone operation to
64.56 MJ, resulting in a 48.6% energy savings and a total cost savings of 24.4%. For
the case of simultaneous optimization of energy and water, a significant total cost savings was obtained compared to energy integration alone and water integration alone.
A total cost savings of 29.4% was obtained, compared to the stand-alone operation.
The performance of the proposed model was also compared with the technique by
Halim and Srinivasan (2011), a total cost of $103 was used in their technique, which
is significantly higher than $94.3 obtained using the proposed model. Furthermore,
the proposed technique is very easy to adopt as opposed to their approach, which
required solving the 3500 MILP scheduling problem to find the best schedule compared to only 3 MILP major iterations of the MINLP problem. Each MILP problem
is solved in a specified CPU time of 2000 s. This complex case study was solved in
a reasonable CPU time of 6074 s, which is less than 2 h, using the proposed model.
When this work was compared to the model by Adekola et al. (2013), the number of
event points required reduced considerably from 17 to 11, which have a direct effect
on reducing CPU time required. In addition, the usage of hot and cold utilities, freshwater, and wastewater are also improved.

Figure 15.10 shows the Gantt chart related to the optimal usage of energy and
water. It also indicates the types of tasks performed in the equipment, the starting
and finishing times of the processes and washing tasks, and the amount of material
processed in each batch.

15.5   CONCLUSIONS

In the method presented in this chapter, wastewater minimization and heat integration
are both embedded within the scheduling framework and solved simultaneously, thus
leading to a truly flexible process schedule. Results from case studies show that addressing profit maximization together with heat integration and wastewater minimization
gives a better overall economic performance. From the case studies, a better objective
value was achieved using the proposed model compared to previous literature models.

NOMENCLATURE

SETS

C $\{c|c \text{ contaminant}\}$

P $\{p|p \text{ time point}\}$

S_{inj} $\{s_{inj}|s_{inj} \text{ any task}\}$

S_{inj_c} $\{s_{inj_c}|s_{inj_c} \text{ task which needs heating}\}$

S_{inj_h} $\{s_{inj_h}|s_{inj_h} \text{ task which needs cooling}\}$

S_{inj_w} $\{s_{inj_w}|s_{inj_w} \text{ task which needs washing afterwards}\}$

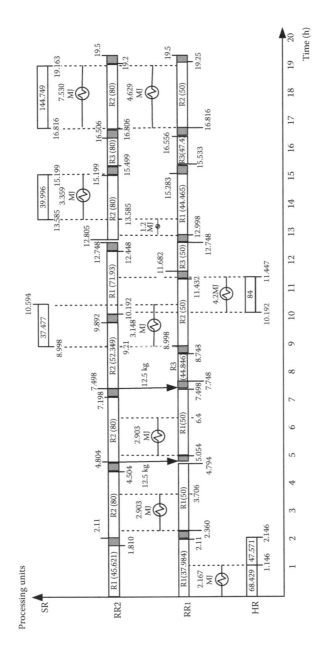

FIGURE 15.10 Resulting production schedule for case study II with direct heat integration and direct water reuse.

PARAMETERS

ΔT	Minimum thermal driving force
ΔT^U	Maximum thermal driving force
$\Delta T^U(s_{\text{inj}})$	Maximum temperature change for a task
$\tau w(s_{\text{inj}})$	Minimum duration required for a washing task
$cin^U(s_{\text{inj}},c)$	Maximum inlet contaminant concentration allowed for contaminant c
$cout^U(s_{\text{inj}},c)$	Maximum outlet contaminant concentration allowed for contaminant c
costcw	Cost of cooling water
costew	Cost of effluent water
costfw	Cost of freshwater
costst	Cost of steam
$cp(s_{\text{inj}_h})$	Specific heat capacity for the heating task
$cp(s_{\text{inj}_c})$	Specific heat capacity for the cooling task
$d(s^p)$	Amount of product produced at the end of the time horizon
H	Time horizon of interest
M	Big-M mostly equivalent to the time horizon
$price(s^p)$	Price of a product
Q^U	Maximum heat requirement from the heating and cooling task
$SMC(s_{\text{inj}})$	Specific contaminant load produced by a task
$T^{\text{in}}_{s_{\text{inj}_h}}$	Inlet temperature of the heating task
$T^{\text{out}}_{s_{\text{inj}_h}}$	Outlet temperature of the heating task
$T^{\text{in}}_{s_{\text{inj}_c}}$	Inlet temperature of the cooling task
$T^{\text{out}}_{s_{\text{inj}_c}}$	Outlet temperature of the cooling task
$W^U_{\text{in}}(s_{\text{inj}})$	Maximum water inlet to a processing task

VARIABLES

$avcl(s_{\text{inj}_h},p)$	Average cooling load required by the hot task at time point p using TAM
$avhl(s_{\text{inj}_c},p)$	Average heating load required by the cold task at time point p using TAM
$cin(s_{\text{inj}}, c, p)$	Inlet contaminant concentration at time point p
$cout(s_{\text{inj}}, c, p)$	Outlet contaminant concentration at time point p
$cl(s_{\text{inj}_h},p)$	Cooling load required by the hot task at time point p
$clp(s_{\text{inj}_h},p,pp)$	Cooling load required by the hot task active at time point p and extended time point pp
$cw\left(s_{\text{inj}_h},p\right)$	External cooling water used by the hot task
$hl(s_{\text{inj}_c},p)$	Heating load required by the cold task at time point p
$hlp(s_{\text{inj}_c},p,pp)$	Heating load required by the cold task active at time point p and extended time point pp
$mew(s_{\text{inj}},p)$	Mass of water entering effluent treatment produced from washing
$mfw(s_{\text{inj}},p)$	Mass of freshwater entering a unit
$mrw(s_{\text{inj}},s_{\text{inj}'},p)$	Mass of water recycled from unit j to another unit j'
$mu(s_{\text{inj}_h},p)$	Amount of material processed by the hot task

$mu(s_{inj_c}, p)$	Amount of material processed by the cold task
$mwin(s_{inj}, p)$	Mass of water entering to wash a unit after a task is performed
$mwout(s_{inj}, p)$	Mass of water leaving after washing
$qe(s_{inj_c}, s_{inj_h}, p, pp)$	Amount of heat load exchanged by the hot and cold unit active at time point p and extended time point pp
$st(s_{inj_c}, p)$	External heating used by the cold task
$tp(s_{inj}, p)$	End time of a heat flow for a task
$tpp(s_{inj}, p, pp)$	Finishing time of a heat flow for a task active at time point p and extended time point pp
$tpw(s_{inj}, p)$	Finishing time of washing operation for unit j
$tu(s_{inj}, p)$	Starting time of a heat flow for a task
$tuu(s_{inj}, p, pp)$	Starting time of a heat flow for a task active at time point p and extended time point pp
$tuw(s_{inj}, p)$	Starting time of washing operation for unit j
$T^{in}(s_{inj}, p, pp)$	Inlet temperature of a task active at time point p and extended time point pp
$T^{out}(s_{inj}, p, pp)$	Outlet temperature of a task active at time point p and extended time point pp
$x(s_{inj_c}, s_{inj_h}, p, pp)$	Binary variable signifying whether heat integration occurs between the hot and cold unit
$y(s_{inj_c}, p)$	Binary variable associated to whether the cold state is active at time point p or not
$y(s_{inj_h}, p)$	Binary variable associated to whether the hot state is active at time point p or not
$y_{int}(s_{inj}, p, pp)$	Binary variable associated to whether the hot and cold states are active at time point p and extended time point pp
$y_{re}(s_{inj}, s_{inj'}, p)$	Binary variable associated with reuse of water from unit j to j' at time point p

REFERENCES

Adekola, O., Stamp, J., Majozi, T., Garg, A., Bandyopadhyay, S., 2013. Unified approach for the optimization of energy and water in multipurpose batch plants using a flexible scheduling framework. *Ind. Eng. Chem Res.* 52, 8488–8506.

Halim, I., Srinivasan, R., 2011. Sequential methodology for integrated optimization of energy and water use during batch process scheduling. *Comput. Chem. Eng.* 35, 1575–1597.

Kondili, E., Pantelides, C.C., Sargent, R.W.H., 1993. A general algorithm for short-term scheduling of batch operations. I. MILP formulation. *Comp. Chem. Eng.* 17, 211–227.

Seid, R., Majozi, T., 2012. Arobust mathematical formulation for multipurpose batch plants. *Chem. Eng. Sci.* 68, 36–53.

16 Retrofit of Industrial Water Network with Mixed Batch and Semicontinuous Processes

Economic versus Environmental Impact

Hella Tokos

CONTENTS

16.1   INTRODUCTION

Increased awareness among public regarding the effects of industrial activities on the environment led to increasing environmental control costs and more stringent environmental regulations. The integration of water networks enables minimization of the flowrate and cost of freshwater for the plant's water supply system by maximizing water reuse and regeneration reuse.

In the literature, studies on the design of water reuse and wastewater treatment networks in industry are mainly focused on continuous processes (Bagajewicz 2000; Karuppiah and Grossmann 2006), while less attention has been directed toward the development of water conservation strategies for batch or mixed batch-continuous processes. The complexity of water network integration for batch process industries lies in the fact that production processes are operating under time-varying conditions and resource demand. In general, two main approaches are utilized for minimization of freshwater consumption, that is, the graphical and mathematically based optimization approach. Graphical design methods are mostly two-stage procedures, where the time dimension is taken as a primary constraint, and concentration as a secondary one, or/and vice versa (Wang and Smith 1995; Foo et al. 2005; Majozi et al. 2006; Chen and Lee 2008; Chan et al. 2008; Kim 2011). Mathematical methods are based on superstructure presentation, and optimize water reuse/recycle by installation of central storage tank for wastewater with acceptable purity (Almato et al. 1999; Majozi 2005; Chen et al. 2007; Gouwsa et al. 2008; Shoaib et al. 2008; Chen at al. 2009), or by providing a storage tank after each water-using operation (Kim and Smith 2004; Tokos and Novak Pintarič 2009). Mathematical models also enable optimization of several subproblems within a single model, for example, the batch scheduling, the water reuse network, and the wastewater treatment network, generating an integrated water network in batch processes (Cheng and Chang 2007).

Water network integration has been successfully applied to some of the process industries by means of water pinch and mathematical optimization approaches. This includes chemical industry (Forstmeier et al. 2005; Feng et al. 2006; Ku-Pineda and Tan 2006; Zheng et al. 2006), oil refineries (Wang and Smith 1994; Ulson de Souzaa et al. 2009), pulp and paper mills (Foo et al. 2006; Lovelady et al. 2007; Manan et al. 2007; Tan et al. 2007; Campos de Faria et al. 2009), textile industry (Ujang et al 2002; Wenzel et al. 2002; Hul et al. 2007a), and food and beverage plants (Thevendiraraj et al. 2003; Matsumura and Mierzwa 2008; Feng et al. 2009; Tokos and Novak Pintarič 2009).

Simultaneous optimization of the water distribution and wastewater treatment system (total water network) makes possible reduction of freshwater consumption and contaminant load of the wastewater discharged in order to meet the environmental regulations at minimum total cost. The water system may discharge wastewater into the central treatment unit or use a local treatment unit to purify wastewater at the production site. This makes possible water regeneration reuse, and a reduction in the contaminant load of wastewater discharged into the central treatment unit.

Several authors have addressed the problem of total water network design. Takama et al. (1980) developed a mathematical program for optimal water allocation within a petroleum refinery based on superstructure representation incorporating reuse and regeneration reuse possibilities. A graphical technique was presented by Kuo and Smith (1998) to explore the interactions between water reuse and wastewater treatment units. The developed graphical methods are mainly based on the water-pinch approach for single (Ng and Foo 2009) and multiple contaminants (Liu Z et al. 2009) or water sources diagram (de Souza et al. 2009; Gomes et al. 2007). A graphical sequential targeting process for a single contaminant water system, considering the minimum freshwater consumption, the minimum regenerated water flow rate, and the optimal regeneration concentration, was explored by Bai et al. (2007) and Feng et al. (2007). Existing design approaches based on mathematical optimization can be classified as sequential or simultaneous. In the case of sequential procedures, the design of water network is decomposed into a series of simpler steps, and the acceptable solution is obtained gradually (Xu et al. 2003; Liu Y et al. 2009). A two-step optimization approach, which generates multiple optimum solutions for the total water network by utilizing mixed integer linear programming (MILP) and nonlinear programming (NLP) subsequently, was proposed by Putra and Amminudin (2008). Simultaneous methods identify the optimal water network based on the interval elimination procedure (Faria and Bagajewicz 2008), deterministic spatial branch and contract algorithm (Karuppiah and Grossmann 2006), stochastic- (Poplewski and Jezowski 2005; Luo and Yuan 2008), superstructure-based mathematical programming, or genetic algorithms (Hul et al. 2007b; Tan et al. 2008; Iancu et al. 2009; Tudor and Lavric 2010; Khor et al. 2012; Liu et al. 2012). The superstructure-based mathematical models are formulated as NLP model (Huang et al. 1999; Chang and Li 2005; de Faria et al. 2009; Tan et al. 2009), MILP model (Jödicke et al. 2001; Matijašević et al. 2010; Poplewski et al. 2010; Ponce-Ortega et al. 2012), or mixed integer nonlinear programming (MINLP) model (Gunaratnam et al. 2005; Chen et al. 2009; Tokos and Novak Pintarič 2009). The water network complexity in case of superstructural approach is controlled by specifying the minimum permissible flow rates in the network, the maximum number of streams allowed at mixing junctions, including the piping costs in the problem formulation (Gunaratnam et al. 2005) or placing additional weight on the piping cost in the objective function (Chang and Li 2005). A design method for an industrial water system, which combines the principles of water-pinch with mathematical programming, was developed by Alva-Argaez et al. (2007) and Bai et al. (2010). Li et al. (2010) introduced a state-time-space superstructure for the optimization of a batch water network in both time and space dimensions. A hybrid optimization strategy, where deterministic and stochastic searching techniques are combined, is suggested to deal with the resulting MINLP model.

In general, process design and optimization problems are based on economic objectives, such as capital investment, net present value, operating costs, and the payback period. The environmental aspect of water networks has recently become very important in design and optimization, in an effort to enhance their environmental performances. With an aim to evaluate the environmental impact of the obtained water network design, the process flow diagram (PFD)-based life cycle assessment (LCA) was used by Lim and Park (2007), Erol and Thoming (2005) used elements of LCA, while Ku-Pineda and Tan (2006) applied sustainable process index and Tokos et al. (2013) the environmental sustainability index.

This chapter presents an MINLP model for water network retrofit estimating both economic and environmental impacts of the design for systems involving batch and semicontinuous processes.

16.2 PROBLEM STATEMENT

The main goal is to obtain the optimum water network with identified reuse and regeneration reuse options between water consumers within the production sections and among them by exploring the trade-off between economic and environmental impacts of the design.

The following is given:

   i. Maximum allowed inlet and outlet contaminant concentrations, and limiting water mass for batch and semicontinuous water-using operations
   ii. Production scheduling data, duration of batch operations, time horizon of interest
   iii. Operation and design parameters for on-site (local) wastewater treatment units operating in batch and/or semicontinuous manner
   iv. Freshwater cost, wastewater treatment cost for on-site and central treatment unit, distance between the water-using operations, distance between the water consumers and on-site treatment units, annual investment cost of the storage tank, piping and local treatment units
   v. Value of selected environmental indicators for current production, benchmarks and weights of the indicators

Determine:

   i. Water reuse and regenerating reuse connection between water consumers,
   ii. The optimal size of on-site wastewater treatment unit (if selected)
   iii. Operating schedule of on-site batch treatment unit depending on the fixed schedule of batch water consumers (if selected)
   iv. Trade-off between the economic (total cost of the water network) and environmental (environmental sustainability index) impact of the obtained water network designs

The water demands can be satisfied by freshwater, wastewater from semicontinuous operations, and wastewater reuse and regeneration reuse between batch processes. Freshwater is defined as an unlimited water source, while wastewater from semicontinuous operations is considered as a water source with limited availability. To avoid possible reduction in extent of water reuse for each reused wastewater stream, a separate storage tank should be installed avoiding mixing of wastewater with different contaminant concentrations. Water streams treated in the semicontinuous treatment units are available for reuse immediately, while those streams purified in batch treatment units are available after a specific treatment time. The operating durations of batch treatment units are significantly shorter than the overall batch time interval.

16.3 MATHEMATICAL MODEL

In this section, only a brief description of the mathematical model is given. The formulation is based on the design method developed by Kim and Smith (2004) modified for specific industrial circumstances of mixed batch and semicontinuous production enabling investigation of the following integration options: (a) direct water reuse between batch and semicontinuous water consumers operating within the same time interval, (b) indirect water reuse between batch and semicontinuous processes operating in different time intervals via storage tank, and (c) regeneration reuse options by designing and scheduling an on-site wastewater treatment system. More detail can be found in Tokos and Novak Pintarič (2009).

With an aim to explore the trade-off between the total cost and environmental sustainability index of the water network designs, bi-objective optimization method was applied. In general, a bi-objective optimization problem can be defined as

$$
\begin{aligned}
&\text{Max} && f_1(\mathbf{x}) \\
&\text{Min} && f_2(\mathbf{x}) \\
&\text{s.t.} && \mathbf{g}(\mathbf{x}) \le 0 \\
&&& \mathbf{h}(\mathbf{x}) = 0 \\
&&& \mathbf{x} = \left[x_1 \cdots x_n \right]^T \\
&&& \mathbf{g} = \left[g_1(\mathbf{x}) \cdots g_{m_1}(\mathbf{x}) \right]^T \\
&&& \mathbf{h} = \left[h_1(\mathbf{x}) \cdots h_{m_2}(\mathbf{x}) \right]^T
\end{aligned}
\tag{16.1}
$$

where $f_j(\mathbf{x})$ is the objective function ($j = 1, 2$), $\mathbf{x}$ is a vector of decision variables, $\mathbf{g}$ is an inequality constraint vector, and $\mathbf{h}$ is an equality constraint. Note that maximization of $f_1(\mathbf{x})$ corresponding to minimization of $-f_1(\mathbf{x})$. Within this formulation, the normal-boundary intersection (NBI) method is used for generation of the trade-off curve (Pareto frontier) between the objectives. This method solves a bi-objective optimization problem by constructing subproblems with an aggregated objective function using the solution obtained in the previous subproblem (Das and Dennis 1998). The original objective functions were normalized according to Equation 16.2 as they have a very different range (Koski and Silvennoinen 1987; Rao and Freiheit 1991):

$$
f_j^{\text{Trans}}(\mathbf{x}) = \frac{f_j(\mathbf{x}) - f_j^{U}(\mathbf{x})}{f_j^{N}(\mathbf{x}) - f_j^{U}(\mathbf{x})} \quad \forall j \in J
\tag{16.2}
$$

where $f_j^{\text{Trans}}(\mathbf{x})$ is the transformed objective function j, $f_j^{N}(\mathbf{x})$ is the nadir point of the objective function j, and $f_j^{U}(\mathbf{x})$ is the utopia point of the objective function j. The utopia objective vector contains the optimum values for the objectives, when each of them is optimized individually by disregarding the other objectives, that is, the most optimistic values. The nadir objective vector consists from the most pessimistic values of the objectives, that is, the value of the second objective function when the first objective function is optimized, individually. The transformed objective function is nondimensional.

16.3.1   ECONOMIC OBJECTIVE FUNCTION

The economic objective function, $F_{\text{Obj}}^{\text{Economic}}/(\text{€}/a)$, is the overall cost of the water network that involves the freshwater cost, cost of electricity needed for pumping water from the wells to/within production site, annual investment costs for the storage tank and piping installations, and the wastewater treatment costs:

$$\min F_{\text{Obj}}^{\text{Economic}} = f_1 + f_2 + f_3 + f_4 + f_5 + f_6 \tag{16.3}$$

The annual freshwater cost, $f_1/(\text{€}/a)$, is given as a product of water consumption for production of one batch, freshwater price, and the total number of batches produced within the overall time horizon:

$$f_1 = \left(\sum_{fw} \sum_n m_{fw,n}^{\text{W}} + \sum_{ww} \sum_j m_{ww,j}^{\text{C}} \right) \cdot \frac{P^{\text{W}} \cdot \lambda_{\text{OHY}}}{\Delta t^{\text{ALL}}} \tag{16.4}$$

where $m_{fw,n}^{\text{W}}/t$ (tons) is the freshwater mass fw to operation n, $m_{ww,j}^{\text{C}}/t$ is the available water mass from semicontinuous operation ww over time interval j, $P^{\text{W}}/(\text{€}/t)$ is the freshwater price, $\lambda_{\text{OHY}}/(h/a)$ is the annual operating time, and $\Delta t^{\text{ALL}}/h$ is the overall processing time for one batch.

The annual investment cost of storage tank installation, $f_2/(\text{€}/a)$, is

$$f_2 = \left(\sum_n CT_n + \sum_{ww} CT_{ww}^{\text{C}} + \sum_{tr} CT_{tr}^{\text{TR,IN}} + \sum_{tr} CT_{tr}^{\text{TR,OUT}} \right) \cdot F_{\text{AN}} \tag{16.5}$$

where $CT_n/\text{€}$ is storage tank investment cost of operation n, $CT_{ww}^{\text{C}}/\text{€}$ is storage tank investment cost of semicontinuous operation ww, $CT_{tr}^{\text{TR,IN}}/\text{€}$ is storage tank investment cost of wastewater before treatment unit tr, $CT_{tr}^{\text{TR,OUT}}/\text{€}$ is storage tank investment cost of purified water after treatment unit tr, and $F_{\text{AN}}/(a^{-1})$ is the annualization factor.

Annual piping cost, $f_3/(\text{€}/a)$, is only evaluated for newly installed pipelines that connect semicontinuous water sources with batch operations, batch operations with treatment units, and integrated batch operations among each other. It is assumed that pipelines connecting process operations with freshwater sources, as well as with the discharge point, already exist at the plant and are, thus, not included in the cost expression. The piping cost is correlated with the cross-sectional area of piping, which is determined based on the water mass:

$$f_3 = \begin{pmatrix} \sum_w \sum_n D_{w,n}^{W} \cdot \left(p \cdot \left(\dfrac{m_{w,n}^{W}}{3600 \cdot \rho \cdot v_{w,n}^{W} \cdot \Delta t_n} \right) + q \cdot Y_{w,n}^{W} \right) \\[2mm] + \sum_n \sum_{nc} D_{n,nc}^{PP} \cdot \left(p \cdot \left(\dfrac{m_{n,nc}^{PP}}{3600 \cdot \rho \cdot v_{n,nc}^{PP} \cdot \Delta t_{nc}} \right) + q \cdot Y_{n,nc}^{PP} \right) \\[2mm] + \sum_n \sum_{nc} \sum_{tr \in batch} D_{n,nc,tr}^{TR} \cdot \left(p \cdot \left(\dfrac{m_{n,nc,tr}^{TR}}{3600 \cdot \rho \cdot v_{n,nc,tr}^{TR} \cdot \Delta t_{tr}^{TR}} \right) + q \cdot Y_{n,nc,tr}^{TR} \right) \\[2mm] + \sum_n \sum_{nc} \sum_{tr \in continuous} D_{n,nc,tr}^{TR} \cdot \left(p \cdot \left(\dfrac{m_{n,nc,tr}^{TR}}{3600 \cdot \rho \cdot v_{n,nc,tr}^{TR} \cdot \Delta t_n} \right) + q \cdot Y_{n,nc,tr}^{TR} \right) \end{pmatrix} \cdot F_{AN}$$

(16.6)

where $m_{w,n}^{W}/t$ is the water mass from water source w to operation n, $m_{nc,n}^{PP}/t$ is the water mass reused from operation nc to operation n, $m_{nc,n,tr}^{TR}/t$ is the water mass reused from operation nc to operation n purified in local treatment unit tr, $D_{w,n}^{W}/m$ is the distance between the water source w and operation n, $D_{n,nc}^{PP}/m$ is the distance between operations n and nc, $D_{n,nc,tr}^{TR}/m$ is the distance between operations n and nc, including the connection with treatment unit tr, $\rho/(t/m^3)$ is density, $v_{w,n}^{W}/(m/s)$ is the velocity of water flow from the water source w to operation n, $v_{n,nc}^{PP}/(m/s)$ is the velocity of the reused water flow from operation n to operation nc, $v_{n,nc,tr}^{TR}/(m/s)$ is the velocity of the reused water flow from operation n to operation nc purified in treatment unit tr, p and q are the variable and fixed parameters, respectively, regarding piping investment costs, $\Delta t_n/h$ is the processing time of operation n while $\Delta t_{tr}^{TR}/h$ is treatment time in local treatment unit tr. Decision variables in Equation 16.6 are: $Y_{w,n}^{W}$ binary variable denoting water mass reused from water source w in operation n, $Y_{n,nc}^{PP}$ binary variable associated with the reused water mass from operation n to operation nc, and $Y_{nc,n,tr}^{TR}$ binary variable associated with the reused water mass from operation nc to operation n purified in local treatment unit t. Pipeline distances in the upper equation are assumed to be the shortest possible distances between the units within the existing pipeline system.

If the processes operate over several time intervals, Equation 16.6 cannot be applied directly, and additional constraints need to be added into the model in order to prevent multiple summations of identical pipeline connections over different time intervals. Assume that process k is divided over several time intervals into parts n. As the reuse water mass between semicontinuous water source and divided process could vary over time intervals, the cross-sectional area of pipeline is determined by means of the largest water mass among the processes n:

$$m_{ww,k}^{W,E} \geq m_{ww,n}^{W}$$

(16.7)

where $m_{ww,k}^{W,E}/t$ is the largest reused water mass from the semicontinuous water source ww to batch operations n, that belongs to the same process k. The latter is associated with the annual piping cost by the logical constraint:

$$Y_{ww,n}^{W} - Y_{ww,k}^{W,E} \leq 0$$

(16.8)

where $Y_{ww,k}^{W,E}$ is the binary variable that denotes the link between water source ww and process k.

For reuse connections between process kc divided into processes nc, and process k divided into processes n, the largest reuse mass is determined by the expressions:

$$m_{kc,k}^{PP,E} \geq m_{nc,n}^{PP} \tag{16.9}$$

$$Y_{nc,n}^{PP} - Y_{kc,k}^{PP,E} \leq 0 \tag{16.10}$$

where $m_{kc,k}^{PP,E}$ /t represents the largest reuse water mass from process kc to process k and $Y_{kc,k}^{PP,E}$ is the binary variable that corresponds to this link.

Similar expressions are added for regeneration reuse connections between divided processes kc and k:

$$m_{kc,k,tr}^{TR,E} \geq m_{nc,n,tr}^{TR} \tag{16.11}$$

$$Y_{nc,n,tr}^{TR} - Y_{kc,k,nc}^{TR,E} \leq 0 \tag{16.12}$$

where $m_{kc,k,tr}^{TR,E}$ /t represents the largest reused water mass from process kc to process k purified in local treatment unit tr, and $Y_{kc,k,tr}^{TR,E}$ is the binary variable of this connection.

In those cases where several processes are divided over the time intervals, Equation 16.6 for the annual piping cost is modified as follows:

$$f_3 = \left[\begin{array}{l} \displaystyle\sum_{ww}\sum_{k} D_{ww,k}^{W,E} \cdot \left(p \cdot \left(\frac{m_{ww,k}^{W,E}}{3600 \cdot \rho \cdot v_{ww,k}^{W,E} \cdot \Delta t_k^E} \right) + q \cdot Y_{ww,k}^{W,E} \right) \\[2ex] + \displaystyle\sum_{k}\sum_{kc} D_{k,kc}^{PP,E} \cdot \left(p \cdot \left(\frac{m_{k,kc}^{PP,E}}{3600 \cdot \rho \cdot v_{k,kc}^{PP,E} \cdot \Delta t_{kc}^E} \right) + q \cdot Y_{k,kc}^{PP,E} \right) \\[2ex] + \displaystyle\sum_{k}\sum_{kc}\sum_{tr \in \text{batch}} D_{k,kc,tr}^{TR,E} \cdot \left(p \cdot \left(\frac{m_{k,kc,tr}^{TR,E}}{3600 \cdot \rho \cdot v_{k,kc,tr}^{TR,E} \cdot \Delta t_{tr}^{TR}} \right) + q \cdot Y_{k,kc,tr}^{TR,E} \right) \\[2ex] + \displaystyle\sum_{k}\sum_{kc}\sum_{tr \in \text{continuous}} D_{k,kc,tr}^{TR,E} \cdot \left(p \cdot \left(\frac{m_{k,kc,tr}^{TR,E}}{3600 \cdot \rho \cdot v_{k,kc,tr}^{TR,E} \cdot \Delta t_k^E} \right) + q \cdot Y_{k,kc,tr}^{TR,E} \right) \end{array} \right] \cdot F_{AN} \tag{16.13}$$

where D/m are distances, v/(m/s) velocities, Δt/h processing times, while k and kc denote the processes divided over several time intervals into processes n and nc, respectively.

The annual wastewater treatment cost, f_4/(€/a), depends on the mass and concentration of wastewater discharged:

$$f_4 = \left(\sum_n \sum_c \frac{0.001 \cdot m_n^{OUT} \cdot C_{c,n}^{OUT}}{m_c^E} + \sum_{ww} \sum_c \frac{0.001 \cdot m_{ww}^{C,\,FOUT} \cdot C_{c,ww}^{W}}{m_c^E} \right) \cdot P^E \cdot \frac{\lambda_{OHY}}{\Delta t^{ALL}}$$

$$+ \left(\sum_c \sum_n \sum_{nc} \sum_{tr \in batch} \frac{0.001 \cdot m_{n,nc,tr}^{TR} \cdot C_{c,n}^{OUT}}{m_c^E} \right) \cdot P^{E,LB} \cdot \frac{\lambda_{OHY}}{\Delta t^{ALL}}$$

$$+ \left(\sum_c \sum_n \sum_{nc} \sum_{tr \in continuous} \frac{0.001 \cdot m_{n,nc,tr}^{TR} \cdot C_{c,n}^{OUT}}{m_c^E} \right) \cdot P^{E,LC} \cdot \frac{\lambda_{OHY}}{\Delta t^{ALL}}$$

$$(16.14)$$

where $C_{c,n}^{OUT}$ /(g /m)^3 is the outlet water mass concentration of operation n, $C_{c,ww}^{W}$ /(g /m)^3 is the mass concentration of water source ww, $m_{ww}^{C,FOUT}$ /t is the mass of wastewater from semicontinuous operation ww to be discharged into the central treatment unit, m_c^E /kg is the equivalent mass load unit of contaminant c, P^E /(€/load unit) is the price of the wastewater treatment, $P^{E,LB}$ /(€/load unit) is the price of wastewater treatment in the local batch treatment unit, and $P^{E,LC}$ /(€/load unit) is the price of wastewater treatment in a continuous local treatment unit.

Investment for the installation of local treatment units, f_5/(€/a), is defined by the equation:

$$f_5 = \left(\sum_{tr \in batch} K_{tr} \cdot \left(m_{tr}^{TR} \right)^{n_{tr}^{TR}} + \sum_{tr \in continuous} K_{tr} \cdot \left(\frac{m_{tr}^{TR}}{J \cdot \rho \cdot \Delta t_n} \right)^{n_{tr}^{TR}} \right) \cdot F_{AN} \quad (16.15)$$

where m_{tr}^{TRC} /t is the capacity of the local treatment unit tr, K_{tr} is the investment parameter, n_{tr}^{TR} is the capacity exponent of local treatment unit tr, and J/(m³/(m² h)) is the average filtrate flux.

The annual cost of electricity needed for pumping water from the wells to/within production site, f_6/(€/a), is estimated from the available industrial data and expressed as a product of consumption ratio and electricity price.

16.3.2 Environmental Objective Function

The environmental objective function, $F_{Obj}^{Environmental}$, is the environmental sustainability index calculated as a sum of weighted and normalized individual indicators:

$$F_{Obj}^{Environmental} = \sum_i I_{NR_i}^{+} \cdot w_i + \sum_i I_{NR_i}^{-} \cdot w_i$$

$$\sum_i w_i = 1, \quad\quad\quad (16.16)$$

$$w_i \geq 0$$

where $I_{NR_i}^+$ and $I_{NR_i}^-$ are the normalized indicators with positive and negative impacts on sustainable development and w_i is a weight of the environmental indicator i. Selection of a suitable set of indicators should always be performed in close cooperation with the industry, using, for example, Global Reporting Initiative (GRI 2002) guidelines. The weights (relative importance) of selected indicators have a significant effect on the environmental sustainability index. A number of weighting techniques exist, some are derived from statistical models such as factor analysis, data envelopment analysis, and unobserved components models, or from participatory methods such as budget allocation processes, analytic hierarchy processes, and conjoint analysis (OECD 2008). Regardless of which method is used, weights are essentially valued judgments and the main source of disagreement, as the decision-makers of companies have different views, and are interested in different indicators.

Normalization is necessary in integration of the selected indicators into a composite index, as they are usually expressed in different units. Before normalization the indicators are divided in two groups regarding their influence on the sustainable development: indicators with positive and negative impact. Increasing value of indicators with positive impact on the sustainable development will bring the company closer to sustainable production, for example, water reuse, while increasing value of indicators with negative impact will lower the environmental, social, and/or economic sustainability of the company, for example, toxic releases into water. Common normalization methods are: minimum-maximum, distance to a reference, and the percentage of annual differences over consecutive years (OECD 2008). In the proposed method, the distance to a reference method is utilized where the normalized value is calculated as a ratio between the indicator and the corresponding benchmark:

$$I_{NR_i}^+ = \frac{I_i^+}{I_i^{\text{Benchmark}}} \quad \forall \, i \in I \tag{16.17}$$

$$I_{NR_i}^- = \frac{I_i^{\text{Benchmark}}}{I_i^-} \quad \forall \, i \in I \tag{16.18}$$

where I_i^+ is indicator i with positive impact on sustainable development, I_i^- is indicator i with negative impact on sustainable development, and $I_i^{\text{Benchmark}}$ is benchmark for indicator i. The benchmark values of the selected indicators should be determined based on the values of the best available techniques (BAT), measurements and standards within the company, local legal regulations, GRI reports for specific production sectors, and other relevant documents. By this benchmarking enables the company to compare the actual business operations against the best-practices and high performances within a particular production sector (Tokos et al. 2013). In other words, a set of environmental indicators provides no information on a company's progress toward sustainable development compared to other companies within the sector. Such information could be obtained only if a reference value, such as a benchmark, is assigned to each indicator (Altham 2007). By using benchmarking, the decision maker could have an insight not only into the environmental impacts of certain designs belonging to the Pareto optimal solutions, but also into the competitiveness of the designs within the particular production sector and identify the need for investment in a new technology.

16.3.3 CONSTRAINTS FOR SEMICONTINUOUS WATER-USING OPERATIONS (WATER REUSE BETWEEN SEMICONTINUOUS AND BATCH WATER CONSUMERS)

The limiting water mass of the semicontinuous stream ww over time interval j is defined by equation:

$$m^{\mathrm{C}}_{ww,j} = q_{m,ww} \cdot \left(t^{\mathrm{E}}_j - t^{\mathrm{S}}_j \right) \quad j = j^0_{ww}, j^0_{ww} + 1, \dots, J_{ww} \tag{16.19}$$

where set j represents time intervals, j^0_{ww} and J_{ww} are the first and last time intervals, respectively, in which continuous stream ww is available, $q_{m,ww}$/(t/h) is the mass flow rate of the continuous stream ww, t^{S}_j/h and t^{E}_j/h are the starting and ending times of the time interval j.

The outlet water mass from the semicontinuous water stream over time interval j is given by equation:

$$m^{\mathrm{C,OUT}}_{ww,j} = m^{\mathrm{C}}_{ww,j} - \sum_n m^{\mathrm{W}}_{ww,n} \quad j = j^0_{ww}, j^0_{ww} + 1, \dots, J_{ww}, \quad \forall n: t^{\mathrm{S}}_n = t^{\mathrm{S}}_j \tag{16.20}$$

where $m^{\mathrm{C,OUT}}_{ww,j}$/t is the unused water mass from the semicontinuous unit to be discharged over the interval j, $m^{\mathrm{W}}_{ww,n}$/t is the water mass from water source ww to operation n and t^{S}_n/h is the starting time of operation n.

Collecting the unused wastewater, $m^{\mathrm{C,OUT}}_{ww,j}$, in a storage tank would enable water reuse over the subsequent time intervals. The mass of wastewater from the semicontinuous operation ww to discharge, $m^{\mathrm{C,FOUT}}_{ww}$/t, is:

$$m^{\mathrm{C,FOUT}}_{ww} = \sum_j m^{\mathrm{C,OUT}}_{ww,j} - \sum_n m^{\mathrm{W}}_{ww,n} \quad j = j^0_{ww}, j^0_{ww} + 1, \dots, J_{ww}, \quad \forall n: t^{\mathrm{S}}_n \geq t^{\mathrm{E}}_{J_{ww}}$$

$$\tag{16.21}$$

The first term on the right-hand side of Equation 16.21 represents the surplus mass of wastewater over those time intervals in which the semicontinuous stream is present, while the second term represents the mass of wastewater used over the subsequent intervals.

A logic constraint is used to identify the existence or nonexistence of a storage tank for semicontinuous operation ww:

$$Y^{\mathrm{W}}_{ww,n} - Y^{\mathrm{C,ST}}_{ww} \leq 0 \quad \forall n: t^{\mathrm{S}}_n \geq t^{\mathrm{E}}_{J_{ww}} \tag{16.22}$$

where $Y^{\mathrm{C,ST}}_{ww}$ is the binary variable for a storage tank of semicontinuous wastewater source ww. This variable obtains value 1 only if the semicontinuous stream is integrated with batch operations after semicontinuous operation ww is finished.

The capacity of a storage tank for the semicontinuous wastewater stream, $m_{ww}^{C,ST}$ /t, is defined by equation:

$$m_{ww}^{C,ST} = \sum_n m_{ww,n}^{W} \quad \forall n : t_n^S \geq t_{J_{ww}}^E \tag{16.23}$$

Investment cost of storage tank for continuous operation, CT_{ww}^C /€, is:

$$CT_{ww}^C = r \cdot m_{ww}^{C,ST} + s \cdot Y_{ww}^{C,ST} \tag{16.24}$$

where r and s are the variable and fixed parameters, respectively, for the storage tank investment cost.

16.3.4 CONSTRAINTS FOR WATER REGENERATION REUSE BETWEEN BATCH WATER-USING OPERATIONS

By water regeneration reuse the overall water consumption can be additionally reduced, as well as the contaminant load of the discharged water. The model explores the possibility of installation of local treatment units operating in batch and continuous mode, simultaneously optimizing the operating schedule in order to suit the production schedules of batch processes. The option of storage tank installation before and after treatment is included in the model, in order to store wastewater and/or purified water if the time schedule of the treatment unit requires the storage.

The mass load balance, mass balance for each operation, and total water mass of the water-using operation given in the original model (Kim and Smith 2004) are extended by $m_{nc,n,tr}^{TR}$ /t, which represents the reused water mass from operation nc to operation n purified in a local treatment unit tr.

The feasibility constraint on the inlet concentration is:

$$\left[\sum_w \left(m_{w,n}^W \cdot C_{c,w}^W \right) + \sum_{nc} \left(m_{nc,n}^{PP} \cdot C_{c,nc}^{OUT} \right) + \sum_{nc} \sum_{tr} \left(m_{nc,n,tr}^{TR} \cdot C_{c,n,tr}^{TR} \right) \right] - m_n^{OP} \cdot C_{c,n}^{IN,MAX} \leq 0 \tag{16.25}$$

where $C_{c,n,tr}^{TR}$ /(g/m)^3 is the mass concentration of water purified in the local treatment unit tr before entering operation n, $C_{c,n}^{IN,MAX}$ /(g/m)^3 is the maximum inlet mass concentration of operation n, and m_n^{OP} /t is the water mass inside operation n.

Additional equations for upper and lower bounds of water mass purified in local treatment units are:

$$m_{nc,n,tr}^{TR} - m_{nc,n,tr}^{UB,TR} \cdot Y_{nc,n,tr}^{TR} \leq 0 \tag{16.26}$$

$$m_{nc,n,tr}^{TR} - m_{nc,n,tr}^{LB,TR} \cdot Y_{nc,n,tr}^{TR} \geq 0 \tag{16.27}$$

where $m_{nc,n,tr}^{UB,TR}, m_{nc,n,tr}^{LB,TR}$ /t are the upper and lower bounds for the reused water mass from operation nc to operation n purified in local treatment unit tr.

The outlet concentration from local treatment unit, $C_{c,n,tr}^{TR}$ /(g/m)^3, is defined by equation:

$$C_{c,n,tr}^{TR} = \left(1 - r_{c,tr}^{TR}\right) \cdot \frac{\sum_{nc}\left(m_{nc,n,tr}^{TR} \cdot C_{c,nc}^{OUT}\right)}{\sum_{nc} m_{nc,n,tr}^{TR}} \tag{16.28}$$

where $r_{c,tr}^{TR}$ is the constant removal ratio of contaminant c in the treatment unit tr.

The capacity of the local treatment unit, m_{tr}^{TRC} /t, is obtained by the inequality:

$$m_{tr}^{TRC} \geq \sum_{nc}\sum_{n} m_{nc,n,tr}^{TR} \quad \forall n \wedge \forall j : t_{nc}^{S} = t_{j}^{S} \tag{16.29}$$

The right-hand side of Equation 16.29 represents the sum of the reused masses from all operations nc within the same interval j to operations n.

The purification of wastewater from operation nc in treatment unit tr can start anytime after the termination of process nc. Time constraints on the treatment unit are expressed by disjunctive expressions which hold for selected connections, while for unselected are redundant:

$$t_{nc,tr}^{S,TR} - t_{nc}^{E} \geq t^{LB} - M \cdot \left(1 - Y_{nc,n,tr}^{TR}\right) \tag{16.30}$$

$$t_{nc,tr}^{S,TR} - t_{nc}^{E} \leq t^{UB} + M \cdot \left(1 - Y_{nc,n,tr}^{TR}\right) \tag{16.31}$$

where $t_{nc,tr}^{S,TR}$ /h is the starting time of the wastewater treatment from operation nc in local treatment unit tr, t_{nc}^{E} /h is the ending time of operation nc, t^{LB}, t^{UB} are the lower and upper bounds of time difference, where $t^{LB} = 0$, and M is a large positive constant. For processes operating within the same time interval j, the starting time is defined by equations:

$$t_{nc,tr}^{S,TR} = t_{nc-1,tr}^{S,TR} \quad \forall nc, \ t_{nc}^{S} = t_{nc-1}^{S}, \ nc = 2,...,N \tag{16.32}$$

$$t_{nc,tr}^{S,TR} \geq t_{nc-1,tr}^{E,TR} \quad \forall nc, \ t_{nc}^{S} > t_{nc-1}^{S}, \ nc = 2,...,N \tag{16.33}$$

The ending time of the purification, $t_{nc,tr}^{E,TR}$, is defined by

$$t_{nc,tr}^{E,TR} = t_{nc,tr}^{S,TR} + \Delta t_{tr}^{TR} \tag{16.34}$$

The purification of wastewater from process nc in treatment unit tr has to be completed before process n starts:

$$t_{n}^{S} - t_{nc,tr}^{E,TR} \geq t^{LB} - M \cdot \left(1 - Y_{nc,n,tr}^{TR}\right) \tag{16.35}$$

$$t_{n}^{S} - t_{nc,tr}^{E,TR} \leq t^{UB} + M \cdot \left(1 - Y_{nc,n,tr}^{TR}\right) \tag{16.36}$$

If the treatment time is shorter than the difference between the ending time of process nc and the starting time of process n, the waiting times before and after treatment are defined by equations:

$$t_n^S - t_{nc}^E - t_{nc,n,tr}^{B,TR} - \Delta t_{tr}^{TR} - t_{nc,n,tr}^{A,TR} \leq M \cdot \left(1 - Y_{nc,n,tr}^{TR}\right) \tag{16.37}$$

$$t_n^S - t_{nc}^E - t_{nc,n,tr}^{B,TR} - \Delta t_{tr}^{TR} - t_{nc,n,tr}^{A,TR} \geq -M \cdot \left(1 - Y_{nc,n,tr}^{TR}\right) \tag{16.38}$$

where $t_{nc,n,tr}^{B,TR}$ /h is the waiting time before treatment, and $t_{nc,n,tr}^{A,TR}$ /h is the waiting time after treatment in unit tr. The upper inequalities become redundant for unselected connections $\left(Y_{nc,n,tr}^{TR} = 0\right)$, while for $Y_{nc,n,tr}^{TR} = 1$ they define the correct time expressions.

Additional constraints for determining waiting time after wastewater treatment are

$$t_n^S - t_{nc,tr}^{E,TR} - t_{nc,n,tr}^{A,TR} \leq M \cdot \left(1 - Y_{nc,n,tr}^{TR}\right) \tag{16.39}$$

$$t_n^S - t_{nc,tr}^{E,TR} - t_{nc,n,tr}^{A,TR} \geq -M \cdot \left(1 - Y_{nc,n,tr}^{TR}\right) \tag{16.40}$$

Additional constraints for determining waiting time before wastewater treatment are

$$t_{nc,tr}^{S,TR} - t_{nc}^E - t_{nc,n,tr}^{B,TR} \leq M \cdot \left(1 - Y_{nc,n,tr}^{TR}\right) \tag{16.41}$$

$$t_{nc,tr}^{S,TR} - t_{nc}^E - t_{nc,n,tr}^{B,TR} \geq -M \cdot \left(1 - Y_{nc,n,tr}^{TR}\right) \tag{16.42}$$

The waiting times of nonselected treatment connections $\left(Y_{nc,n,tr}^{TR} = 0\right)$ are forced to zero by the following constraints:

$$t_{nc,n,tr}^{A,TR} \leq M \cdot Y_{nc,n,tr}^{TR} \tag{16.43}$$

$$t_{nc,n,tr}^{B,TR} \leq M \cdot Y_{nc,n,tr}^{TR} \tag{16.44}$$

The following constraints are used to identify those processes nc that need the installation of a storage tank for purified water after treatment:

$$t_{nc,n,tr}^{A,TR} \leq t^{UB} \cdot Y_{tr,nc}^{ST,TR,OUT} \tag{16.45}$$

$$t_{nc,n,tr}^{A,TR} \geq t^{LB} \cdot Y_{tr,nc}^{ST,TR,OUT} \tag{16.46}$$

where $Y_{tr,nc}^{ST,TR,OUT}$ is a binary variable for the storage tank after treatment unit tr, which holds the purified water from process nc.

The required storage tank capacity after purification, $m_{tr}^{ST,TR,OUT}$ /t, is defined by the constraint:

$$m_{tr}^{ST,TR,OUT} \geq \sum_{nc} \sum_{n} \left(m_{nc,n,tr}^{TR} \cdot Y_{tr,nc}^{ST,TR,OUT} \right) \quad \forall n \wedge \forall j : t_{nc}^{S} = t_{j}^{S} \quad (16.47)$$

The right-hand side of Equation 16.47 represents the sum of purified streams, from all operations nc within the same interval j, that need storage after treatment before entering operations n.

Storage tank cost, $CT_{tr}^{TR,OUT}$ /€, is:

$$CT_{tr}^{TR,OUT} = r \cdot m_{tr}^{ST,TR,OUT} + s \cdot Y_{tr}^{ST,OUT} \quad (16.48)$$

where $Y_{tr}^{ST,OUT}$ is the binary variable for the storage tank after treatment unit tr, as defined by equation:

$$Y_{tr,nc}^{ST,TR,OUT} \leq Y_{tr}^{ST,OUT} \quad (16.49)$$

Those processes that require the installation of a storage tank for wastewater before treatment are identified by the constraints:

$$t_{nc,n,tr}^{B,TR} \leq t^{UB} \cdot Y_{tr,nc}^{ST,TR,IN} \quad (16.50)$$

$$t_{nc,n,tr}^{B,TR} \geq t^{LB} \cdot Y_{tr,nc}^{ST,TR,IN} \quad (16.51)$$

where $Y_{tr,nc}^{ST,TR,IN}$ is the binary variable for the storage tank before treatment unit tr, to hold the wastewater from process nc.

The required storage tank capacity before the treatment unit, $m_{tr}^{ST,TR,IN}$ /t, is defined by the constraint:

$$m_{tr}^{ST,TR,IN} \geq \sum_{nc} \sum_{n} \left(m_{nc,n,tr}^{TR} \cdot Y_{tr,nc}^{ST,TR,IN} \right) \quad \forall n \wedge \forall j : t_{nc}^{S} = t_{j}^{S} \quad (16.52)$$

Storage tank cost, $CT_{tr}^{TR,IN}$ /€, is:

$$CT_{tr}^{TR,IN} = r \cdot m_{tr}^{ST,TR,IN} + s \cdot Y_{tr}^{ST,IN} \quad (16.53)$$

where $Y_{tr}^{ST,IN}$ is a binary variable for the storage tank before treatment unit tr, defined by the equation:

$$Y_{tr,nc}^{ST,TR,IN} \leq Y_{tr}^{ST,IN} \quad (16.54)$$

Under assumption that the treatment time for the local batch treatment unit is shorter than the time intervals defined by the operating schedule for production, the treatment unit is always available for purification of the generated wastewater:

$$t_{nc,tr}^{S,TR} = t_{nc}^{E} \quad \Delta t_{tr}^{TR} \leq t_{nc}^{E} - t_{nc}^{S} \tag{16.55}$$

In this specific case, the time scheduling for the batch treatment unit is obtained by Equations 16.32, 16.34, and 16.55. A storage tank before treatment is not needed, while the purified water is stored after treatment in the tank until the starting time of operation n, in which water will be reused. The existence or nonexistence of a storage tank after the treatment unit is determined by a logical constraint:

$$Y_{nc,n,tr}^{TR} - Y_{tr,nc}^{ST,TR,OUT} \leq 0 \quad \forall n : t_{nc,tr}^{E,TR} \leq t_{n}^{S} \tag{16.56}$$

The required storage tank capacity after purification is defined by Equation 16.47 and the storage tank cost by Equation 16.48.

The scheduling of the continuous treatment unit only differs from that of the batch treatment unit, when defining the treatment's ending time:

$$t_{nc,tr}^{E,TR} = t_{nc,tr}^{S,TR} \tag{16.57}$$

In case of industrial application of the described model, the number of decision variables which need to be determined in large (around 50,200 constraints, 23,300 continuous variables, and 10,800 binary variables) results in a complex problem with high computational cost. To overcome this drawback, a two-level solution procedure is proposed based on a temporal decomposition. Taking in consideration that the daily production schedule changes over the working week, and the daily schedules repeat weekly, additional constraint will be needed to prevent multiple summations of the installation cost of the storage tanks, similar to Equation 16.13. Accordingly, for a process integrated with other processes over several time intervals different storage capacities could be determined in different intervals. The largest value is considered as the final storage capacity of process k:

$$m_{k}^{ST,E} \geq m_{n}^{ST} \quad \forall n \in N \wedge \forall k \in K \tag{16.58}$$

$$Y_{n}^{ST} - Y_{k}^{ST,E} \leq 0 \quad \forall n \in N \wedge \forall k \in K \tag{16.59}$$

where m_{n}^{ST} / t is the storage tank capacity of operation n, $m_{k}^{ST,E} / t$ represents the largest reuse water mass of process k, Y_{n}^{ST} is the binary variable for storage tank to operation n, and $Y_{k}^{ST,E}$ is the binary variable for installation of a storage tank for process k. The storage tank investment cost of process k, $CT_{k} / \text{€}$, is:

$$CT_{k} = r \cdot m_{k}^{ST,E} + s \cdot Y_{k}^{ST,E} \quad \forall k \in K \tag{16.60}$$

At the first step of the proposed method, integration of water networks is performed for each day of a week according to daily production schedules in order to identify possible intra-daily connections between production sections. At the second step, the identified connections are fixed and the integration is performed for each section separately for the weekly operation schedule. The water network considered for integration is extended with the processes involved in the fixed connection from the other sections. Further modifications included in the mathematical model to facilitate the solution process are

i. The reused water mass between operations located in different sections is fixed to the value obtained in the first step. The corresponding binary variables are set to 1.

ii. The expression for annual freshwater cost is modified in order to exclude freshwater cost of those operations from other sections that supplies water to the section considered for integration, because freshwater cost of these operations is considered during optimization of their home sections.

iii. The expression for annual wastewater treatment cost is modified in order to exclude the treatment cost of those operations from other sections that use wastewater from the section, under integration, because the reused water of these operations will be discharged and accounted during optimization of their home sections.

iv. As the investment for the fixed intra-daily connections is included in objective function of all sections, it should be subtracted from the total investment required for the joint integration of the sections.

More detail about the decomposition method can be found in Tokos et al. (2012).

16.4 INDUSTRIAL CASE STUDY

The case-studied brewery has two brewhouses, operating interchangeably depending on the demand, cellar, and packaging area, with four filling lines. The production section consists of processes operating in batch mode, while the packing area operating in mixed batch and semicontinuous mode. Semicontinuous processes present in the packaging area are the rinsers for nonreturnable glass bottles and cans.

The annual beer production in the brewery is 114,200 m^3, with freshwater consumption of 653,300 m^3/a. The volume ratio of water consumption to beer sold is determined to be 6.04 m^3/m^3, which is considerably higher than the range defined in BREF (2006), 3.7–4.7 m^3/m^3. Based on the obtained water balance, the ratio between freshwater consumption and beer production is calculated and compared with the relevant benchmarks for each department within the brewery (Table 16.1). This analysis identified the cellar with filters as one of the critical points in the brewery. The specific water consumption of packing area was also higher than the specified BAT value. Higher water consumption was observed in the case of bottle washing (returnable glass bottles), CIP system, and pasteurization.

The total annual wastewater discharge from the brewery is 437,690 m^3/a, from which 429,600 m^3/a is discharged from the production area and 8090 m^3/a from the

TABLE 16.1

Water Consumption in Different Production Departments

			EBC (m³/m³)		
Department	Case Study (m³/m³)	BREF (m³/m³)	Good Practice	Best Practice	PPAH (m³/m³)
Brewhouse to wort cooling and CIP	1.72	1.74–2.60	1.75	1.48	1.8–2.2
Wort cooling			–	–	0.0–2.4
Fermentation and yeast handling and CIP	0.8	0.32–0.53	0.09	0.05	0.5–0.8
Maturation and CIP		0.24–0.67	0.09	0.05	0.3–0.6
Filtration and BBT and CIP	0.37	0.31–1.09	0.28	0.06	0.1–0.5
Keg washing	0.067	0.13–0.61	0.23	0.13	0.1–0.2
Bottle washing	1.15	0.59–1.63	0.34	0.17	0.9–2.1
Bottle and can pasteurization	0.29	2.00–2.04	0.16	0.08	–
General CIP	0.41		0.42	0.20	–
Wastewater treatment conversion	1.23		0.20	0.16	–
Boilers			0.36	0.16	0.1–0.3
Evaporator cooling tower			0.55	0.38	–
Air compressors and CO₂			0.06	0.04	0.1–0.5
Domestic			0.07	0.04	–

Source: Adapted from EBC. 1990. *Manual of Good Practice: Water in Brewing. European Brewery Convention*, Nürnberg: Fachverlag Hans Carl; PPAH. 1998. *Pollution Prevention and Abatement Handbook: Toward Cleaner Production*. Washington: World Bank Group.

evaporative cooling towers. The wastewater from the production and boiler house is discharged into a central wastewater treatment unit whilst, from the cooling system, into the local river in accordance with the environmental regulations. The volume ratio of total wastewater discharged to beer produced is 3.83 m³/m³, which is slightly higher than the lower bound given in BREF (2006), 3–9 m³/m³.

As a result of this preliminary analysis, an initiative has emerged in the company for reconstructing the existing water network. The set goal is reduction of the fresh-water consumption and the contaminant load of wastewater discharged into the central wastewater treatment system by retrofitting the total water network taking in the consideration the environmental impact and competitiveness of the obtained network.

Further to the water balance, the fluctuation of contaminants (chemical oxygen demand—COD, pH, and conductivity) was monitored in the inlet and outlet water streams. The measurements were carried out for 3 months, during the highest production rate. Based on these measurements, the maximal inlet values of contaminants were determined for each water consumer. The results of laboratory measurements have confirmed a strong influence of production intensity and packaging material types (returnable or nonreturnable glass bottles) on outlet water quality. The average COD value from the washer for nonreturnable glass bottles was 34 g/m³. In the case of returnable glass bottles, the COD value was within the range from 200 to 460 g/m³. Any variability in the dirt on the returnable packaging material and craters can result in extremely high COD values. For example, the average COD value from the crater washer was 50 g/m³, but in case of rather dirty craters the COD value can reach 1600 g/m³. Taking in consideration the limiting inlet and outlet concentrations of the processes in the packaging area, the water reuse opportunities were analyzed, as several semicontinuous wastewater streams with low contaminant concentrations are available for reuse in batch operations with lower purity requirements. In case of the brewhoues and the cellar, the possibilities for regeneration reuse were analyzed, as the concentration of wastewaters is unacceptable for direct reuse.

Five indicators were selected for construction of the environmental sustainability index used as environmental objective function Equation 16.16: volume ratio of freshwater consumed to beer production, volume ratio of wastewater discharged from production to beer produced, indirect emission of CO_2 by electricity generation per volume of beer produced, indirect emission of CO by electricity generation per volume of beer produced, and indirect emission of NO_x by electricity generation per volume of beer produced. The benchmarks for the selected indicators are 3.7 m³/m³, 2.89 m³/m³, 2.05 kg/m³, 0.019 kg/m³, and 0.0067 kg/m³, respectively. The benchmark for volume ratio of wastewater discharged from production to beer produced is set to 2.89 m³/m³, which is the lowest value of wastewater volume discharged from the brewery between 2005 and 2007. It should be noticed that this value is slightly lower than the lower limit of wastewater discharge specified in BREF (2006). The benchmarks for indirect emission were calculated based on lower bound for electricity consumption, 75 kW h/m³, and emissions by electricity production in power plant, which amounts to 0.594 kg/ kW h for CO_2, 0.00055 kg/kW h for CO, 0.00195 kg/kW h for NO_x. The model calculates the indirect emissions based on electricity needed for pumping water from the wells to/within production site. The indicators' weights determined by budget allocation method are 0.63 for freshwater consumption, 0.1 for wastewater discharge, and 0.09 for greenhouse gas emissions. This method calculates the indicator weights based on expert opinion. In order to establish a proper weighting system, it is essential to bring together experts representing a wide spectrum of knowledge and experience, for example, experts from different production sectors and management within the company. The economic criterion used is the total cost of the water network that involves the freshwater cost, cost of electricity needed for pumping water from the wells to/ within production site wastewater treatment cost, and annual investment costs of the storage tank, piping and local treatment unit installation, Equation 16.3.

16.4.1 SEPARATE INTEGRATION OF WATER USE IN THE PRODUCTION AND THE PACKAGING SECTIONS

In the packaging area by increasing the weight of the environmental criterion, the number of water reuse opportunities identified within the optimal water network for the weekly schedule increases. The possibility of wastewater reuse between the can rinser and the pasteurization process was found for the lowest value of the environmental criterion weight, 0. The optimal water network does not contain water reuse options between batch processes. By increasing the environmental criterion weight on 0.2, an additional water reuse connection was identified between the rinser for nonreturnable glass bottles and the washer for returnable glass bottles. Wastewater reuse between batch processes operating within the same time interval, the pasteurization process and the washer for returnable glass bottles was found at environmental criterion weight of 0.4. With further increase of the environmental criterion weight, two more opportunities were identified for reduction of freshwater consumption: wastewater reuse from the washer for returnable bottles and the crater washer, and water reuse between the rinser for nonreturnable glass bottles and the crater washer. The generated Pareto front is given in Figure 16.1. The weekly freshwater consumption could be reduced from 2233 t to 1493 t. The environmental sustainability index improved from 0.911 to 0.952. The total annual network cost was between 249,499 €/a (at the environmental criterion weight of 0), and 499,822 €/a (at the environmental criterion weight of 1). Information about the total savings of freshwater and wastewater treatment cost, the piping investment cost for new connections, net present values, and the payback periods are given in Table 16.2. It should be noted that several water networks can exist with the same sustainability index but different total costs because the water consumptions of the operations can vary between the lower and upper bounds in the model formulation.

In similar manner, by increasing the weight of the environmental criterion for the weekly schedule in the production area, the number of water reuse and regeneration reuse opportunities identified within the optimal water network increases. The optimal water network for the weekly schedule in the production area at an increasing weight of environmental criterion would include: reuse of wastewater from filtration as water for pouring the batch material in the cellar or in the brewhouse, regeneration reuse of wastewater from pouring operation in CIP systems. The wastewater purification is carried out by nanofiltration technology, with membrane area of 28 m³. The generated Pareto front is given in Figure 16.2. The weekly freshwater consumption could be reduced from

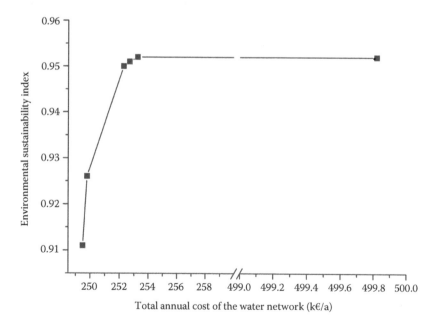

FIGURE 16.1   Pareto front for the packaging area.

TABLE 16.2
Results for the Packaging Area

w_E	$1-w_E$	Environmental Sustainability Index (–)	Freshwater Cost (€/y)	Wastewater Treatment Cost (€/y)	Storage Tank Cost (€/y)	Piping Cost (€/y)	Total Cost (€/y)	Net Present Value (€)	Payback Period (y)
0.0	1.0	0.911	210,773	4589	0	7069	249,499	331,687	0.29
0.2	0.8	0.926	207,273	4545	0	11,345	249,781	331,687	0.45
0.4	0.6	0.95	202,104	4502	0	19,730	252,291	323,677	0.75
0.6	0.4	0.951	201,866	4499	0	20,411	252,702	322,187	0.78
0.8	0.2	0.952	201,618	4498	0	21,263	253,271	320,084	0.81
1.0	0.0	0.952	201,618	4512	131,737	136,062	499,822	621,594	–
Current		0.667	287,956	5238	–	–	330,175	–	–

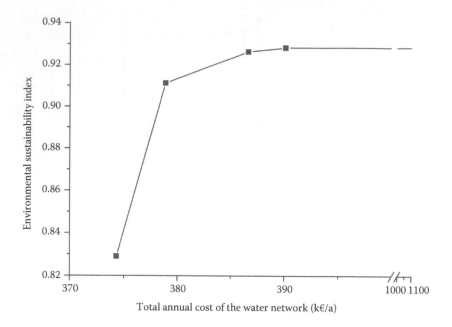

FIGURE 16.2   Pareto front for the production area.

2,592 t to 1861 t. The environmental sustainability index improved from 0.829 to 0.928. The total annual network cost is between 374,316 €/a (at the environmental criterion weight of 0) and 1,199,092 €/a (at the environmental criterion weight of 1). The economic results are given in Table 16.3. In case of the highest total annual network cost, both batch and semicontinuous local treatment units are included in the water network.

The environmental sustainability index of all networks in the packaging and production sections is lower than 1. This is an indication that investment in a new technology is needed with lower freshwater consumption to keep the brewery competitive on the market. For example, new washers for returnable glass bottles have lower freshwater consumption, and/or the tunnel pasteurizers could be replaced by flash pasteurizers. Another possibility is joint integration of the production and packaging sections.

The size of the MINLP model was approximately 6600 constraints, 3300 continuous variables, and 1500 binary variables in the case of the packaging area, and approximately 21,500 constraints, 10,500 continuous variables, and 4600 binary variables for the production area. The model was solved by the GAMS/DICOPT solver (GAMS 2007) using 218 s of CPU time at the PC (P4, 2.6 GHz and 512 MB RAM) in the case of the packaging area, and 3884 s in the case of the production area.

16.4.2   Joint Integration of Production and Packing Sections

In the first step of joint integration, the water networks of the production and the packaging sections are simultaneously integrated for each working day by

Retrofit of Industrial Water Network

TABLE 16.3
Results for the Production Area

w_E	$1-w_E$	Environmental Sustainability Index (–)	Freshwater Cost (€/y)	Wastewater Treatment Cost (€/y)	Storage Tank Cost (€/y)	Piping Cost (€/y)	LT Investment (€/y)	Total Cost (€/y)	Net Present Value (€)	Payback Period (y)
0.0	1.0	0.829	281,528	58,379	0	4658	0	374,316	300,163	0.21
0.2	0.8	0.911	256,008	55,610	21,823	4958	13,446	378,898	294,686	1.46
0.4	0.6	0.911	256,008	55,610	21,823	4958	13,446	378,898	294,686	1.46
0.6	0.4	0.926	252,035	55,479	21,915	17,069	13,528	386,660	266,674	1.91
0.8	0.2	0.928	251,328	55,473	21,915	21,342	13,528	390,146	253,637	2.08
1.0	0.0	0.928	251,272	48,334	337,003	436,317	99,612	1,199,092	-2,830,369	–
Current		0.667	349,947	58,954	0	0	0	445,882	–	–

identifying the daily water reuse and regeneration reuse options. By changing the weight coefficient of the environmental criterion between 0.0 and 0.6, one connection is identifiable between the sections, that is, wastewater from pasteurization in the packaging area is reused as pouring water for batch material in the brewhouse. A storage tank would be needed for wastewater reuse between the sections. The capacity of the storage tank would be 105 t. Two connections between the sections can be found at higher weight coefficients (0.8 and 1.0): reuse of wastewater from pasteurization in the packaging area as pouring water for batch material in the brewhouse and the cellar. The required storage tank capacity is 210 t. The generated Pareto fronts are given in Figure 16.3. It should be noted that the daily schedule is changing during the week, that is, on day 1 all four filling lines operate, on day 4 one, while in day 5 two filling lines are shut down, giving the curve distribution seen in Figure 16.3. The environmental sustainability index of all obtained networks is above 1.

Within the second step two analyses were performed, as one or both reuse options identified in the first step can be fixed. A water reuse connection between pasteurization in the packaging area and batch material pouring in brewhouse is fixed in the first evaluation. The optimal water network in the production and the packaging areas contains water reuse and regeneration reuse options as in the case of separated integration. The weekly freshwater consumption could be reduced from 4725 t to 2882 t. The environmental sustainability index improved from 0.993 to 1.093. The total annual network cost is between 699,282 €/a (at the environmental criterion weight of 0) and 3,906,295 €/a (at the environmental criterion weight of 1)

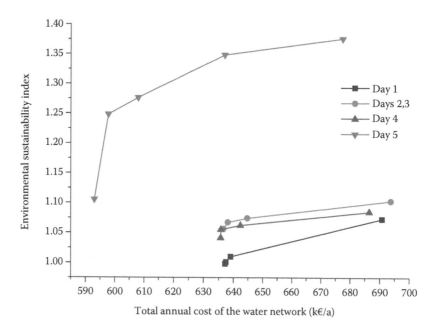

FIGURE 16.3 Pareto fronts for the each working day, first step of the proposed temporal decomposition method.

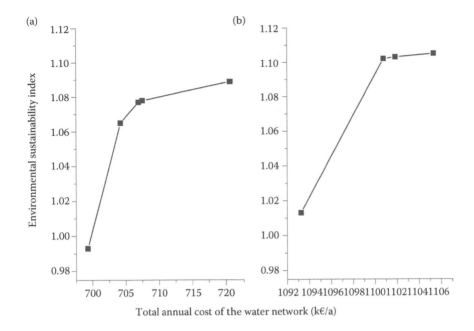

FIGURE 16.4 Pareto fronts obtained in the second step for (a) one and (b) two fixed connection indentified in the first step.

(Figure 16.4a). The economic results are given in Table 16.4. Both of the identified connections are fixed in the second evaluation. The optimal water networks include the same connections as in case of separated integration of the production sections. The weekly freshwater consumption could be reduced from 4725 t to 2847 t. The environmental sustainability index increases from 1.013 to 1.106. The total annual network cost is between 1,093,237 €/a and 5,749,971 €/a (Figure 16.4b). The net present value for all the obtained optimal water networks is negative due to high investment required (Table 16.5).

According to the results, the environmental sustainability index is slightly higher than 1, except in the optimal water network obtained by fixing only one connection at the environmental criterion weight equal to 0. This means that the company can perform better than the assigned benchmarks by reusing water between pasteurization in the packaging area and batch material pouring in the brewhouse. Currently, the savings of freshwater and wastewater treatment cost are not high enough to justify application of both reuse opportunities between the production and packaging area.

The size of the MINLP model was up to 16,900 constraints, 8300 continuous variables, and 3800 binary variables in the case of the first step and the model was solved by the GAMS/DICOPT solver (GAMS 2007) using up to 2245 s of CPU time at the PC (P4, 2.6 GHz and 512 MB RAM). In the case of the second step, the model had up to 39,000 constraints, 18,600 continuous variables, and 8400 binary variables, which were solved using up to 10,907 s of CPU time.

TABLE 16.4

Results for Joint Integration of Production and Packaging Area, One Water Reuse Connection Fixed[a]

w_E	$1-w_E$	Environmental Sustainability Index (–)	Freshwater Cost (€/y)	Wastewater Treatment Cost (€/y)	Storage Tank Cost (€/y)	Piping Cost (€/y)	LT Investment (€/y)	Total Cost (€/y)	Net Present Value (€)	Payback Period (y)
0.0	1.0	0.933	428,193	62,429	19,051	139,577	0	699,282	118,948	3.04
0.2	0.8	1.065	399,172	59,598	40,905	144,156	13,473	704,189	132,861	3.41
0.4	0.6	1.077	394,576	59,559	40,905	152,114	13,473	706,923	134,347	3.51
0.6	0.4	1.078	394,327	59,557	40,905	152,966	13,473	707,492	351,939	3.52
0.8	0.2	1.089	390,500	59,459	40,966	170,329	13,528	720,638	303,267	3.86
1.0	0.0	1.093	389,055	53,840	510,946	2,795,468	111,291	3,906,295	–11,858,967	–
Current		–	637,903	64,192	0	0	0	776,057	–	–

[a] Fixed connection between pasteurization in the packaging area and batch material pouring in the brewhouse.

TABLE 16.5

Results for Joint Integration of Production and Packaging Area, Two Water Reuse Connections Fixed[a]

w_E	$1-w_E$	Environmental Sustainability Index (–)	Freshwater Cost (€/y)	Wastewater Treatment Cost (€/y)	Storage Tank Cost (€/y)	Piping Cost (€/y)	LT Investment (€/y)	Total Cost (€/y)	Net Present Value (€)	Payback Period (y)
0.0	1.0	1.013	419,668	62,357	38,102	523,974	0	1,093,237	−1,132,602	–
0.2	0.8	1.102	385,904	59,421	60,017	536,512	13,528	1,100,762	−1,146,728	–
0.4	0.6	1.102	385,904	59,421	60,017	536,512	13,528	1,100,762	−1,146,728	–
0.6	0.4	1.103	385,513	59,418	60,017	538,045	13,528	1,101,850	−1,115,763	–
0.8	0.2	1.105	384,806	59,412	60,017	542,319	13,528	1,105,335	−1,163,800	–
1.0	0.0	1.106	384,356	51,987	561,830	4,631,134	75,466	5,749,971	−18,897,367	–
Current		–	637,903	64,192	0	0	0	–	–	–

[a] Fixed connections are wastewater reuse from pasteurization in the packaging area as pouring water for batch material in the brewhouse and the cellar.

16.5 CONCLUSION

This chapter presented a bi-objective optimization method for retrofitting industrial water networks, which simultaneously evaluates the economic and environmental impacts of the obtained optimal network design. The environmental impact is evaluated via benchmarking. The economic criterion used is the total cost of the water network. The approach uses an MINLP model which enables water reuse and regeneration reuse in batch and semicontinuous processes. The Pareto front is generated using the NBI method. The proposed approach was applied to an industrial case study in a brewery. The results obtained show that the benchmark cannot be reached by the individual integrations of two sections, and investment in new technologies with lower freshwater consumption would be needed. The environmental sustainability index rose to slightly above 1 by integrating water networks of the production and the packaging sections, which means that the brewery can achieve better performance than its competitors by using the results of process integration.

SYMBOLS USED

SETS

c	a set of contaminant
fw	a set of freshwater sources
i	set of the selected environmental indicators
j	a set of objective functions or time intervals
k	a set of processes divided over time intervals
n	a set of water-using operation
w	a common set of water sources, $w = fw \cup ww$
ww	a set of semicontinuous water sources

PARAMETERS

Δt_n	processing time of operation n, h
Δt^{ALL}	overall time interval, h
$\Delta t_{tr}^{\text{TR}}$	treatment time in local treatment unit tr, h
λ_{OHY}	annual operating time, h/a
$v_{n,nc}^{\text{PP}}$	velocity of reuse water flow from operation n to operation nc, m/s
$v_{n,nc,tr}^{\text{TR}}$	velocity of reuse water flow from operation n to operation nc purified in treatment unit tr, m/s
$v_{w,n}^{\text{W}}$	velocity of water flow from water source w to operation n, m/s
ρ	density, t/m³
$C_{c,w}^{\text{W}}$	mass concentration of water source w, g/m³
$C_{c,n}^{\text{IN,MAX}}$	maximum inlet mass concentration of operation n, g/m³
$D_{n,nc}^{\text{PP}}$	distance between operation n and nc, m
$D_{k,kc}^{\text{PP,E}}$	distance between processes k and kc, m
$D_{n,nc,tr}^{\text{TR}}$	distance between operations n and nc including the connection with treatment unit tr, m
$D_{k,kc,tr}^{\text{TR,E}}$	distance between processes k and kc through treatment unit tr, m

$D_{w,n}^{\mathrm{W}}$	distance between the water source w and operation n, m
$D_{w,k}^{\mathrm{W}}$	distance between water source w and process k, m
F_{AN}	annualization factor, a^{-1}
I_i^+	indicator i with positive impact on sustainable development
I_i^-	indicator i with negative impact on sustainable development
$I_i^{\mathrm{Benchmark}}$	benchmark for indicator i
J	average filtrate flux, $\mathrm{m}^3/(\mathrm{m}^2 \cdot \mathrm{h})$
K_{tr}	investment parameter for local treatment unit tr
M	large number
$m_{ww,j}^{\mathrm{C}}$	limiting water mass of the continuous stream ww in time interval j, t
m_c^{E}	equivalent mass load unit of contaminant c, kg
$m_{nc,n,tr}^{\mathrm{LB,TR}}$	lower bound for reused water mass from operation nc to operation n purified in local treatment unit tr, t
$m_{nc,n,tr}^{\mathrm{UB,TR}}$	upper bound for reused water mass from operation nc to operation n purified in local treatment unit tr, t
n_{tr}^{TR}	capacity exponent of local treatment unit tr
p	variable parameter of piping investment cost
P^{E}	price of wastewater treatment, €/load unit
$P^{\mathrm{E,LB}}$	price of wastewater treatment in the local batch treatment unit, €/load unit
$P^{\mathrm{E,LC}}$	price of wastewater treatment in continuous local treatment unit, €/load unit
P^{W}	freshwater price, €/t
r	variable parameter of storage tank investment cost
$r_{c,tr}^{\mathrm{TR}}$	removal ratio of contaminant c in treatment unit tr
s	fixed parameter of storage tank investment cost
t_j^{E}	ending time of interval j, h
t_n^{E}	ending time of operation n, h
t^{LB}	lower bound of time difference, h
t_j^{S}	starting time of time interval j, h
t_n^{S}	starting time of operation n, h
t^{UB}	upper bound of time difference, h
q	fixed parameter of piping investment cost
$q_{m,ww}$	mass flow rate of the continuous stream ww, kg/h
w_i	weight of the selected environmental indicator i

VARIABLES

$C_{c,n}^{\mathrm{OUT}}$	outlet water mass concentration of operation n, g/m^3
$C_{c,nc,tr}^{\mathrm{TR}}$	mass concentration of water purified in local treatment unit tr from operation nc, g/m^3
CT_k	storage tank investment cost of process k, /€
CT_n	storage tank investment cost of operation n, €
CT_{ww}^{C}	storage tank investment cost of continuous operation ww, €
$CT_{tr}^{\mathrm{TR,IN}}$	storage tank investment cost of wastewater before treatment unit tr, €
$CT_{tr}^{\mathrm{TR,OUT}}$	storage tank investment cost of purified water after treatment unit tr, €

f_1	freshwater cost, €/a
f_2	annual investment cost of storage tank installation, €/a
f_3	annual investment cost of piping, €/a
f_4	wastewater treatment cost, €/a
f_5	annual investment costs of the local treatment units, €/a
f_6	annual cost of electricity needed for pumping water from the wells to/within production site, €/a
$f_j(\mathbf{x})$	objective function j
$f_j^N(\mathbf{x})$	nadir point of objective function j
$f_j^{Trans}(\mathbf{x})$	transformed objective function j
$f_j^U(\mathbf{x})$	utopia point of objective function j
$F_{Obj}^{Economic}$	economic objective function, €/a
$F_{Obj}^{Environmental}$	environmental objective function
$\mathbf{g}$	inequality constraint vector
$\mathbf{h}$	equality constraint vector
$I_{NR_i}^+$	normalized indicator i with positive impact on sustainable development
$I_{NR_i}^-$	normalized indicator i with negative impact on sustainable development
$m_{ww}^{C,FOUT}$	mass of wastewater from semicontinuous operation ww to discharge, t
$m_{ww,j}^{C,OUT}$	water mass from continuous unit to discharge in the interval j, t
$m_{ww}^{C,ST}$	capacity of storage tank for continuous water stream ww, t
m_n^{OP}	water mass inside operation n, t
$m_{nc,n}^{PP}$	reuse water mass from operation nc to operation n, t
$m_{kc,k}^{PP,E}$	reuse water mass from process kc to process k, t
m_n^{ST}	storage tank capacity of operation n, t
$m_k^{ST,E}$	reuse water mass from process k, t
$m_{tr}^{ST,TR,IN}$	capacity of storage tank placed before treatment unit tr, t
$m_{tr}^{ST,TR,OUT}$	capacity of storage tank placed after treatment unit tr, t
$m_{nc,n,tr}^{TR}$	reuse water mass from operation nc to operation n purified in local treatment unit tr, t
m_{tr}^{TRC}	capacity of local treatment unit tr, t
$m_{kc,k,tr}^{TR,E}$	reuse water mass from process kc to process k purified in local treatment unit tr, t
$m_{w,n}^W$	water mass from water source w to operation n, t
$m_{ww,k}^{W,E}$	water mass from water source ww to process k, t
$t_{n,nc,tr}^{A,TR}$	waiting time after treatment unit tr, h
$t_{n,nc,tr}^{B,TR}$	waiting time before treatment unit tr, h
$t_{nc,tr}^{E,TR}$	ending time of wastewater treatment from operation nc in treatment unit tr, h
$t_{nc,tr}^{S,TR}$	starting time of wastewater treatment from operation nc in treatment unit tr, h
$\mathbf{x}$	vector of decision variables

Binary Variables

$Y_{nc,n,tr}^{TR}$	binary variable for reused water mass from operation nc to operation n purified in local treatment unit tr

$Y_{kc,k,tr}^{TR,E}$	binary variable for reused water mass from process kc to process k purified in local treatment unit tr
$Y_{nc,n}^{PP}$	binary variable for reused water mass from operation nc to operation n
$Y_{kc,k}^{PP,E}$	binary variable for reused water mass from process kc to process k
Y_n^{ST}	binary variable for storage tank to operation n
$Y_k^{ST,E}$	binary variable for storage tank to process k
$Y_{ww,n}^{W}$	binary variable for water mass from water source w to operation n
$Y_{ww,k}^{W,E}$	binary variable for water mass from water source ww to process k
$Y_{ww}^{C,ST}$	binary variable for a central storage tank for continuous water source ww
$Y_{tr}^{ST,IN}$	binary variable for storage tank before treatment unit tr
$Y_{tr}^{ST,OUT}$	binary variable for storage tank after treatment unit tr
$Y_{tr,nc}^{ST,TR,IN}$	binary variable for storage tank before treatment unit tr
$Y_{tr,nc}^{ST,TR,OUT}$	binary variable for storage tank after treatment unit tr

REFERENCES

Almato M, Espuna A, Puigjaner L. 1999. Optimisation of water use in batch process industries. *Computers and Chemical Engineering* 23(10): 1427–37.

Altham W. 2007. Benchmarking to trigger cleaner production in small businesses: Dry cleaning case study. *Journal of Cleaner Production* 15(8–9): 798–813.

Alva-Argaez A, Kokossis AC, Smith R. 2007. The design of water-using systems in petroleum refining using a water-pinch decomposition. *Chemical Engineering Journal* 128(1): 33–46.

Bagajewicz M. 2000. A review of recent design procedures for water networks in refineries and process plants. *Computers and Chemical Engineering* 24(9–10): 2093–113.

Bai J, Feng X, Deng C. 2007. Graphically based optimization of single-contaminant regeneration reuse water systems. *Chemical Engineering Research and Design* 85(8): 1178–87.

Bai J, Feng X, Deng Ch. 2010. Optimal design of single-contaminant regeneration reuse water networks with process decomposition. *Process Systems Engineering* 56(4): 915–29.

BREF, European Commission, Reference Document on Best Available Techniques in Food, Drink and Milk Industries, Seville, Spain, 2006. http://eippcb.jrc.ec.europa.eu/reference/BREF/fdm_bref_0806.pdf. Accessed: 25. 07. 2014.

Campos de Faria D, Ulson de Souza AA, Guelli U, Souzaa SMA. 2009. Optimization of water networks in industrial processes. *Journal of Cleaner Production* 17(9): 857–62.

Chan JH, Foo DCY, Kumaresan S, Aziz RA, Hassan MAA. 2008. An integrated approach for water minimisation in a PVC manufacturing process. *Clean Technologies and Environmental Policy* 10(1): 67–79.

Chang CT, Li BH. 2005. Improved optimization strategies for generating practical water-usage and -treatment network structures. *Industrial & Engineering Chemistry Research* 44(10): 3607–18.

Chen ChL, Lee JY. 2008. A graphical technique for the design of water-using networks in batch processes. *Chemical Engineering Science* 63(14): 3740–54.

Chen ChL, Lee JY, Tang JW, Ciou YJ. 2009. Synthesis of water-using network with central reusable storage in batch processes. *Computers and Chemical Engineering* 33(1): 267–76.

Chen CL, Tang JW, Ciou YJ, Chang CY. 2007. Synthesis of water-using network with central reusable storage in batch processes. *Chemical Engineering Transactions* 12: 409–14.

Cheng KF, Chang CT. 2007. Integrated water network designs for batch processes. *Industrial and Engineering Chemistry Research* 46(4): 1241–53.

Das I, Dennis J. 1998. Normal-boundary intersection: A new method for generating the Pareto surface in nonlinear multicriteria optimization problems. *SIAM Journal on Optimization* 8(3): 631–57.

de Faria DC, de Souza AAU, de Souza SMAGU. 2009. Optimization of water networks in industrial processes. *Journal of Cleaner Production* 17(9): 857–62.

de Souza AAU, Forgiarini E, Brandão HL, Xavier MF, Pessoa FLP, de Souza SMAGU. 2009. Application of Water Source Diagram (WSD) method for the reduction of water consumption in petroleum refineries. *Resources, Conservation and Recycling* 53(3): 149–54.

EBC. 1990. *Manual of Good Practice: Water in Brewing. European Brewery Convention*, Nürnberg: Fachverlag Hans Carl.

Erol P, Thoming J. 2005. ECO-design of reuse and recycling networks by multi-objective optimization. *Journal of Cleaner Production* 13(15): 1492–1503.

Faria DC, Bagajewicz MJ. 2008. A new approach for design of multicomponenet water/wastewater networks. *Computer Aided Chemical Engineering* 25: 43–8.

Feng X, Bai J, Zheng X. 2007. On the use of graphical method to determine the targets of single-contaminant regeneration recycling water systems. *Chemical Engineering Science* 62(8): 2127–38.

Feng X, Huang L, Zhang X, Liu Y. 2009. Water system integration of a brewhouse. *Energy Conversion and Management* 50(2): 352–9.

Feng X, Wang N, Chen E. 2006. Water system integration in a catalyst plant. *Chemical Engineering Research and Design* 84(8): 645–51.

Foo DCY, Manan ZA, Tan YL. 2005. Synthesis of maximum water recovery network for batch process systems. *Journal of Cleaner Production* 13(15): 1381–94.

Foo DCY, Manan ZA, Tan YL. 2006. Use cascade analysis to optimize water networks. *Chemical Engineering Progress* 102(7): 45–52.

Forstmeier M, Goers B, Wozny G. 2005. Water network optimisation in the process industry-Case study of a liquid detergent plant. *Journal of Cleaner Production* 13(5): 495–98.

GAMS Beta 22.4. 2007. *The Solver Manuals*. Washington, USA: GAMS Development Corporation.

Global Reporting Initiative. 2002. *GRI—Sustainability Reporting Guidelines 2002 on Economic, Environmental and Social Performance*. Boston, USA: Global Reporting Initiative (GRI).

Gomes JFS, Queiroz EM, Pessoa FLP. 2007. Design procedure for water/wastewater minimization: single contaminant. *Journal of Cleaner Production* 15(5): 474–85.

Gouwsa JF, Majozi T, Gadalla M. 2008. Flexible mass transfer model for water minimization in batch plants. *Chemical Engineering and Processing* 47(12): 2323–35.

Gunaratnam M, Alva-Argez A, Kokossis A, Kim JK, Smith R. 2005. Automated design of total water systems. *Industrial & Engineering Chemistry Research* 44(3): 588–99.

Huang C, Chang C, Ling H, Chang C. 1999. A mathematical programming model for water usage and treatment network design. *Industrial & Engineering Chemistry Research* 38(7): 2666–79.

Hul S, Ng DKS, Tan RR, Chiang ChL, Foo DCY. 2007a. Crisp and fuzzy optimization approaches for water network retrofit. *Chemical Product and Process Modeling* 2(3): 1934–2659.

Hul S, Tan RR, Auresenia J, Fuchino T, Foo DCY. 2007b. Water network synthesis using mutation-enhanced particle swarm optimization. *Process Safety and Environmental Protection* 86(6): 507–14.

Iancu P, Plesu V, Lavric V. 2009. Waste water network retrofitting through optimal placement of regeneration unit. *Chemical Engineering Transactions* 18: 851–856. doi: 10.3303/CET0918139.

Jödicke G, Fischer U, Hungerbühler K. 2001. Wastewater reuse: A new approach to screen for designs with minimal total costs. *Computers and Chemical Engineering* 25(2–3): 203–15.

Karuppiah R, Grossmann IE. 2006. Global optimization for the synthesis of integrated water systems in chemical processes. *Computers and Chemical Engineering* 30(4): 650–73.

Khor CS, Chachuat B, Shah N. 2012. A superstructure optimization approach for water network synthesis with membrane separation-based regenerators. *Computers and Chemical Engineering* 42(11): 48–63.

Kim JK. 2011. Design of discontinuous water-using systems with a graphical method. *Chemical Engineering Journal* 172(2–3): 799–810.

Kim JK, Smith R. 2004. Automated design of discontinuous water systems. *Trans IChemE* 82(B3): 238–48.

Koski J, Silvennoinen R. 1987. Norm methods and partial weighting in multicriterion optimization of structures. International *Journal for Numerical Methods in Engineering* 24(6): 1101–21.

Kuo W, Smith R. 1998. Designing for the interactions between water use and effluent treatment. *Transactions of the Institution of Chemical Engineers* 76(3): 287–301.

Ku-Pineda V, Tan R. 2006. Environmental performance optimization using process water integration and sustainable process index. *Journal of Cleaner Production* 14(18): 1586–92.

Li LJ, Zhou RJ, Dong HG. 2010. State-time-space superstructure-based MINLP formulation for batch water-allocation network design. *Industrial & Engineering Chemistry Research* 49(1): 236–51.

Lim S-R, Park JM. 2007. Environmental and economic analysis of a water network system using LCA and LCC. *AIChE Journal* 53(12): 3253–62.

Liu Y, Feng X, Chu KH, Deng Ch. 2012. Optimization of water network with single and two outflow water-using processes. *Chemical Engineering Journal* 192(1): 49–57.

Liu Y, Li G, Wang L, Zhang J, Shams K. 2009. Optimal design of an integrated discontinuous water-using network coordinating with a central continuous regeneration unit. *Industrial & Engineering Chemistry Research* 48(24): 10924–40.

Liu Z, Li Y, Liu Z, Wang Y. 2009. A simple method for design of water-using networks with multiple contaminants involving regeneration reuse. *AIChE Journal* 55(6): 1628–33.

Lovelady E, El-Halwagi MM, Krishnagopalan G. 2007. An integrated approach to the optimization of water usage and discharge in pulp and paper plants. *Journal of Environment and Pollution* 29(1–3): 274–307.

Luo Y, Yuan X. 2008. Global optimization for the synthesis of integrated water systems with particle swarm optimization algorithm. *Chinese Journal of Chemical Engineering* 16(1): 11–5.

Majozi T. 2005. Wastewater minimisation using central reusable water storage in batch processes, *Computers and Chemical Engineering Journal*, 29(7): 1631–1646.

Majozi T, Brouckaert CJ, Buckley CA. 2006. A graphical technique for wastewater minimisation in batch processes. *Journal of Environmental Management* 78(4): 317–29.

Manan ZA, Tan YL, Foo DCY, Tea SY. 2007. Application of water cascade analysis technique for water minimisation in a paper mill plant. *International Journal of Environment and Pollution* 29(1–3): 90–103.

Matijašević Lj, Dejanović I, Spoja D. 2010. A water network optimization using MATLAB-A case study. *Resources, Conservation and Recycling* 54(12): 1362–67.

Matsumura EM, Mierzwa JC. 2008.Water conservation and reuse in poultry processing plant-A case study. *Resources, Conservation and Recycling* 52(6): 835–42.

Ng DKS, Foo DCY. 2009. Automated targeting technique for single-impurity resource conservation networks—Part 1 and 2. *Industrial & Engineering Chemistry Research* 48(16): 7637–61.

OECD. 2008. *Handbook on Constructing Composite Indicators: Methodology and User Guide*. Paris: OECD—Organisation for Economic Co-operation and Development.

Ponce-Ortega JM, Rivera FN, El-Halwagi MM, Jiménez-Gutiérrez A. 2012. An optimization approach for the synthesis of recycle and reuse water integration networks. *Clean Technologies and Environmental Policy* 14(1): 133–51.

Poplewski G, Jezowski JM. 2005. Stochastic optimization based approach for designing cost optimal water networks. *Computer Aided Chemical Engineering* 20(A): 727–32.

Poplewski G, Walczyk K, Jezowski J. 2010. Optimization-based method for calculating water networks with user specified characteristics. *Chemical Engineering Research & Design* 88(1A): 109–20.

PPAH. 1998. *Pollution Prevention and Abatement Handbook: Toward Cleaner Production.* Washington: World Bank Group.

Putra ZA, Amminudin KA. 2008. Two-step optimization approach for design of a total water system. *Industrial & Engineering Chemistry Research* 47(16): 6045–57.

Rao SS, Freiheit TI. 1991. A modified game theory approach to multiobjective optimization. *Journal of Mechanical Design* 113(3): 286–91.

Shoaib AM, Aly SM, Awad ME, Foo DCY, El-Halwagi MM. 2008. A hierarchical approach for the synthesis of batch water network. *Computers and Chemical Engineering* 32(3): 530–9.

Takama N, Kuriyama T, Shiroko K, Umeda T. 1980. Optimal water allocation in a Petroleum Refinery. *Computers and Chemical Engineering* 4(4): 251–8.

Tan RR, Col-long KJD, Foo DCY, Hul S, Ng DKS. 2008. A methodology for the design of efficient resource conservation networks using adaptive swarm intelligence. *Journal of Cleaner Production* 16(7): 822–32.

Tan RR, Ng DKS, Foo DCY, Aviso KB. 2009. A superstructure model for the synthesis of single-contaminant water networks with partitioning regenerators. *Process Safety and Environmental Protection* 87(3): 197–205.

Tan YL, Manan ZA, Foo DCY. 2007. Retrofit of water network with regeneration using water pinch analysis. *Process Safety and Environmental Protection* 85(4): 305–17.

Thevendiraraj S, Klemeš J, Paz D, Aso G, Cardenas GJ. 2003. Water and wastewater minimisation study of a citrus plant. *Resources, Conservation and Recycling* 37(3): 227–50.

Tokos H, Novak Pintarič Z. 2009. Synthesis of batch water network for a brewery plant. *Journal of Cleaner Production* 17(16): 1465–79.

Tokos H, Novak-Pintarič Z, Yang Y. 2013. Bi-objective optimization of water network via benchmarking. *Journal of Cleaner Production* 39: 168–79.

Tokos H, Novak Pintarič Z, Yang Y, Kravanja Z. 2012. Multilevel strategies for the retrofit of large-scale industrial water system: A brewery case study. *AIChE Journal* 58(3): 884–98.

Tudor R, Lavric V. 2010. Optimization of total networks of water using and treatment units by genetic algorithms. *Industrial & Engineering Chemistry Research* 49(8): 3715–31.

Ujang Z, Wong CL, Manan ZA. 2002. Industrial wastewater minimization using water pinch analysis: A case study of an old textile plant. *Water Science and Technology* 46(11–12): 77–84.

Ulson de Souzaa AA, Forgiarini E, Brandăoa HL, Xavier MF, Pessoab FLP, Guelli U, Souzaa SMA. 2009. Application of water source diagram (WSD) method for the reduction of water consumption in petroleum refineries. *Resources, Conservation and Recycling* 53(3): 149–54.

Wang YP, Smith R. 1994. Wastewater minimisation. *Chemical Engineering Science* 49(7): 981–1006.

Wang YP, Smith R. 1995. Time pinch analysis. *Trans IChemE* A73: 905–14.

Wenzel H, Dunn RF, Gottrup L, Kringelum J. 2002. Process integration design methods for water conservation and wastewater reduction in industry. Part 3: experience of industrial application. *Clean Technologies and Environmental Policy* 4(1): 16–25.

Xu D, Hu Y, Hau B, Wang X. 2003. Minimization of the flowrate of fresh water and corresponding regenerated water in water-using system with regeneration re-use. *Chinese Journal of Chemical Engineering* 11(3): 257–63.

Zheng P, Feng X, Qian F, Cao D. 2006. Water system integration of a chemical plant. *Energy Conversion and Management* 47(15–16): 2470–8.

Index